STUDENT'S SOLUTIONS MANUAL

EDGAR N. REYES
Southeastern Louisiana University

COLLEGE ALGEBRA

FOURTH EDITION

Mark Dugopolski
Southeastern Louisiana University

PEARSON

Addison
Wesley

Boston   San Francisco   New York
London Toronto Sydney Tokyo Singapore Madrid
Mexico City Munich Paris Cape Town Hong Kong Montreal

PEARSON

Addison
Wesley

This Student's Solutions Manual includes worked out solutions to all odd-numbered exercises in the exercise sets, and all odd-numbered exercises in the end of the chapter "Review Exercises". Also, included are solutions to the "For Thought", "Chapter Test", and "Tying It All Together" exercises.

TABLE OF CONTENTS

For Thought

1. False, 0 is not an irrational number.

2. True

3. True

4. False, 0 does not have a multiplicative inverse.

5. True

6. False, the case $a = w = z$ satisfies $a \leq w$ and $w \leq z$ but it does not satisfy $a < z$.

7. False, since $a - (b - c) = a - b + c$.

8. False, the distance is $|a - b|$.

9. False, a TI-89 calculator can give the irrational number $\sqrt{2}$.

10. False, the opposite is $-(a + b)$, which is $-a - b$.

P.1 Exercises

1. e, true

3. h, true

5. g, true

7. c, true

9. All

11. $\{-\sqrt{2}, \sqrt{3}, \pi, 5.090090009...\}$

13. $\{0, 1\}$ **15.** $x + 7$

17. $5x + 15$

19. $5(x + 1)$ **21.** $(-13 + 4) + x$

23. $\dfrac{1}{0.125}$ or 8 **25.** $\sqrt{3}$

27. $y^2 - x^2$ **29.** 7.2

31. $\sqrt{5}$

33. $d(13, 8) = |13 - 8| = 5$

35. $d(-5, 17) = |-5 - 17| = |-22| = 22$

37. $d(-6, -18) = |-6 - (-18)| = |12| = 12$

39. $d\left(-\dfrac{1}{2}, \dfrac{1}{4}\right) = \left|-\dfrac{1}{2} - \dfrac{1}{4}\right| = \left|-\dfrac{3}{4}\right| = \dfrac{3}{4}$

41. 8 **43.** -49

45. $(-8)(-8) = 16$ **47.** $-\dfrac{1}{64}$

49. $2 \cdot 5 - 3 \cdot 6 = 10 - 18 = -8$

51. $|3 - 4 \cdot 5| - 5 = |3 - 20| - 5 = 12$

53. $|-4 \cdot 3| - |-3 \cdot 5| = 12 - 15 = -3$

55. $\dfrac{4}{4} = 1$

57. $4 - 5 \cdot 3^2 = 4 - 5 \cdot 9 = 4 - 45 = -41$

59. -7

61. $3 \cdot 6 + 2 \cdot 4 = 18 + 8 = 26$

63. $\dfrac{26}{5} \div \dfrac{1}{2} \cdot 5 = \dfrac{52}{5} \cdot 5 = 52$

65. $(12 - 1)(1 + 8) = 11 \cdot 9 = 99$

67. $2 - 3|3 - 24| = 2 - 3 \cdot 21 = -61$

69. $49 - 36 = 13$

71. $3 - 4(-1)^2 = 3 - 4(1) = -1$

73. $\dfrac{2(9)}{25 - 16} = \dfrac{18}{9} = 2$

75. $b^2 - 4ac = 3^2 - 4(-2)(4) = 9 - (-32) = 41$

77. $\dfrac{a - c}{b - c} = \dfrac{-2 - 4}{3 - 4} = \dfrac{-6}{-1} = 6$

79. $a^2 - b^2 = (-2)^2 - (3)^2 = 4 - 9 = -5$

81. $(a - b)(a + b) = (-2 - 3)(-2 + 3) = (-5)(1) = -5$

83. $(a - b)(a^2 + ab + b^2) =$ $(-2 - 3)((-2)^2 + (-2)(3) + (3)^2) =$ $(-5)(4 - 6 + 9) = (-5)(7) = -35$

85. $a^b + c^b = (-2)^3 + (4)^3 = -8 + 64 = 56$

87. $-2x$ **89.** $0.85x$

91. $-6xy$

93. $3 - 2x$

95. $\dfrac{6x}{2} - \dfrac{2y}{2} = 3x - y$

97. $-3x - 6$

99. $-8 + 2x - 9 + 9x = 11x - 17$

101. a)

$$0.60(220 - a - r) + r =$$
$$132 - 0.60a - 0.60r + r =$$
$$0.40r - 0.60a + 132$$

b) If $a = 20$ and $r = 70$, then the target heart rate is

$$0.40r - 0.60a + 132 =$$
$$0.40(70) - 0.60(20) + 132 =$$
$$148 \text{ beats/min}$$

c)

$$0.60\left(205 - \dfrac{a}{2} - r\right) + r =$$
$$123 - 0.30a - 0.60r + r =$$
$$0.40r - 0.30a + 123$$

103. Since $\dfrac{1}{3} = \dfrac{4}{12}$, we find

$$-\dfrac{1}{2} < -\dfrac{5}{12} < -\dfrac{1}{3} < 0 < \dfrac{1}{3} < \dfrac{5}{12} < \dfrac{1}{2}.$$

For Thought

1. True, since $\dfrac{1}{2} + \dfrac{1}{2} = 1$.

2. True, since $2^{100} = (2^2)^{50} = 4^{50}$.

3. True, since $9^8 9^8 = (9 \cdot 9)^8 = 81^8$.

4. True, since $(0.25)^{-1} = (1/4)^{-1} = 4$.

5. False, since $\dfrac{5^{10}}{5^{-12}} = 5^{10-(-12)} = 5^{22}$.

6. False, since $2^3 \cdot 2^{-1} = 2^2 = 4$.

7. True, since $-\dfrac{1}{3^3} = -\dfrac{1}{27}$.

8. True, since $a^{-2} = (1/a)^2$ provided $a \neq 0$.

9. False, since $10^{-4} = 0.0001$.

10. False, since $98.6 \times 10^8 = 9.86 \times 10^9$.

P.2 Exercises

1. $3^{-1} = \dfrac{1}{3}$

3. $-4^{-2} = -\dfrac{1}{16}$

5. $\dfrac{1}{2^{-3}} = (2)^3 = 8$

7. $\left(\dfrac{2}{3}\right)^3 = \dfrac{8}{27}$

9. $\left(-\dfrac{2}{1}\right)^2 = (-2)^2 = 4$

11. $\dfrac{1}{2} \cdot 16 \cdot \dfrac{1}{10} = \dfrac{8}{10} = \dfrac{4}{5}$

13. $\dfrac{6^3}{3^2} = 24$

15. $1 + \dfrac{1}{2} = \dfrac{3}{2}$

17. $-\dfrac{2}{1000} = -\dfrac{1}{500}$

19. $-6x^{11}y^{11}$

21. $x^6 + x^6 = 2x^6$

23. $-6b^5 - 15b^5 = 21b^5$

25. $-2a^{11} + 4a^{11} = 2a^{11}$

27. $m^3 + m^2 + 3m^3 = 4m^3 + m^2$

29. $25 \cdot 9 = 225$

31. $\dfrac{1}{6}x^0 y^{-3} = \dfrac{1}{6y^3}$

33. $3x^4$

35. $\dfrac{-3}{-6} \cdot \dfrac{m^{-1}}{m^{-1}} \cdot \dfrac{n}{n^{-1}} = \dfrac{n^2}{2}$

37. $8a^6 + 9a^6 = 17a^6$

39. $-4x^6$

41. $\dfrac{(-2)^3(x^2)^3}{27} = \dfrac{-8x^6}{27}$

43. $\dfrac{25}{y^4}$

45. $\left(\dfrac{3x^5}{4y}\right)^{-3} = \dfrac{64y^3}{27x^{15}}$

47. $(x^{3b-3})(x^{8-2b}) = x^{b+5}$

49. $-125a^{6t}b^{-9t} = -\dfrac{125a^{6t}}{b^{9t}}$

51. $\dfrac{-3y^{6v}}{2x^{5w}}$

53. $\left(a^{-s+5}\right)^4 = a^{-4s+20}$

55. $43{,}000$ **57.** 0.0000356

59. 5×10^6

61. 6.72×10^{-5}

63. 0.000000007 **65.** 2×10^{10}

67. $20 \times 10^{15} = 2 \times 10^{16}$

69. $2 \times 10^{-6+3} = 2 \times 10^{-3}$

71. $40 \times 10^{-30} = 4 \times 10^{-29}$

73. $\dfrac{(8 \times 10^{18})(5 \times 10^{-6})}{(4 \times 10^{-10})} = \dfrac{40 \times 10^{12}}{4 \times 10^{-10}} = 1 \times 10^{23}$

75. 9.936×10^{-5}

77. Using $\pi \approx 3.14$, we find
$$\dfrac{(178.45 \times 10^{-18})(0.00666 \times 10^{-28})}{3.14(79.21 \times 10^{-8})} \approx$$
$$\dfrac{1.188 \times 10^{-46}}{248.72 \times 10^{-8}} \approx 4.78 \times 10^{-41}$$

79. BMI$= \dfrac{703(335)}{79^2} \approx 37.7$.

81. Since 1 ton = 2000 pounds, the ratio is
$$D = \dfrac{2200(2000)175^{-3}10^6}{2240} \approx 366.5$$

83. Each person owes $\dfrac{7.634 \times 10^{12}}{2.956 \times 10^8} = \$25{,}825$

85. The number of seconds in 10 billion years is $60^2 \cdot 24 \cdot 365 \cdot 10^{10} = 3.1536 \times 10^{17}$. So, the amount of mass transformed by the sun in 10 billion years is 5 million $\times 3.1536 \times 10^{17} \approx 1.577 \times 10^{24}$ tons

87. The earth's radius is $\dfrac{6.9599 \times 10^5}{109.2} \approx 6379$ km.

89. $\dfrac{1.989 \times 10^{30}}{5.976 \times 10^{24}} \approx 3.3 \times 10^5$.

91. **a)** 5365.4

b) 9702.38

c) Let a, b, c, d, e, f be whole numbers. Then
$$a\cdot10^3 + b\cdot10^2 + c\cdot10^1 + d\cdot10^0 + e\cdot10^{-1} + f\cdot10^{-2}$$
is the number $abcdef$ obtained by juxtaposition.

d) $9\cdot10^3 + 6\cdot10^1 + 3\cdot10^0 + 2\cdot10^{-1} + 4\cdot10^{-2} + 1\cdot10^{-3}$
Not unique for $9063.241 = 90 \cdot 10^2 + 63 \cdot 10^1 + 241 \cdot 10^{-3}$.

e) $4\cdot10^4 + 3\cdot10^3 + 2\cdot10^0 + 1\cdot10^{-1} + 9\cdot10^{-2}$

For Thought

1. False, since $8^{-\frac{1}{3}} = \dfrac{1}{2}$.

2. True, since $16^{\frac{1}{4}} = 2 = 4^{\frac{1}{2}}$.

3. False, since $\sqrt{\dfrac{4}{6}} = \dfrac{2}{\sqrt{6}}$.   **4.**   True

5. False, since $(-1)^{\frac{1}{2}}$ is not a real number.

6. False, since $\sqrt[3]{7^2} = 7^{\frac{2}{3}}$.   **7.**   False, since $9^{\frac{1}{2}} = 3$.

8. True, since $\dfrac{1}{\sqrt{3}} \cdot \dfrac{\sqrt{3}}{\sqrt{3}} = \dfrac{\sqrt{3}}{3}$.

9. True, since $\dfrac{2^{\frac{1}{2}}}{2^{\frac{1}{3}}} = 2^{\frac{1}{2}-\frac{1}{3}} = 2^{\frac{1}{6}} = \sqrt[6]{2}$.

10. True, since $\sqrt[3]{7^5} = \sqrt[3]{7^3 \cdot 7^2} = 7\sqrt[3]{49}$.

P.3 Exercises

1. -3 **3.** 8

5. -4 **7.** 81

9. $(8^{1/3})^{-4} = \dfrac{1}{16}$

11. $\left(\dfrac{1}{4}\right)^{1/2} = \dfrac{1}{2}$

13. $\left(\dfrac{4}{9}\right)^{3/2} = \left(\left(\dfrac{4}{9}\right)^{1/2}\right)^3 = \left(\dfrac{2}{3}\right)^3 = \dfrac{8}{27}$

15. $|x|$ **17.** a^3

19. a^2

21. $x^{3\cdot(1/3)}y^{6\cdot(1/3)} = xy^2$

23. $y^{2/3+7/3} = y^{9/3} = y^3$

25. $x^2 y^{1/2}$

27. $6a^{1/2+1} = 6a^{3/2}$

29. $3a^{1/2-1/3} = 3a^{1/6}$

31. $a^{2+1/3}b^{1/2-1/2} = a^{7/3}$

33. $\dfrac{x^2 y}{z^3}$

35. 30 **37.** -2

39. -2

41. $\dfrac{2}{3}$ **43.** 0.1

45. $\dfrac{\sqrt[3]{-8}}{\sqrt[3]{1000}} = \dfrac{-2}{10} = -\dfrac{1}{5}$

47. $\sqrt[4]{(2^4)^3} = \sqrt[4]{2^{12}} = 2^{12\cdot(1/4)} = 2^3 = 8$

49. $10^{2/3} = \sqrt[3]{10^2}$

51. $\dfrac{3}{y^{3/5}} = \dfrac{3}{\sqrt[5]{y^3}}$

53. $x^{-1/2}$

55. $(x^3)^{1/5} = x^{3/5}$

57. $4x$

59. $2y^3$

61. $\dfrac{\sqrt{xy}}{10}$

63. $\dfrac{-2a}{b^{15\cdot(1/3)}} = -\dfrac{2a}{b^5}$

65. $\sqrt{4(7)} = 2\sqrt{7}$

67. $\dfrac{1}{\sqrt{5}} \cdot \dfrac{\sqrt{5}}{\sqrt{5}} = \dfrac{\sqrt{5}}{5}$

69. $\dfrac{\sqrt{x}}{\sqrt{8}} \cdot \dfrac{\sqrt{2}}{\sqrt{2}} = \dfrac{\sqrt{2x}}{\sqrt{16}} = \dfrac{\sqrt{2x}}{4}$

71. $\sqrt[3]{8 \cdot 5} = 2\sqrt[3]{5}$

73. $\sqrt[3]{-250x^4} = \sqrt[3]{(-125)(2)x^3 x} = -5x\sqrt[3]{2x}$

75. $\sqrt[3]{\dfrac{1}{2}} \cdot \sqrt[3]{\dfrac{4}{4}} = \sqrt[3]{\dfrac{4}{8}} = \dfrac{\sqrt[3]{4}}{2}$

77. $2\sqrt{2} + 2\sqrt{5} - 2\sqrt{3}$

79. $-10\sqrt{18} = -30\sqrt{2}$

81. $12(5a) = 60a$

83. $(-5)^2(3) = 75$

85. $\sqrt{\dfrac{9}{a^3}} = \dfrac{3}{a\sqrt{a}} \cdot \dfrac{\sqrt{a}}{\sqrt{a}} = \dfrac{3\sqrt{a}}{a^2}$

87. $\dfrac{5}{\sqrt{x}} \cdot \dfrac{\sqrt{x}}{\sqrt{x}} = \dfrac{5\sqrt{x}}{x}$

89. $2x\sqrt{5x} + 3x\sqrt{5x} = 5x\sqrt{5x}$

91. $\sqrt[6]{3^2}\,\sqrt[6]{2^3} = \sqrt[6]{72}$

93. $\sqrt[12]{3^4}\,\sqrt[12]{x^3} = \sqrt[12]{81x^3}$

95. $\sqrt[6]{(xy)^3}\,\sqrt[6]{(2xy)^2} = \sqrt[6]{x^3 y^3}\,\sqrt[6]{4x^2 y^2} = \sqrt[6]{4x^5 y^5}$

97. $(7^{1/2})^{1/3} = \sqrt[6]{7}$

99. Since $E = \sqrt{\dfrac{2(25)(6000)}{140}} \approx 46.3$, the most economic order quantity is $E = 46$.

101. $S = \dfrac{16(42,700)}{[(2200)(2000)]^{2/3}} \approx 25.4$

103. a) 50%, 30%

   b) If $n = 5$, the depreciation rate is

$$r = 1 - \left(\dfrac{200}{5000}\right)^{1/5} \approx 0.47469 \text{ or } 47.5\%.$$

   If $n = 10$, the depreciation rate is

$$r = 1 - \left(\dfrac{200}{5000}\right)^{1/10} \approx 0.2752 \text{ or } 27.5\%.$$

105. $D = \sqrt{4^2 + 6^2 + 12^2} = 14$ in.

107. $A = \sqrt{12(6)(5)(1)} = \sqrt{360} \approx 19.0$ ft^2

109. Each expression is equivalent to $t^{4/5}$ except for $\sqrt[4]{t^5}$ which is $t^{5/4}$, the expression in (b).

111. No, since $\sqrt{9+16} = 5 \neq 7 = \sqrt{9} + \sqrt{16}$.

For Thought

1. True, since $i \cdot (-i) = 1$.

2. True, since $\overline{0+i} = 0 - i = -i$.

3. False, the set of real numbers is a subset of the complex numbers.

4. True, $(\sqrt{3} - i\sqrt{2})(\sqrt{3} + i\sqrt{2}) = 3 + 2 = 5$.

5. False, since $(2 + 5i)(2 + 5i) =$
$4 + 20i + 25i^2 = 4 + 20i - 25 = -21 + 20i$.

6. False, $5 - \sqrt{-9} = 5 - 3i$.

7. True, since $(3i)^2 + 9 = (-9) + 9 = 0$.

8. True, since $(-3i)^2 + 9 = (-9) + 9 = 0$.

9. True, since $i^4 = i^2 \cdot i^2 = (-1)(-1) = 1$.

10. False, $i^{18} = (i^4)^4 i^2 = (1)^4(-1) = -1$.

P.4 Exercises

1. $0 + 6i$, imaginary

3. $\dfrac{1}{3} + \dfrac{1}{3}i$, imaginary

5. $\sqrt{7} + 0i$, real

7. $\dfrac{\pi}{2} + 0i$, real

9. $7 + 2i$

11. $1 - i - 3 - 2i = -2 - 3i$

13. $1 + 3 - i\sqrt{2} + 2i\sqrt{2} = 4 + i\sqrt{2}$

15. $5 - \dfrac{1}{2} + \dfrac{1}{3}i + \dfrac{1}{2}i = \dfrac{9}{2} + \dfrac{5}{6}i$

17. $-18i + 12i^2 = -12 - 18i$

19. $8 + 12i - 12i - 18i^2 = 26 + 0i$

21. $(4 - 5i)(6 + 2i) = 24 + 8i - 30i + 10 = 34 - 22i$

23. $(5 - 2i)(5 + 2i) = 25 - 4i^2 = 25 - 4(-1) = 29$

25. $(\sqrt{3} - i)(\sqrt{3} + i) = 3 - i^2 = 3 - (-1) = 4$

27. $9 + 24i + 16i^2 = -7 + 24i$

29. $5 - 4i\sqrt{5} + 4i^2 = 1 - 4i\sqrt{5}$

31. $\left(i^4\right)^4 \cdot i = (1)^4 \cdot i = i$

33. $\left(i^4\right)^{24} i^2 = 1^{24}(-1) = -1$

35. Since $i^4 = 1$, we get $i^{-1} = i^{-1}i^4 = i^3 = -i$.

37. Since $i^4 = 1$, we get $i^{-3} = i^{-3}i^4 = i^1 = i$.

39. Since $i^{16} = 1$, we get $i^{-13} = i^{-13}i^{16} = i^3 = -i$.

41. Since $i^{-4} = 1$, we get $i^{-38} = i^2i^{-40} =$
$i^2(i^{-4})^{10} = i^2(1) = -1$.

43. $(3 - 9i)(3 + 9i) = 9 - 81i^2 = 90$

45. $\left(\dfrac{1}{2} + 2i\right)\left(\dfrac{1}{2} - 2i\right) = \dfrac{1}{4} - 4i^2 = \dfrac{1}{4} + 4 = \dfrac{17}{4}$

47. $i(-i) = -i^2 = 1$

49. $(3 - i\sqrt{3})(3 + i\sqrt{3}) = 9 - 3i^2 = 9 - 3(-1) = 12$

51. $\dfrac{1}{2-i} \cdot \dfrac{2+i}{2+i} = \dfrac{2+i}{5} = \dfrac{2}{5} + \dfrac{1}{5}i$

53. $\dfrac{-3i}{1-i} \cdot \dfrac{1+i}{1+i} = \dfrac{-3i+3}{2} = \dfrac{3}{2} - \dfrac{3}{2}i$

55.
$\dfrac{-3+3i}{i} \cdot \dfrac{-i}{-i} = \dfrac{3i - 3i^2}{1} = 3i - 3(-1) = 3 + 3i$

57.
$\dfrac{1-i}{3+2i} \cdot \dfrac{3-2i}{3-2i} = \dfrac{3 - 5i - 2}{13} = \dfrac{1}{13} - \dfrac{5}{13}i$

59.
$\dfrac{2-i}{3+5i} \cdot \dfrac{3-5i}{3-5i} = \dfrac{6 - 10i - 3i - 5}{34} = \dfrac{1}{34} - \dfrac{13}{34}i$

61. $2i - 3i = -i$

63. $-4 + 2i$

65. $\left(i\sqrt{6}\right)^2 = -6$

67. $(i\sqrt{2})(i\sqrt{50}) = i^2\sqrt{2}\cdot 5\sqrt{2} = (-1)(2)(5) = -10$

69.
$$\frac{-2}{2} + \frac{i\sqrt{20}}{2} = -1 + i\frac{2\sqrt{5}}{2} = -1 + \sqrt{5}i$$

71. $-3 + \sqrt{9 - 20} = -3 + i\sqrt{11}$

73. $2i\sqrt{2}\left(i\sqrt{2} + 2\sqrt{2}\right) = 4i^2 + 8i = -4 + 8i$

75.
$$\frac{-2 + \sqrt{-16}}{2} = \frac{-2 + 4i}{2} = -1 + 2i$$

77.
$$\frac{-4 + \sqrt{16 - 24}}{4} = \frac{-4 + 2\sqrt{2}i}{4} = \frac{-2 + i\sqrt{2}}{2}$$

79.
$$\frac{-6 - \sqrt{-32}}{2} = \frac{-6 - 4i\sqrt{2}}{2} = -3 - 2i\sqrt{2}$$

81.
$$\frac{-6 - \sqrt{36 + 48}}{-4} = \frac{-6 - 2\sqrt{21}}{-4} = \frac{3 + \sqrt{21}}{2}$$

83. $(3 - 5i)(3 + 5i) = 3^2 + 5^2 = 34$

85. $(3 - 5i) + (3 + 5i) = 6$

87.
$$\frac{3 - 5i}{3 + 5i} \cdot \frac{3 - 5i}{3 - 5i} = \frac{9 - 15i - 15i - 25}{34} =$$
$$-\frac{16}{34} - \frac{30}{34}i = -\frac{8}{17} - \frac{15}{17}i$$

89. $(6 - 2i) - (7 - 3i) = 6 - 7 - 2i + 3i = -1 + i$

91. $i^5(i^2 - 3i) = i(-1 - 3i) = -i + 3 = 3 - i$

93. If r is the remainder when n is divided by 4, then $i^n = i^r$. The possible values of r are $0, 1, 2, 3$ and for i^r they are $1, i, -1, -i$, respectively.

95. Note, $w + \overline{w} = (a + bi) + (a - bi) = 2a$ is a real number and $w - \overline{w} = (a + bi) - (a - bi) = 2bi$ is an imaginary number.

When a complex number is added to its complex conjugate the sum is twice the real part of the complex number. When the complex conjugate of a complex number is subtracted from the complex number, the difference is an imaginary number.

97. Its reciprocal is $\dfrac{1}{a + bi} = \dfrac{a - bi}{(a + bi)(a - bi)}$

$$= \frac{a - bi}{a^2 + b^2} = \frac{a}{a^2 + b^2} - \frac{b}{a^2 + b^2}i.$$

99. Note, $i^{n!} = 1$ for $n \geq 4$. Then we obtain

$$i^{4!} + \ldots + i^{100!} = 97.$$

Since $i^{0!} = i$, $i^{1!} = i$, $i^{2!} = -1$, and $i^{3!} = -1$, we conclude that

$$i^{0!} + \ldots + i^{3!} + (i^{4!} + \ldots + i^{100!}) = -2 + 2i + 97 = 95 + 2i.$$

For Thought

1. False, since polynomials do not have negative exponents.

2. False, since the degree is 4.

3. False, since $(8)^2 \neq 9 + 25$.

4. True, since $50^2 - 1^2 = 2499$.

5. False, since $(x + 3)^2 = x^2 + 6x + 9$.

6. False, since the left-hand side is $-x - 6$.

7. False, since $(a + b)^3 = a^3 + 3a^2b + 3ab^2 + b^3$.

8. False, since divisor times quotient plus the remainder is equal to the dividend.

9. False, since the equation is not defined at $x = -2$.

10. True, since $P(17) + Q(17) = 2310 = S(17)$.

P.5 Exercises

1. Degree 3, leading coefficient 1, trinomial

3. Degree 2, leading coefficient -3, binomial

5. Degree 0, leading coefficient 79, monomial

7. $P(-2) = 4 + 6 + 2 = 12$

9. $M(-3) = 27 + 45 + 3 + 2 = 77$

11. $8x^2 + 3x - 1$

13. $4x^2 - 3x - 9x^2 + 4x - 3 = -5x^2 + x - 3$

15. $4ax^3 - a^2x - 5a^2x^3 + 3a^2x - 3 =$
$(-5a^2 + 4a)x^3 + 2a^2x - 3$

17. $2x - 1$

19. $3x^2 - 3x - 6$

21. $-18a^5 + 15a^4 - 6a^3$

23. $(3b^2 - 5b + 2)b - (3b^2 - 5b + 2)3 =$
$3b^3 - 5b^2 + 2b - 9b^2 + 15b - 6 =$
$3b^3 - 14b^2 + 17b - 6$

25. $2x(4x^2 + 2x + 1) - (4x^2 + 2x + 1) =$
$8x^3 + 4x^2 + 2x - 4x^2 - 2x - 1 =$
$8x^3 - 1$

27. $(x - 4)z + (x - 4)3 = xz - 4z + 3x - 12$

29. $a(a^2 + ab + b^2) - b(a^2 + ab + b^2) =$
$a^3 + a^2b + ab^2 - a^2b - ab^2 - b^3 = a^3 - b^3$

31. $a^2 - 2a + 9a - 18 = a^2 + 7a - 18$

33. $2y^2 + 18y - 3y - 27 = 2y^2 + 15y - 27$

35. $4x^2 + 18x - 18x - 81 = 4x^2 - 81$

37. $4x^2 + 10x + 10x + 25 = 4x^2 + 20x + 25$

39. $6x^4 + 10x^2 + 12x^2 + 20 = 6x^4 + 22x^2 + 20$

41. $(1 + \sqrt{2})(3 + \sqrt{2}) = 3 + \sqrt{2} + 3\sqrt{2} + 2 = 5 + 4\sqrt{2}$

43. $20 - 15\sqrt{2} + 4\sqrt{2} - 3\sqrt{2}\sqrt{2} = 14 - 11\sqrt{2}$

45. $6\sqrt{2}\sqrt{2} - 3\sqrt{6} + 2\sqrt{6} - \sqrt{3}\sqrt{3} = 9 - \sqrt{6}$

47. $\sqrt{5}\sqrt{5} + \sqrt{15} + \sqrt{15} + \sqrt{3}\sqrt{3} = 8 + 2\sqrt{15}$

49. $(3x)^2 + 2(3x)(5) + (5)^2 = 9x^2 + 30x + 25$

51. $(x^n)^2 - 3^2 = x^{2n} - 9$

53. $(\sqrt{2})^2 - 5^2 = -23$

55. $(3\sqrt{6})^2 - 2(3\sqrt{6})(1) + 1 = 55 - 6\sqrt{6}$

57. $(3x^3)^2 - 2(3x^3)(4) + 4^2 = 9x^6 - 24x^3 + 16$

59. $\dfrac{5}{1 - \sqrt{7}} \cdot \dfrac{1 + \sqrt{7}}{1 + \sqrt{7}} = \dfrac{5 + 5\sqrt{7}}{1 - 7} = \dfrac{-5 - 5\sqrt{7}}{6}$

61. $\dfrac{\sqrt{10}}{\sqrt{5} - 2} \cdot \dfrac{\sqrt{5} + 2}{\sqrt{5} + 2} = \dfrac{\sqrt{50} + 2\sqrt{10}}{5 - 4} = 5\sqrt{2} + 2\sqrt{10}$

63. $\dfrac{\sqrt{6}}{6 + \sqrt{3}} \cdot \dfrac{6 - \sqrt{3}}{6 - \sqrt{3}} = \dfrac{6\sqrt{6} - 3\sqrt{2}}{33} = \dfrac{2\sqrt{6} - \sqrt{2}}{11}$

65.
$$\frac{1 + \sqrt{2}}{2 - \sqrt{3}} \cdot \frac{2 + \sqrt{3}}{2 + \sqrt{3}} = \frac{2 + 2\sqrt{2} + \sqrt{3} + \sqrt{6}}{4 - 3} =$$
$2 + 2\sqrt{2} + \sqrt{3} + \sqrt{6}$

67.
$$\frac{\sqrt{2} + \sqrt{3}}{\sqrt{6}} \cdot \frac{\sqrt{6}}{\sqrt{6}} = \frac{\sqrt{12} + \sqrt{18}}{6} =$$
$$\frac{2\sqrt{3} + 3\sqrt{2}}{6}$$

69. $\dfrac{36x^6}{-4x^3} = -9x^3$

71. $\dfrac{3x^2}{-3x} + \dfrac{6x}{3x} = -x + 2$

73. Factor & simplify: $\dfrac{(x + 3)^2}{x + 3} = x + 3$

75. Factor & simplify: $\dfrac{(a - 1)(a^2 + a + 1)}{a - 1} =$
$a^2 + a + 1$

77. Quotient $x + 5$, remainder 13 since

$$\begin{array}{r} x + 5 \\ x - 2 \enclose{longdiv}{x^2 + 3x + 3} \\ \underline{x^2 - 2x} \\ 5x + 3 \\ \underline{5x - 10} \\ 13 \end{array}$$

79. Quotient $2x - 6$, remainder 13 since

$$\begin{array}{r} 2x - 6 \\ x + 3 \enclose{longdiv}{2x^2 + 0x - 5} \\ \underline{2x^2 + 6x} \\ -6x - 5 \\ \underline{-6x - 18} \\ 13 \end{array}$$

81. Quotient $x^2 + x + 1$, remainder 0 since

$$\begin{array}{r} x^2 + x + 1 \\ x - 3 \enclose{longdiv}{x^3 - 2x^2 - 2x - 3} \\ \underline{x^3 - 3x^2} \\ x^2 - 2x \\ \underline{x^2 - 3x} \\ x - 3 \\ \underline{x - 3} \\ 0 \end{array}$$

83. $x + 1 + \dfrac{-1}{x-1}$ since

$$\begin{array}{r} x + 1 \\ x - 1 \;\overline{)\,x^2 + 0x - 2} \\ \underline{x^2 - x} \\ x - 2 \\ \underline{x - 1} \\ -1 \end{array}$$

85. $\dfrac{2x^2}{x} - \dfrac{3x}{x} + \dfrac{1}{x} = 2x - 3 + \dfrac{1}{x}$

87. $\dfrac{x^2}{x} + \dfrac{x}{x} + \dfrac{1}{x} = x + 1 + \dfrac{1}{x}$

89. $x + \dfrac{1}{x-2}$ since

$$\begin{array}{r} x \\ x - 2 \;\overline{)\,x^2 - 2x + 1} \\ \underline{x^2 - 2x} \\ 1 \end{array}$$

91. $x^2 + 2x - 24$

93. $2a^{10} - 3a^5 - 27$

95. $-y - 9$ **97.** $6a - 6$

99. $w^2 + 8w + 16$

101. $16x^2 - 81$

103. $3y^5 - 9xy^2$

105. $2b - 1$

107. $x^6 - 64$

109. $9w^4 - 12w^2n + 4n^2$

111. 1

113. The area is $(x + 3)(2x - 1) = 2x^2 + 5x - 3$.

115. Divide the volume by the height.

$$\begin{array}{r} x^2 + 5x + 6 \\ x + 4 \;\overline{)\,x^3 + 9x^2 + 26x + 24} \\ \underline{x^3 + 4x^2} \\ 5x^2 + 26x \\ \underline{5x^2 + 20x} \\ 6x + 24 \\ \underline{6x + 24} \\ 0 \end{array}$$

Then the area of the bottom is $x^2 + 5x + 6$ cm^2.

117. The dimensions of the rectangular habitat are $6 - 2x$ and $4 - 2x$. The area is the product of the dimensions which is $4x^2 - 20x + 24$ square miles.

119. a) Approximately \$0.20

    b) $A(.1160) - A(.0569) \approx \$0.21$

121. The fine is $F(25) = 500 + 200(25) = \$5,500$.

123. Let x be the length of a side of the square. The areas of the square and rectangle are x^2 and $(x + 10)(x - 10) = x^2 - 100$, respectively. Wilson had 100 ft^2 less area than he thought.

125. Let $x \gg 0$ represent a sufficiently large number. If $x \gg 0$, then the sign of $P(x)$ is the same as the sign of the leading coefficient. If $x < 0$, $|x| \gg 0$, and the degree is even, then the sign of $P(x)$ is the same as the sign of the leading coefficient. If $x < 0$, $|x| \gg 0$, and the degree is odd, then the sign of $P(x)$ is opposite the sign of the leading coefficient.

For Thought

1. False, since it factors as $(x + 8)(x - 2)$.

2. True   **3.**   True

4. False, since $a^2 - 1 = (a - 1)(a + 1)$.

5. False, since $a^3 - 1 = (a - 1)(a^2 + a + 1)$.

6. False, since $x^3 - y^3 = (x - y)(x^2 + xy + y^2)$.

7. False, it factors as $(2a + 3b)(4a^2 - 6ab + 9b^2)$.

8. False, since $(x + 2)(x - 6) = x^2 - 4x - 12$.

9. True   **10.**   True

P.6 Exercises

1. $6x^2(x - 2),\ -6x^2(-x + 2)$

3. $4a(1 - 2b),\ -4a(2b - 1)$

5. $ax(-x^2 + 5x - 5),\ -ax(x^2 - 5x + 5)$

7. $1(m - n),\ -1(n - m)$

9. $x^2(x + 2) + 5(x + 2) = (x^2 + 5)(x + 2)$

11. $y^2(y - 1) - 3(y - 1) = (y^2 - 3)(y - 1)$

13. $ady + d - awy - w = d(ay + 1) - w(ay + 1) = (d - w)(ay + 1)$

15. $x^2y^2 - ay^2 - (bx^2 - ab) = y^2(x^2 - a) - b(x^2 - a) = (y^2 - b)(x^2 - a)$

17. $(x + 2)(x + 8)$

19. $(x - 6)(x + 2)$

21. $(m - 2)(m - 10)$

23. $(t - 7)(t + 12)$

25. $(2x + 1)(x - 4)$

27. $(4x + 1)(2x - 3)$

29. $(3y + 5)(2y - 1)$

31. $(t - u)(t + u)$

33. $(t + 1)^2$ **35.** $(2w - 1)^2$

37. $(y^{2t} - 5)(y^{2t} + 5)$

39. $(3zx + 4)^2$

41. $(t - u)(t^2 + tu + u^2)$

43. $(a - 2)(a^2 + 2a + 4)$

45. $(3y + 2)(9y^2 - 6y + 4)$

47. $(3xy^2)^3 - (2z^3)^3 = (3xy^2 - 2z^3)(9x^2y^4 + 6xy^2z^3 + 4z^6)$

49. $(x^n)^3 - 2^3 = (x^n - 2)(x^{2n} + 2x^n + 4)$

51. Replace y^3 by w and y^6 by w^2.
Then $w^2 + 10w + 25 = (w + 5)^2 = (y^3 + 5)^2$.

53. Replace $2a^2b^4$ by w and $4a^4b^8$ by w^2.
So, $w^2 - 4w - 5 = (w + 1)(w - 5) = (2a^2b^4 + 1)(2a^2b^4 - 5)$.

55. Replace $(2a + 1)$ by w and $(2a + 1)^2$ by w^2.
Then $w^2 + 2w - 24 = (w + 6)(w - 4) = ((2a + 1) + 6)((2a + 1) - 4) = (2a + 7)(2a - 3)$.

57. Replace $(b^2 + 2)$ by w and $(b^2 + 2)^2$ by w^2.
We find $w^2 - 5w + 4 = (w - 4)(w - 1) = ((b^2 + 2) - 4)((b^2 + 2) - 1) = (b^2 - 2)(b^2 + 1)$.

59. $-3x^3 + 27x = -3x(x^2 - 9) = -3x(x - 3)(x + 3)$

61. $2t(8t^3 + 27w^3) = 2t(2t + 3w)(4t^2 - 6tw + 9w^2)$

63. $a^3 + a^2 - 4a - 4 = a^2(a + 1) - 4(a + 1) = (a^2 - 4)(a + 1) = (a - 2)(a + 2)(a + 1)$

65. $x^4 - 2x^3 - 8x + 16 = x^3(x - 2) - 8(x - 2) = (x - 2)(x^3 - 8) = (x - 2)^2(x^2 + 2x + 4)$

67. $-2x(18x^2 - 9x - 2) = -2x(6x + 1)(3x - 2)$

69. $a^7 - a^6 - 64a + 64 = a^6(a - 1) - 64(a - 1) = (a^6 - 64)(a - 1) = (a^3 - 8)(a^3 + 8)(a - 1) = (a - 2)(a^2 + 2a + 4)(a + 2)(a^2 - 2a + 4)(a - 1)$

71. $-(3x + 5)(2x - 3)$

73. Replace $(a^2 + 2)$ by w and $(a^2 + 2)^2$ by w^2.
$w^2 - 4w + 3 = (w - 3)(w - 1) = ((a^2 + 2) - 3)((a^2 + 2) - 1) = (a^2 - 1)(a^2 + 1) = (a - 1)(a + 1)(a^2 + 1)$

75. Yes, since

$$
\require{enclose}
\begin{array}{r}
x^2 + x + 1 \\
x + 3 \enclose{longdiv}{x^3 + 4x^2 + 4x + 3} \\
\underline{x^3 + 3x^2} \\
x^2 + 4x \\
\underline{x^2 + 3x} \\
x + 3 \\
\underline{x + 3} \\
0
\end{array}
$$

and $x^3 + 4x^2 + 4x + 3 = (x + 3)(x^2 + x + 1)$.

77. No, since

$$
\begin{array}{r}
3x^2 + 8x - 4 \\
x - 1 \enclose{longdiv}{3x^3 + 5x^2 - 12x - 9} \\
\underline{3x^3 - 3x^2} \\
8x^2 - 12x \\
\underline{8x^2 - 8x} \\
-4x - 9 \\
\underline{-4x + 4} \\
-13
\end{array}
$$

and the remainder is not 0.

79.

$$
\begin{array}{r}
x^2 + 5x + 6 \\
x - 1 \enclose{longdiv}{x^3 + 4x^2 + x - 6} \\
\underline{x^3 - x^2} \\
5x^2 + x \\
\underline{5x^2 - 5x} \\
6x - 6 \\
\underline{6x - 6} \\
0
\end{array}
$$

Then $x^3 + 4x^2 + x - 6 = (x-1)(x^2 + 5x + 6) = (x-1)(x+2)(x+3)$.

81.

$$
\begin{array}{r}
x^2 + 2x + 2 \\
x - 3 \overline{\smash{\big)}\, x^3 - x^2 - 4x - 6} \\
\underline{x^3 - 3x^2} \\
2x^2 - 4x \\
\underline{2x^2 - 6x} \\
2x - 6 \\
\underline{2x - 6} \\
0
\end{array}
$$

So $x^3 - x^2 - 4x - 6 = (x-3)(x^2 + 2x + 2)$.

83.

$$
\begin{array}{r}
x^3 + 3x^2 - x - 3 \\
x + 2 \overline{\smash{\big)}\, x^4 + 5x^3 + 5x^2 - 5x - 6} \\
\underline{x^4 + 2x^3} \\
3x^3 + 5x^2 \\
\underline{3x^3 + 6x^2} \\
-x^2 - 5x \\
\underline{-x^2 - 2x} \\
-3x - 6 \\
\underline{-3x - 6} \\
0
\end{array}
$$

Then we obtain

$$
\begin{aligned}
x^4 + 5x^3 + 5x^2 - 5x - 6 &= \\
(x+2)(x^3 + 3x^2 - x - 3) &= \\
(x+2)(x^2(x+3) - (x+3)) &= \\
(x+2)(x+3)(x^2 - 1) &= \\
(x+2)(x+3)(x-1)(x+1).
\end{aligned}
$$

85. The area of the bottom is the volume divided by the height.

$$
\begin{array}{r}
x^2 + x + 1 \\
x - 1 \overline{\smash{\big)}\, x^3 + 0x^2 + 0x - 1} \\
\underline{x^3 - x^2} \\
x^2 + 0x \\
\underline{x^2 - x} \\
x - 1 \\
\underline{x - 1} \\
0
\end{array}
$$

Then $x^2 + x + 1$ ft^2 is the area of the bottom.

87. The volume is $V = x(6 - 2x)(7 - 2x)$.

From the tabulated values below

x	0.5	1	2
Volume	15 in.^{3}	20 in.^{3}	12 in.^{3}

we see that $x = 1$ produces the largest volume.

89. **b** is not a perfect square trinomial since $(\sqrt{1000a} - b)^2 \neq 1000a^2 - 200b + b^2$. The others are perfect squares.

91. No, since $1^3 + 1^3 = 2 \neq n^3$ for any integer n.

For Thought

1. False, since 2 is not factor. **2.** True

3. False, since the first expression is not defined at $x = 0$.

4. False, since $\dfrac{(a-b)(a+b)}{a-b} = a + b$.

5. False, since the LCD is $x(x+1)$. **6.** True

7. True **8.** True

9. True, since $\dfrac{2(500) + 1}{500 - 3} = \dfrac{1001}{497} \approx 2.014$

10. True, since $\dfrac{5x + 1}{x} \approx \dfrac{5x}{x} = 5$ when $|x|$ is large.

P.7 Exercises

1. $\{x \mid x \neq -2\}$

3. $\{x \mid x \neq 4, -2\}$

5. $\{x \mid x \neq \pm 3\}$

7. All real numbers since $x^2 + 3 \neq 0$ for any real number x.

9. $\dfrac{3(x-3)}{(x-3)(x+2)} = \dfrac{3}{x+2}$

11. $\dfrac{10a - 8b}{12b - 15a} = \dfrac{2(5a - 4b)}{-3(5a - 4b)} = -\dfrac{2}{3}$

13. $\dfrac{a^3 b^6}{a^2 b^3 - a^4 b^2} = \dfrac{a^3 b^6}{a^2 b^2(b - a^2)} = \dfrac{ab^4}{b - a^2}$

15. $\dfrac{y^2 z}{x^3}$

17. $\dfrac{a^3 - b^3}{a^2 - b^2} = \dfrac{(a-b)(a^2+ab+b^2)}{(a-b)(a+b)} =$

$\dfrac{a^2 + ab + b^2}{a + b}$

19. $\dfrac{(3y-1)^2}{(1-3y)(1+3y)} = \dfrac{-(3y-1)}{1+3y}$

21. $\dfrac{2a}{3b^2} \cdot \dfrac{9b}{14a^2} = \dfrac{1}{b} \cdot \dfrac{3}{7a} = \dfrac{3}{7ab}$

23. $\dfrac{12a}{7} \cdot \dfrac{49}{2a^3} = \dfrac{42}{a^2}$

25. $\dfrac{(a-3)(a+3)}{3(a-2)} \cdot \dfrac{(a-2)(a+2)}{(a-3)(a+2)} = \dfrac{a+3}{3}$

27. $\dfrac{(x-y)(x+y)}{9} \cdot \dfrac{9(2)}{(x+y)^2} = \dfrac{2x-2y}{x+y}$

29. $\dfrac{(x-y)(x+y)}{-3xy} \cdot \dfrac{3xy(2xy^2)}{-2(x-y)} = x^2y^2 + xy^3$

31. $\dfrac{(b-3)(b+2)}{(3-b)(3+b)} \cdot \dfrac{2(b+3)}{(b+4)(b+2)} =$

$\dfrac{-2}{b+4}$

33. $\dfrac{16a}{12a^2}$

35. $\dfrac{x-5}{x+3} \cdot \dfrac{x-3}{x-3} = \dfrac{x^2-8x+15}{x^2-9}$

37. $\dfrac{x}{x+5} \cdot \dfrac{x+1}{x+1} = \dfrac{x^2+x}{x^2+6x+5}$

39. $12a^2b^3$

41. Since $3a + 3b = 3(a+b)$ and $2a + 2b = 2(a+b)$, the LCD is $6(a+b)$.

43. Since $x^2 + 5x + 6 = (x+3)(x+2)$ and $x^2 - x - 6 = (x-3)(x+2)$, the LCD is $(x+3)(x-3)(x+2)$.

45. $\dfrac{3(3)}{2x(3)} + \dfrac{x}{6x} = \dfrac{9+x}{6x}$

47. $\dfrac{(x+3)(x+1)}{(x-1)(x+1)} - \dfrac{(x+4)(x-1)}{(x+1)(x-1)} =$

$\dfrac{x^2+4x+3}{(x-1)(x+1)} - \dfrac{x^2+3x-4}{(x+1)(x-1)} =$

$\dfrac{x+7}{(x+1)(x-1)}$

49. $\dfrac{3a}{a} + \dfrac{1}{a} = \dfrac{3a+1}{a}$

51. $\dfrac{(t-1)(t+1)}{t+1} - \dfrac{1}{t+1} = \dfrac{t^2-2}{t+1}$

53. $\dfrac{x}{(x+2)(x+1)} + \dfrac{x-1}{(x+3)(x+2)} =$

$\dfrac{x(x+3)}{(x+2)(x+1)(x+3)} +$

$\dfrac{(x-1)(x+1)}{(x+3)(x+2)(x+1)} =$

$\dfrac{2x^2+3x-1}{(x+1)(x+2)(x+3)}$

55. $\dfrac{1}{x-3} - \dfrac{5}{-2(x-3)} = \dfrac{2}{2(x-3)} - \dfrac{-5}{2(x-3)} =$

$\dfrac{7}{2x-6}$

57. $\dfrac{y^2}{(x-y)(x^2+xy+y^2)} +$

$\dfrac{(x+y)(x-y)}{(x^2+xy+y^2)(x-y)} =$

$\dfrac{y^2}{(x-y)(x^2+xy+y^2)} +$

$\dfrac{x^2-y^2}{(x^2+xy+y^2)(x-y)} =$

$\dfrac{x^2}{x^3-y^3}$

59. $\dfrac{(x+1)(x-1)}{x(x+1)(x-1)} + \dfrac{x(x+1)}{(x-1)(x)(x+1)} -$

$\dfrac{x(x-1)}{(x+1)(x)(x-1)} =$

$\dfrac{x^2-1}{x(x+1)(x-1)} + \dfrac{x^2+x}{(x-1)(x)(x+1)} -$

$\dfrac{x^2-x}{(x+1)(x)(x-1)} =$

$\dfrac{x^2+2x-1}{x(x^2-1)}$

61.
$$\frac{5(x+1)(x-3)+2x(x-3)-6x(x+1)}{x(x+1)(x-3)}=$$
$$\frac{5(x^2-2x-3)+2x^2-6x-6x^2-6x}{x(x+1)(x-3)}=$$
$$\frac{x^2-22x-15}{x(x+1)(x-3)}$$

63. $\dfrac{25}{36a}\cdot\dfrac{27}{10a}=\dfrac{5^2\cdot3\cdot9}{4\cdot9\cdot5\cdot2a^2}=\dfrac{15}{8a^2}$

65.
$$\frac{\left(\dfrac{4}{a}-\dfrac{3}{b}\right)(ab^2)}{\left(\dfrac{1}{ab}+\dfrac{2}{b^2}\right)(ab^2)}=\frac{4b^2-3ab}{b+2a}$$

67.
$$\frac{\dfrac{1}{b^2}-\dfrac{1}{ab^2}}{\dfrac{3}{a^2}+\dfrac{1}{a^2b}}=\frac{\left(\dfrac{1}{b^2}-\dfrac{1}{ab^2}\right)(a^2b^2)}{\left(\dfrac{3}{a^2}+\dfrac{1}{a^2b}\right)(a^2b^2)}=\frac{a^2-a}{3b^2+b}$$

69.
$$\frac{\left(a+\dfrac{4}{a+4}\right)(a+4)}{\left(a-\dfrac{4a+4}{a+4}\right)(a+4)}=$$
$$\frac{a^2+4a+4}{a^2+4a-(4a+4)}=$$
$$\frac{(a+2)^2}{(a+2)(a-2)}=\frac{a+2}{a-2}$$

71.
$$\frac{\left(\dfrac{t+2}{t-1}-\dfrac{t-3}{t}\right)((t-1)t)}{\left(\dfrac{t+4}{t}+\dfrac{t-2}{t-1}\right)((t-1)t)}=$$
$$\frac{(t^2+2t)-(t-3)(t-1)}{(t+4)(t-1)+(t^2-2t)}=$$
$$\frac{(t^2+2t)-(t^2-4t+3)}{(t^2+3t-4)+(t^2-2t)}=\frac{6t-3}{2t^2+t-4}$$

73. $\dfrac{(x^{-1}+1)x}{(x^{-1}-1)x}=\dfrac{1+x}{1-x}$

75. $\dfrac{(a^2+a^{-1}b^{-3})ab^3}{ab^3}=\dfrac{a^3b^3+1}{ab^3}$

77.
$$\frac{(x^2-y^2)xy}{(x^{-1}-y^{-1})xy}=\frac{(x-y)(x+y)xy}{y-x}=$$
$$-(x+y)xy=-x^2y-xy^2$$

79.
$$\left(\frac{1}{m}-\frac{1}{n}\right)^{-2}=\left(\frac{n-m}{mn}\right)^{-2}=$$
$$\left(\frac{mn}{n-m}\right)^2=\frac{m^2n^2}{n^2-2mn+m^2}$$

81. $\dfrac{3}{7}$ **83.** $\dfrac{3}{506}$

85. $S(2)=\dfrac{4-5}{4-9}=\dfrac{1}{5}$

87. $\dfrac{1200-5}{1200-9}=\dfrac{1195}{1191}$

89. $\dfrac{9(16)-1}{3(16)-2}\approx3.1087$

91. $T(-400)=\dfrac{9(160,000)-1}{3(160,000)-2}\approx3.00001$

93 (a) Average cost decreases as capacity increases

(b) The average costs are

n	7	12	22
$A(n)$	\$27.14	\$24.17	\$22.27

95. a) The costs are tabulated below.

p	$C(p)$
50%	\$6 million
75%	\$18 million
99%	\$594 million

b) The cost of cleaning up goes up without bound. A 100% clean up is impossible.

c) domain is $\{p|0\le p<100\}$

97. The portion of the invoices Gina and Bert can file in one hour are $\dfrac{1}{4}$ and $\dfrac{1}{6}$, respectively. The part they can file together in one hour is $\dfrac{1}{4}+\dfrac{1}{6}=\dfrac{5}{12}$.

99. Let d be the distance between the restaurant and his home. The number of hours dashing home and returning to the restaurant are $\dfrac{d}{250}$ and $\dfrac{d}{300}$, respectively. His average speed is
$$\frac{2d}{\dfrac{d}{250}+\dfrac{d}{300}}=\frac{2}{\dfrac{1}{250}+\dfrac{1}{300}}\approx272.7\text{ mph.}$$

101. If a rational expression simulates an application, then the domain describes the extent to which the simulation applies. This is one reason.

103. (a) Since $1 + \dfrac{1}{2} = \dfrac{3}{2}$, $1 + \dfrac{1}{3/2} = \dfrac{5}{3}$, and so on..., the exact answer is $\dfrac{8}{13}$.

(b) Since $1 - \dfrac{1}{3} = \dfrac{2}{3}$, $1 - \dfrac{1}{2/3} = -\dfrac{1}{2}$, and so on..., the exact answer is $\dfrac{3}{2}$.

Review Exercises

1. False, since $\sqrt{2}$ is an irrational number.

3. False, since -1 is a negative number.

5. False, since terminating decimal numbers are rational numbers.

7. False, since $\{1, 2, 3, ...\}$ is the set of natural numbers.

9. False, since $\dfrac{1}{3} = 0.3333...$

11. False, since the additive inverse of 0.5 is -0.5

13. $-3x - 12 + 20x = 17x - 12$

15. $\dfrac{2x}{10} + \dfrac{x}{10} = \dfrac{3x}{10}$

17. $\dfrac{3(x-2)}{3 \cdot 3} = \dfrac{x-2}{3}$

19. $\dfrac{-6}{8} = -\dfrac{3}{4}$

21. $3 - 5 = -2$ **23.** $|-4| = 4$

25. $8 - 18 \div 3 + 5 = 8 - 6 + 5 = 7$

27. $3 \cdot 3 \div 6 + 3^3 = 9 \div 6 + 27 = 1.5 + 27 = 28.5$

29. 625 **31.** $4 + 40 = 44$

33. $\dfrac{1}{2} + 1 = \dfrac{3}{2}$

35. $\dfrac{-2^3 \cdot 2^1}{3^1} = -\dfrac{16}{3}$

37. $(8^2)^{-1/3} = 64^{-1/3} = \dfrac{1}{4}$

39. $5x^2$

41. 11

43. $\sqrt{4 \cdot 7s^2 \cdot s} = 2s\sqrt{7s}$

45. $\sqrt[3]{-1000(2)} = -10\sqrt[3]{2}$

47. $\sqrt{\dfrac{5}{2a}} \cdot \sqrt{\dfrac{2a}{2a}} = \dfrac{\sqrt{10a}}{2a}$

49. $\sqrt[3]{\dfrac{2}{5}}\sqrt[3]{\dfrac{25}{25}} = \dfrac{\sqrt[3]{50}}{5}$

51. $3n\sqrt{2n} + 5n\sqrt{2n} = 8n\sqrt{2n}$

53. $\dfrac{2\sqrt{3}}{\sqrt{3}-1} \cdot \dfrac{\sqrt{3}+1}{\sqrt{3}+1} = \dfrac{6 + 2\sqrt{3}}{3 - 1} = 3 + \sqrt{3}$

55. $\dfrac{\sqrt{6}}{2\sqrt{2} + 3\sqrt{2}} = \dfrac{\sqrt{6}}{5\sqrt{2}} = \dfrac{\sqrt{12}}{10} = \dfrac{\sqrt{3}}{5}$

57. $320,000,000$

59. 0.000185 **61.** 5.6×10^{-5} **63.** 2.34×10^6

65. $125 \cdot 10^{18} = 1.25 \times 10^{20}$

67. $\dfrac{(8 \times 10^2)^2 (10^{-5})^{-3}}{(2 \times 10^6)^3 (2 \times 10^{-5})} =$

$\dfrac{(64 \times 10^4)(10^{15})}{(8 \times 10^{18})(2 \times 10^{-5})} =$

$\dfrac{64}{8(2)} \times 10^6 = 4 \times 10^6$

69. $3 - 4 - 7i + 6i = -1 - i$

71. $16 - 40i - 25 = -9 - 40i$

73. $2 + 6i - 6i + 18 = 20$

75. $\dfrac{2 - 3i}{i} \cdot \dfrac{-i}{-i} = \dfrac{-2i - 3}{1} = -3 - 2i$

77. $\dfrac{1-i}{2+i} \cdot \dfrac{2-i}{2-i} = \dfrac{1 - 3i}{5} = \dfrac{1}{5} - \dfrac{3}{5}i$

79. $\dfrac{1+i}{2-3i} \cdot \dfrac{2+3i}{2+3i} = \dfrac{-1 + 5i}{13} = -\dfrac{1}{13} + \dfrac{5}{13}i$

81. $\dfrac{6 + 2i\sqrt{2}}{2} = 3 + i\sqrt{2}$

83. $\dfrac{-6 + \sqrt{-20}}{-8} = \dfrac{-6 + 2i\sqrt{5}}{-8} = \dfrac{3}{4} - \dfrac{\sqrt{5}}{4}i$

85. $i^{32}i^2 + i^{16}i^3 = (1)(-1) + (1)(-i) = -1 - i$

87. $3x^2 - x^2 - x + 2x - 2 - 5 = 2x^2 + x - 7$

89. $-4x^4 - 3x^3 + x - x^4 + 6x^3 + 2x =$
$-5x^4 + 3x^3 + 3x$

91. $(3a^2 - 2a + 5)a - (3a^2 - 2a + 5) \cdot 2 =$
$3a^3 - 2a^2 + 5a - 6a^2 + 4a - 10 =$
$3a^3 - 8a^2 + 9a - 10$

93. $b^2 - 6by + 9y^2$

95. $3t^2 - 7t - 6$

97. $-\dfrac{35y^5}{7y^2} = -5y^3$

99. $9 - 2 = 7$

101. $1 + 2\sqrt{3} + \sqrt{3}^2 = 1 + 2\sqrt{3} + 3 = 4 + 2\sqrt{3}$

103. $2^2\sqrt{5}^2 + 4\sqrt{15} + \sqrt{3}^2 = 20 + 4\sqrt{15} + 3 =$
$23 + 4\sqrt{15}$

105. Quotient $x^2 + 4x - 1$, remainder 1, since

$$
\begin{array}{r}
x^2 + 4x - 1 \\
x - 2 \enclose{longdiv}{x^3 + 2x^2 - 9x + 3} \\
\underline{x^3 - 2x^2} \\
4x^2 - 9x \\
\underline{4x^2 - 8x} \\
-x + 3 \\
\underline{-x + 2} \\
1
\end{array}
$$

107. Quotient $3x + 2$, remainder 4, since

$$
\begin{array}{r}
3x + 2 \\
2x - 1 \enclose{longdiv}{6x^2 + x + 2} \\
\underline{6x^2 - 3x} \\
4x + 2 \\
\underline{4x - 2} \\
4
\end{array}
$$

109. $x - 2 + \dfrac{1}{x + 2}$ since

$$
\begin{array}{r}
x - 2 \\
x + 2 \enclose{longdiv}{x^2 + 0x - 3} \\
\underline{x^2 + 2x} \\
-2x - 3 \\
\underline{-2x - 4} \\
1
\end{array}
$$

111. $2 + \dfrac{13}{x - 5}$ since

$$
\begin{array}{r}
2 \\
x - 5 \enclose{longdiv}{2x + 3} \\
\underline{2x - 10} \\
13
\end{array}
$$

113. $6x(x^2 - 1) = 6x(x - 1)(x + 1)$

115. $(3h + 4t)^2$ **117.** $(t + y)(t^2 - ty + y^2)$

119. $x^2(x + 3) - 9(x + 3) = (x^2 - 9)(x + 3) =$
$(x - 3)(x + 3)(x + 3) = (x - 3)(x + 3)^2$

121. $t^6 - 1 = (t^3 - 1)(t^3 + 1) =$
$(t - 1)(t^2 + t + 1)(t + 1)(t^2 - t + 1)$

123. $(6x + 5)(3x - 4)$

125. $ab(a^2 + 3a - 18) = ab(a + 6)(a - 3)$

127. $x^2(2x + y) - (2x + y) = (x^2 - 1)(2xy + 1) =$
$(x - 1)(x + 1)(2x + y)$

129. $\dfrac{2x + 6}{x + 3} = \dfrac{2(x + 3)}{x + 3} = 2$

131.
$$\dfrac{(x - 1)(x + 4)}{(x - 2)(x + 4)} - \dfrac{(x + 3)(x - 2)}{(x + 4)(x - 2)} =$$
$$\dfrac{x^2 + 3x - 4 - (x^2 + x - 6)}{(x - 2)(x + 4)} = \dfrac{2x + 2}{(x - 2)(x + 4)}$$

133. $\dfrac{(x - 3)(x + 3)}{x + 3} \cdot \dfrac{1}{-2(x - 3)} = -\dfrac{1}{2}$

135. $\dfrac{c^7}{a^6b^2} \cdot \dfrac{a^2b^6c^{10}}{a^4b^3} = \dfrac{bc^{17}}{a^8}$

137. $\dfrac{1}{(x - 2)(x + 2)} + \dfrac{3(x + 2)}{(x - 2)(x + 2)} = \dfrac{3x + 7}{x^2 - 4}$

139. $\dfrac{5x}{30x^2} - \dfrac{21}{30x^2} = \dfrac{5x - 21}{30x^2}$

141. $\dfrac{(a - 5)(a + 5)}{(a - 5)(a + 1)} \cdot \dfrac{(a - 1)(a + 1)}{2(a + 5)} = \dfrac{a - 1}{2}$

143.
$$\dfrac{(x - 4)(x + 4)}{(x + 4)(x + 1)} \cdot \dfrac{(x + 1)(x^2 - x + 1)}{-2(x - 4)} =$$
$$\dfrac{-x^2 + x - 1}{2}$$

145.

$$\frac{a-2}{(a+5)(a+1)} + \frac{2a+1}{(a-1)(a+1)} =$$

$$\frac{(a-2)(a-1)}{(a+5)(a+1)(a-1)} +$$

$$\frac{(2a+1)(a+5)}{(a-1)(a+1)(a+5)} =$$

$$\frac{(a^2-3a+2)+(2a^2+11a+5)}{(a+5)(a+1)(a-1)} =$$

$$\frac{3a^2+8a+7}{(a+1)(a-1)(a+5)}$$

147.

$$\frac{\left(\dfrac{5}{2x}-\dfrac{3}{4x}\right)(4x)}{\left(\dfrac{1}{2}-\dfrac{2}{x}\right)(4x)} = \frac{10-3}{2x-8} = \frac{7}{2x-8}$$

149.

$$\frac{\left(\dfrac{1}{y^2-2}-3\right)(y^2-2)}{\left(\dfrac{5}{y^2-2}+4\right)(y^2-2)} = \frac{1-3(y^2-2)}{5+4(y^2-2)} =$$

$$\frac{-3y^2+7}{4y^2-3}$$

151.

$$\frac{\left(a^{-2}-b^{-3}\right)a^2b^3}{\left(a^{-1}b^{-1}\right)a^2b^3} = \frac{b^3-a^2}{ab^2}$$

153. $\left(p^{-1}+pq^{-3}\right)\dfrac{pq^3}{pq^3} = \dfrac{q^3+p^2}{pq^3}$

155. $P(2) = (2)^3-3(2)^2+2-9 = 8-3(4)-7 = -11$

157. $P(0) = -9$

159. $R(-1) = \dfrac{3(-1)-1}{2(-1)-9} = \dfrac{-4}{-11} = \dfrac{4}{11}$

161. $R(50) = \dfrac{3(50)-1}{2(50)-9} = \dfrac{149}{91}$

163. a) 300 ft

   b) $\dfrac{130^2}{32} - \dfrac{125^2}{32} \approx 39.8$ feet

165. The number of hydrogen atoms in one kilogram of hydrogen is

$$\frac{1000}{1.7 \times 10^{-24}} \approx 5.9 \times 10^{26}.$$

167. $|-2.35-8.77| = |-11.12| = 11.12$

169. The parts of the lawn Howard and Will can mow in 2 hours are $\dfrac{2}{6}$ and $\dfrac{2}{4}$, respectively. In 2 hours, the portion they will mow together is $\dfrac{2}{6} + \dfrac{2}{4} = \dfrac{5}{6}$.

Chapter P Test

1. All   **2.** $\{-1.22, -1, 0, 2, 10/3\}$

3. $\{-\pi, -\sqrt{3}, \sqrt{5}, 6.020020002...\}$ **4.** $\{0, 2\}$

5. $|6-25| - 6 = |-19| - 6 = 13$ **6.** $8^{1/3} = 2$

7. $-\dfrac{1}{(27^{1/3})^2} = -\dfrac{1}{3^2} = -\dfrac{1}{9}$

8. $\dfrac{9+12+9}{(-5)(6)} = \dfrac{30}{-30} = -1$

9. $6x^5y^7$ **10.** $2x^2 - 2x^2 = 0$

11. $\dfrac{(ab(b+a))^2}{a^2b^2} = \dfrac{a^2b^2(b+a)^2}{a^2b^2} = a^2+2ab+b^2$

12. $\dfrac{-8a^{-3}b^{18}}{2^{-2}a^{-6}b^8} = -32a^3b^{10}$

13. $3\sqrt{3} - 2\sqrt{2} + 4\sqrt{2} = 3\sqrt{3} + 2\sqrt{2}$

14.

$$\frac{2\sqrt{2}}{\sqrt{6}-\sqrt{2}} \cdot \frac{\sqrt{6}+\sqrt{2}}{\sqrt{6}+\sqrt{2}} = \frac{2\sqrt{12}+4}{4} =$$

$$\frac{2(2\sqrt{3})+4}{4} = \sqrt{3}+1$$

15.

$$\frac{1}{x\sqrt[3]{4x}} = \frac{\sqrt[3]{2x^2}}{x\sqrt[3]{4x}\sqrt[3]{2x^2}} = \frac{\sqrt[3]{2x^2}}{2x^2}$$

16. $\sqrt{4(3)x^2xy^8y\cdot 1} = 2xy^4\sqrt{3xy}$

17. $16 - 24i - 9 = 7 - 24i$

18. $\dfrac{2-i}{3+i} \cdot \dfrac{3-i}{3-i} = \dfrac{5-5i}{10} = \dfrac{1}{2} - \dfrac{1}{2}i$

19. $i^4i^2 - i^{32}i^3 = (1)(-1) - (1)(-i) = -1 + i$

20. $2i\sqrt{2}(i\sqrt{2}+\sqrt{6}) = -4 + 2i\sqrt{12} = -4 + 4i\sqrt{3}$

21. $3x^3 + 3x^2 - 12x$

22. $-x^2 + 3x - 4 - 4x^2 + 6x - 9 = -5x^2 + 9x - 13$

23. $x(x^2 - 2x - 1) + 3(x^2 - 2x - 1) =$
$x^3 - 2x^2 - x + 3x^2 - 6x - 3 = x^3 + x^2 - 7x - 3$

24. $\dfrac{(2h-1)(4h^2 + 2h + 1)}{2h - 1} = 4h^2 + 2h + 1$

25. $x^2 - 6xy - 27y^2$

26.

$$
\begin{array}{r}
x^2 + 2x + 2 \\
x - 2 \overline{)x^3 + 0x^2 - 2x - 4} \\
\underline{x^3 - 2x^2} \\
2x^2 - 2x \\
\underline{2x^2 - 4x} \\
2x - 4 \\
\underline{2x - 4} \\
0
\end{array}
$$

Then $\dfrac{x^3 - 2x - 4}{x - 2} = x^2 + 2x + 2$.

27. $9x^2 - 48x + 64$ **28.** $4t^8 - 1$

29.

$$\frac{x(x^2 - 5x + 6)}{2x(x - 3)} \cdot \frac{4(x^3 + 8)}{2x(x^2 - 4)} =$$

$$\frac{x(x - 3)(x - 2)}{2x(x - 3)} \cdot \frac{4(x + 2)(x^2 - 2x + 4)}{2x(x - 2)(x + 2)} =$$

$$\frac{x^2 - 2x + 4}{x}$$

30.

$$\frac{(x + 5)(x + 4)}{(x - 3)(x - 1)(x + 4)} +$$

$$\frac{(x - 1)^2}{(x + 4)(x - 3)(x - 1)} =$$

$$\frac{(x^2 + 9x + 20) + (x^2 - 2x + 1)}{(x + 4)(x - 3)(x - 1)} =$$

$$\frac{2x^2 + 7x + 21}{(x + 4)(x - 3)(x - 1)}$$

31.

$$\frac{a - 1}{(2a - 3)(2a + 3)} + \frac{a - 2}{2a - 3} =$$

$$\frac{a - 1}{(2a - 3)(2a + 3)} + \frac{(a - 2)(2a + 3)}{(2a - 3)(2a + 3)} =$$

$$\frac{(a - 1) + (2a^2 - a - 6)}{(2a - 3)(2a + 3)} = \frac{2a^2 - 7}{4a^2 - 9}$$

32.

$$\frac{\dfrac{1}{2a^2b} - 2a}{\dfrac{1}{4ab^3} + \dfrac{1}{3b}} \cdot \frac{12a^2b^3}{12a^2b^3} = \frac{6b^2 - 24a^3b^3}{3a + 4a^2b^2}$$

33. $a(x^2 - 11x + 18) = a(x - 9)(x - 2)$

34. $m(m^4 - 1) = m(m^2 - 1)(m^2 + 1) =$
$m(m - 1)(m + 1)(m^2 + 1)$

35. $(3x - 1)(x + 5)$

36. $bx(x - 3) + w(x - 3) = (bx + w)(x - 3)$

37. The per capita income is $\dfrac{9.777 \times 10^{12}}{2.956 \times 10^8} =$ $\$33,075$

38. The annual appreciation rate is
$\left(\dfrac{2.5}{1.3}\right)^{1/3} - 1 \approx 0.24355$ or 24.4%.

39. The altitude is $A(2) = -64 + 240 = 176$ feet.

For Thought

1. True, since $5(1) = 6 - 1$.

2. True, since $x = 3$ is the solution to both equations.

3. False, -2 is not a solution of the first equation since $\sqrt{-2}$ is not a real number.

4. True

5. False, $x = 0$ is the solution.   **6.**   True

7. False, since $|x| = -8$ has no solution.

8. False, $\dfrac{x}{x-5}$ is undefined at $x = 5$.

9. False, since we should multiply by $-\dfrac{3}{2}$.

10. False, $0 \cdot x + 1 = 0$ has no solution.

1.1 Exercises

1. No, since $2(3) - 4 = 2 \neq 9$.

3. Yes, since $(-4)^2 = 16$.

5. Since $3x = 5$, the solution set is $\left\{ \dfrac{5}{3} \right\}$.

7. Since $-3x = 6$, the solution set is $\{-2\}$.

9. Since $14x = 7$, the solution set is $\left\{ \dfrac{1}{2} \right\}$.

11. Since $7 + 3x = 4x - 4$, the solution set is $\{11\}$.

13. Since $x = -\dfrac{4}{3} \cdot 18$, the solution set is $\{-24\}$.

15. Multiplying by 6 we get
$$\begin{aligned} 3x - 30 &= -72 - 4x \\ 7x &= -42. \end{aligned}$$

The solution set is $\{-6\}$.

17. Multiply both sides of the equation by 12.
$$\begin{aligned} 18x + 4 &= 3x - 2 \\ 15x &= -6 \\ x &= -\dfrac{2}{5}. \end{aligned}$$

The solution set is $\left\{ -\dfrac{2}{5} \right\}$.

19. Note, $3(x - 6) = 3x - 18$ is true by the distributive law. It is an identity and the solution set is R.

21. Note, $5x = 4x$ is equivalent to $x = 0$. A conditional equation whose solution set is $\{0\}$.

23. Equivalently, we get $2x + 6 = 3x - 3$ or $9 = x$. A conditional equation whose solution set is $\{9\}$.

25. Using the distributive property, we find
$$\begin{aligned} 3x - 18 &= 3x + 18 \\ -18 &= 18. \end{aligned}$$

The equation is inconsistent and the solution set is $\emptyset$.

27. An identity and the solution set is $\{x | x \neq 0\}$.

29. Multiplying $2(w - 1)$, we get
$$\begin{aligned} \dfrac{1}{w-1} - \dfrac{1}{2w-2} &= \dfrac{1}{2w-2} \\ 2 - 1 &= 1. \end{aligned}$$

An identity and the solution set is $\{w | w \neq 1\}$

31. Multiply by $6x$.
$$\begin{aligned} 6 - 2 &= 3 + 1 \\ 4 &= 4 \end{aligned}$$

An identity with solution set $\{x | x \neq 0\}$.

33. Multiply by $3(z - 3)$.
$$\begin{aligned} 3(z + 2) &= -5(z - 3) \\ 3z + 6 &= -5z + 15 \\ 8z &= 9 \end{aligned}$$

A conditional equation with solution set $\left\{ \dfrac{9}{8} \right\}$.

35. Multiplying by $(x - 3)(x + 3)$.
$$\begin{aligned} (x + 3) - (x - 3) &= 6 \\ 6 &= 6 \end{aligned}$$

An identity with solution set
$$\{x | x \neq 3, x \neq -3\}.$$

37. Multiply by $(y - 3)$.

$$\begin{aligned} 4(y - 3) + 6 &= 2y \\ 4y - 6 &= 2y \\ y &= 3 \end{aligned}$$

Since division by zero is not allowed, $y = 3$ does not satisfy the original equation. We have an inconsistent equation and so the solution set is $\emptyset$.

39. Multiply by $t + 3$.

$$\begin{aligned} t + 4t + 12 &= 2 \\ 5t &= -10 \end{aligned}$$

A conditional equation with solution set $\{-2\}$.

41. Since $-4.19 = 0.21x$ and

$$x = \frac{-4.19}{0.21} \approx -19.952$$

the solution set is approximately $\{-19.952\}$.

43. Divide by 0.06.

$$\begin{aligned} x - 3.78 &= \frac{1.95}{0.06} \\ x &= 32.5 + 3.78 \\ x &= 36.28 \end{aligned}$$

The solution set is $\{36.28\}$.

45.

$$\begin{aligned} 2a &= -1 - \sqrt{17} \\ a &= \frac{-1 - \sqrt{17}}{2} \\ a &\approx \frac{-1 - 4.1231}{2} \\ a &\approx -2.562 \end{aligned}$$

The solution set is approximately $\{-2.562\}$.

47.

$$\begin{aligned} 0.001 &= 3(y - 0.333) \\ 0.001 &= 3y - 0.999 \\ 1 &= 3y \\ \frac{1}{3} &= y \end{aligned}$$

The solution set is $\left\{\frac{1}{3}\right\}$.

49. Factoring x, we get

$$\begin{aligned} x\left(\frac{1}{0.376} + \frac{1}{0.135}\right) &= 2 \\ x(2.6596 + 7.4074) &\approx 2 \\ 10.067x &\approx 2 \\ x &\approx 0.199 \end{aligned}$$

The solution set is approximately $\{0.199\}$.

51.

$$\begin{aligned} x^2 + 6.5x + 3.25^2 &= x^2 - 8.2x + 4.1^2 \\ 14.7x &= 4.1^2 - 3.25^2 \\ 14.7x &= 16.81 - 10.5625 \\ 14.7x &= 6.2475 \\ x &= 0.425 \end{aligned}$$

The solution set is $\{0.425\}$.

53.

$$\begin{aligned} (2.3 \times 10^6)x &= 1.63 \times 10^4 - 8.9 \times 10^5 \\ x &= \frac{1.63 \times 10^4 - 8.9 \times 10^5}{2.3 \times 10^6} \\ x &\approx -0.380 \end{aligned}$$

The solution set is approximately $\{-0.380\}$.

55. Solution set is $\{\pm 8\}$.

57. Since $x - 4 = \pm 8$, we get $x = 4 \pm 8$.
The solution set is $\{-4, 12\}$.

59. Since $x - 6 = 0$, we find $x = 6$.
The solution set is $\{6\}$.

61. Since the absolute value of a real number is not a negative number, the equation

$$|x + 8| = -3$$

has no solution. The solution set is $\emptyset$.

63. Since we have

$$2x - 3 = 7 \ \text{ or } \ 2x - 3 = -7$$

we get $2x = 10$ or $2x = -4$. The solution set is $\{-2, 5\}$.

65. Multiplying $\frac{1}{2}|x - 9| = 16$ by 2 we obtain

$$|x - 9| = 32.$$

Then $x - 9 = 32$ or $x - 9 = -32$.
The solution set is $\{-23, 41\}$.

67. Since $2|x + 5| = 10$, we find

$$|x + 5| = 5.$$

Then $x + 5 = \pm 5$ or $x = \pm 5 - 5$.

The solution set is $\{-10, 0\}$.

69. Dividing $8|3x - 2| = 0$ by 8, we obtain $|3x - 2| = 0$. Then $3x - 2 = 0$ and the solution set is $\{2/3\}$.

71. Subtracting 7, we find $2|x| = -1$ and $|x| = -\frac{1}{2}$. Since an absolute value is not equal to a negative number, the solution set is $\emptyset$.

73. Since $0.95x = 190$, the solution set is $\{200\}$.

75.

$$
\begin{aligned}
0.1x - 0.05x + 1 &= 1.2 \\
0.05x &= 0.2
\end{aligned}
$$

The solution set is $\{4\}$.

77. Simplifying $x^2 + 4x + 4 = x^2 + 4$, we obtain $4x = 0$. The solution set is $\{0\}$.

79. Since $|2x - 3| = |2x + 5|$, we get

$$2x - 3 = 2x + 5 \quad \text{or} \quad 2x - 3 = -2x - 5.$$

Solving for x, we find $-3 = 5$ (an inconsistent equation) or $4x = -2$. The solution set is $\{-1/2\}$.

81. Multiply by 4.

$$
\begin{aligned}
2x + 4 &= x - 6 \\
x &= -10
\end{aligned}
$$

The solution set is $\{-10\}$.

83. Multiply by 30.

$$
\begin{aligned}
15(y - 3) + 6y &= 90 - 5(y + 1) \\
15y - 45 + 6y &= 90 - 5y - 5 \\
26y &= 130
\end{aligned}
$$

The solution set is $\{5\}$.

85. Since $7|x + 6| = 14$, we find

$$|x + 6| = 2.$$

Then $x + 6 = 2$ or $x + 6 = -2$.
The solution set is $\{-4, -8\}$.

87. Since $-4|2x - 3| = 0$, we get

$$|2x - 3| = 0.$$

Then $2x - 3 = 0$ and the solution set is $\{3/2\}$.

89. Since $-5|5x + 1| = 4$, we find

$$|5x + 1| = -4/5.$$

Since the absolute value is not a negative number, the solution set is $\emptyset$.

91. Multiply by $(x - 2)(x + 2)$.

$$
\begin{aligned}
3(x + 2) + 4(x - 2) &= 7x - 2 \\
3x + 6 + 4x - 8 &= 7x - 2 \\
7x - 2 &= 7x - 2
\end{aligned}
$$

An identity with solution set $\{x \mid x \neq 2, x \neq -2\}$.

93. Multiply $(x + 3)(x - 2)$ to

$$\frac{4}{x + 3} + \frac{3}{x - 2} = \frac{7x + 1}{(x + 3)(x - 2)}.$$

Then we find

$$
\begin{aligned}
4(x - 2) + 3(x + 3) &= 7x + 1 \\
4x - 8 + 3x + 9 &= 7x + 1 \\
7x + 1 &= 7x + 1.
\end{aligned}
$$

An identity and the solution set is $\{x \mid x \neq 2 \text{ and } x \neq -3\}$.

95. Multiply by $(x-3)(x-4)$.

$$
\begin{aligned}
(x-4)(x-2) &= (x-3)^2 \\
x^2 - 6x + 8 &= x^2 - 6x + 9 \\
8 &= 9
\end{aligned}
$$

An inconsistent equation and so the solution set is $\emptyset$.

97. **a)** About 1995

   b) Increasing

   c) Let $y = 0.90$. Solving for x, we find

$$
\begin{aligned}
0.90 &= 0.0102x + 0.644 \\
\frac{0.90 - 0.644}{0.0102} &= x \\
25 &\approx x.
\end{aligned}
$$

   In the year 2015 $(= 1990 + 25)$, 90% of mothers will be in the labor force.

99. Since we have

$$
B = 21{,}000 - 0.15B
$$

we obtain

$$
1.15B = 21{,}000
$$

and the bonus is

$$
B = \frac{21{,}000}{1.15} = \$18{,}260.87.
$$

101. Rewrite the left-hand side as a sum.

$$
\begin{aligned}
10{,}000 + \frac{500{,}000{,}000}{x} &= 12{,}000 \\
\frac{500{,}000{,}000}{x} &= 2{,}000 \\
500{,}000{,}000 &= 2000x \\
250{,}000 &= x
\end{aligned}
$$

Thus, $250{,}000$ vehicles must be sold.

For Thought

1. False, $P(1 + rt) = S$ implies $P = \dfrac{S}{1 + rt}$.

2. False, since the perimeter is twice the sum of the length and width.   **3.** False, since $n + 1$ and $n + 3$ are even integers if n is odd.

4. True

5. True, since $x + (-3 - x) = -3$.   **6.**   False

7. False, for if the house sells for x dollars then

$$
\begin{aligned}
x - 0.09x &= 100{,}000 \\
0.91x &= 100{,}000
\end{aligned}
$$

and the house sells for $x = \$109{,}890.11$.

8. True

9. False, a correct equation is $4(x - 2) + 5 = 3x$.

10. False, since 9 and $x + 9$ differ by x.

1.2 Exercises

1. $r = \dfrac{I}{Pt}$

3. Since $F - 32 = \dfrac{9}{5}C$, $C = \dfrac{5}{9}(F - 32)$.

5. Since $2A = bh$, we get $b = \dfrac{2A}{h}$.

7. Since $By = C - Ax$, we obtain $y = \dfrac{C - Ax}{B}$.

9. Multiplying by $RR_1R_2R_3$, we find

$$
R_1R_2R_3 = RR_2R_3 + RR_1R_3 + RR_1R_2
$$

$$
\begin{aligned}
R_1R_2R_3 - RR_1R_3 - RR_1R_2 &= RR_2R_3 \\
R_1(R_2R_3 - RR_3 - RR_2) &= RR_2R_3.
\end{aligned}
$$

   Then $R_1 = \dfrac{RR_2R_3}{R_2R_3 - RR_3 - RR_2}$.

11. Since $a_n - a_1 = (n-1)d$, we obtain

$$
\begin{aligned}
n - 1 &= \frac{a_n - a_1}{d} \\
n &= \frac{a_n - a_1}{d} + 1 \\
n &= \frac{a_n - a_1 + d}{d}.
\end{aligned}
$$

13. Since $S = \dfrac{a_1(1 - r^n)}{1 - r}$, we obtain

$$
\begin{aligned}
a_1(1 - r^n) &= S(1 - r) \\
a_1 &= \frac{S(1 - r)}{(1 - r^n)}.
\end{aligned}
$$

15. Multiplying by 2.37, one obtains

$$2.4(2.37) = L + 2D - F\sqrt{S}$$
$$5.688 - L + F\sqrt{S} = 2D$$

and $D = \dfrac{5.688 - L + F\sqrt{S}}{2}$.

17. By using the formula $I = Prt$, one gets

$$51.30 = 950r \cdot 1$$
$$0.054 = r.$$

The simple interest rate is 5.4%.

19. Since $D = RT$, we find

$$5570 = 2228 \cdot T$$
$$2.5 = T.$$

and the surveillance takes 2.5 hours.

21. Note, that we have

$$C = \frac{5}{9}(F - 32).$$

If $F = 23^o F$, then

$$C = \frac{5}{9}(23 - 32) = -5^oC.$$

23. If x is the cost of the car before taxes, then

$$1.08x = 40,230$$
$$x = \$37,250.$$

25. Let S be the saddle height and let L be the inside measurement.

$$S = 1.09L$$
$$37 = 1.09L$$
$$\frac{37}{1.09} = L$$
$$33.9 \approx L$$

The inside leg measurement is 33.9 inches.

27. Let x be the amount of her game-show winnings.

$$0.14\frac{x}{3} + 0.12\frac{x}{6} = 4000$$
$$6\left(0.14\frac{x}{3} + 0.12\frac{x}{6}\right) = 24000$$
$$0.28x + 0.12x = 24000$$
$$0.40x = 24000$$
$$x = \$60,000.$$

Her winnings is $60,000.

29. If x is the length of the shorter piece in feet, then the length of the longer side is $2x + 2$. Then we obtain

$$x + x + (2x + 2) = 30$$
$$4x = 28$$
$$x = 7.$$

The length of each shorter piece is 7 ft and the longer piece is $(2 \cdot 7 + 2)$ or 16 ft.

31. If x is the length of the side of the larger square lot then $2x$ is the amount of fencing needed to divide the square lot into four smaller lots. The solution to

$$4x + 2x = 480$$

is $x = 80$. The side of the larger square lot is 80 feet and its area is 6400 ft^2.

33. Note, Bobby will complete the remaining 8 laps in $\dfrac{8}{90}$ of an hour. If Ricky is to finish at the same time as Bobby, then Ricky's average speed s over 10 laps must satisfy

$$\frac{10}{s} = \frac{8}{90}.$$

$\left(\text{Note: } time = \dfrac{distance}{speed}\right).$

The above equation is equivalent to

$$900 = 8s.$$

Thus, Ricky's average speed is 112.5 mph.

35. Let d be the halfway distance between San Antonio and El Paso, and let s be the speed in the last half of the trip. Junior took $\dfrac{d}{80}$ hours to get to the halfway point and the last half took $\dfrac{d}{s}$ hours to drive. Since the total distance is $2d$ and $distance = rate \times time$,

$$2d = 60\left(\frac{d}{80} + \frac{d}{s}\right)$$
$$160sd = 60\left(sd + 80d\right)$$
$$160sd = 60sd + 4800d$$
$$100sd = 4800d$$
$$100d(s - 48) = 0.$$

Since $d \neq 0$, the speed for the last half of the trip was $s = 48$ mph.

37. If x is the part of the start-up capital invested at 5% and $x + 10,000$ is the part invested at 6%, then

$$
\begin{aligned}
0.05x + 0.06(x + 10,000) &= 5880 \\
0.11x + 600 &= 5880 \\
0.11x &= 5280 \\
x &= 48,000.
\end{aligned}
$$

Norma invested $48,000 at 5% and $58,000 at 6% for a total start-up capital of $106,000.

39. Let x and $1500 - x$ be the number of employees from the Northside and Southside, respectively. Then

$$
\begin{aligned}
(0.05)x + 0.80(1500 - x) &= 750 \\
0.05x + 1200 - 0.80x &= 750 \\
450 &= 0.75x \\
600 &= x.
\end{aligned}
$$

There were 600 and 900 employees at the Northside and Southside, respectively.

41. Let x be the number of hours it takes both combines working together to harvest an entire wheat crop.

	rate
old	1/72
new	1/48
combined	1/x

Then we obtain

$$
\frac{1}{72} + \frac{1}{48} = \frac{1}{x}.
$$

Multiply both sides by $144x$ and get

$$
2x + 3x = 144.
$$

The solution is $x = 28.8$ hr which is the time it takes both combines to harvest the entire wheat crop.

43. Let t be the number of hours since 8:00 a.m.

	rate	time	work completed
Batman	1/8	$t - 2$	$(t-2)/8$
Robin	1/12	t	$t/12$

$$
\begin{aligned}
\frac{t-2}{8} + \frac{t}{12} &= 1 \\
24\left(\frac{t-2}{8} + \frac{t}{12}\right) &= 24 \\
3(t - 2) + 2t &= 24 \\
5t - 6 &= 24 \\
t &= 6
\end{aligned}
$$

At 2 p.m., all the crime have been cleaned up.

45. Since there are 5280 feet to a mile and the circumference of a circle is $C = 2\pi r$,

the radius r of the race track is

$$
r = \frac{5280}{2\pi}.
$$

Since the length of a side of the square plot is twice the radius, the area of the plot is

$$
\left(2 \cdot \frac{5280}{2\pi}\right)^2 \approx 2,824,677.3 \text{ ft}^2.
$$

Dividing this number by $43,560$ results to

$$
64.85 \quad \text{acres}
$$

which is the acreage of the square lot.

47. The area of a trapezoid is $A = \frac{1}{2}h(b_1 + b_2)$.

$$
\begin{aligned}
90,000 &= \frac{1}{2}h(500 + 300) \\
90,000 &= 400h \\
225 &= h
\end{aligned}
$$

Thus, the streets are 225 ft apart.

49. Since the volume of a circular cylinder is $V = \pi r^2 h$, we have

$$
\frac{22,000}{7.5} = \pi 15^2 \cdot h.
$$

Solving for h, we get $h = 4.15$ ft, the depth of water in the pool.

51. Let r be the radius of the semicircular turns. Since the circumference of a circle is given by $C = 2\pi r$, we have

$$514 = 2\pi r + 200.$$

Solving for r, we get

$$r = \frac{157}{\pi} \approx 49.9747 \text{ m}.$$

Note, the width of the rectangular lot is $2r$. Then the dimension of the rectangular lot is 99.9494 m by 199.9494 m; its area is $19,984.82$ m^2, which is equivalent to 1.998 hectares.

53. Let x be Lorinda's taxable income.

$$
\begin{aligned}
14,325 + 0.28(x - 70,350) &= 17,167 \\
0.28x - 19,698 &= 2,842 \\
0.28x &= 22,540 \\
x &\approx 80,500.
\end{aligned}
$$

Lorinda's taxable income is about \$80,500.

55. Let x be the amount of water to be added. The volume of the resulting solution is $4 + x$ liters and the amount of pure baneberry in it is $0.05(4)$ liters. Since the resulting solution is a 3% extract, we have

$$
\begin{aligned}
0.03(4 + x) &= 0.05(4) \\
0.12 + 0.03x &= 0.20 \\
x &= \frac{0.08}{0.03} \\
x &= \frac{8}{3}.
\end{aligned}
$$

The amount of water to be added is $\frac{8}{3}$ liters.

57. The costs of x pounds of dried apples is $(1.20)4x$ and the cost of $(20 - x)$ pounds of dried apricots is $4(1.80)(20 - x)$. Since the 20 lb-mixture costs \$1.68 per quarter-pound, we obtain

$$
\begin{aligned}
4(1.68)(20) &= (1.20)4x + 4(1.80)(20 - x) \\
134.4 &= 4.80x + 144 - 7.20x \\
2.40x &= 9.6 \\
x &= 4.
\end{aligned}
$$

The mix needs 4 lb of dried apples and 16 lb of dried apricots.

59. Let x and $8 - x$ be the number of dimes and nickels, respectively. Since the candy bar costs 55 cents, we have

$$55 = 10x + 5(8 - x).$$

Solving for x, we find $x = 3$. Thus, Dana has 3 dimes and 5 nickels.

61. Let x be the amount of water needed. The volume of the resulting solution is $200 + x$ liters and the amount of active ingredient in it is $0.4(200)$ or 80 liters. Since the resulting solution is a 25% extract, we have

$$
\begin{aligned}
0.25(200 + x) &= 80 \\
200 + x &= 320 \\
x &= 120.
\end{aligned}
$$

The amount of water to be added is 120 liters.

63. Let x be the number of gallons of the stronger solution. The amount of salt in the new solution is $5(0.2) + x(0.5)$ or $(1 + 0.5x)$ lb, and the volume of new solution is $(5 + x)$ gallons. Since the new solution contains 0.3 lb of salt per gallon, we obtain

$$
\begin{aligned}
0.30(5 + x) &= 1 + 0.5x \\
1.5 + 0.3x &= 1 + 0.5x \\
0.5 &= 0.2x.
\end{aligned}
$$

Then $x = 2.5$ gallons, the required amount of the stronger solution.

65. Let x be the number of hours it takes both pumps to drain the pool simultaneously.

	The part drained in 1 hr
Together	$1/x$
Large pump	$1/5$
Small pump	$1/8$

It follows that

$$\frac{1}{5} + \frac{1}{8} = \frac{1}{x}.$$

Multiplying both sides by $40x$, we find

$$8x + 5x = 40.$$

Then $x = 40/13$ hr, or about 3 hr and 5 min.

67. a) About 1992

b) Since the revenues are equal, we obtain

$$
\begin{aligned}
13.5n + 190 &= 7.5n + 225 \\
6n &= 35 \\
n &\approx 5.8.
\end{aligned}
$$

In the year 1992 (=1986+6), restaurant revenues and supermarket revenues were the same.

69. If h is the number of hours it will take two hikers to pick a gallon of wild berries, then

$$
\begin{aligned}
\frac{1}{2} + \frac{1}{2} &= \frac{1}{h} \\
1 &= \frac{1}{h} \\
1 &= h.
\end{aligned}
$$

Two hikers can pick a gallon of wild berries in 1 hr.

If m is the number of minutes it will take two mechanics to change the oil of a Saturn, then

$$
\begin{aligned}
\frac{1}{6} + \frac{1}{6} &= \frac{1}{m} \\
\frac{1}{3} &= \frac{1}{m} \\
m &= 3.
\end{aligned}
$$

Two mechanics can change the oil in 3 minutes.

If w is the number of minutes it will take 60 mechanics to change the oil, then

$$
\begin{aligned}
60 \cdot \frac{1}{6} &= \frac{1}{w} \\
w &= \frac{1}{10} \text{ min} \\
w &= 6 \text{ sec.}
\end{aligned}
$$

So, 60 mechanics working together can change the oil in 6 sec (an unreasonable situation and answer).

For Thought

1. False, the point $(2, -3)$ is in Quadrant IV.

2. False, the point $(4, 0)$ does not belong to any quadrant.

3. False, since the distance is $\sqrt{(a - c)^2 + (b - d)^2}$.

4. False, since $Ax + By = C$ is a linear equation.

5. True, since the x-intercept can be obtained by replacing y by 0.

6. False, since $\sqrt{7^2 + 9^2} = \sqrt{130} \approx 11.4$

7. True

8. True

9. True

10. False, it is a circle of radius $\sqrt{5}$.

1.3 Exercises

1. $(4, 1)$, Quadrant I

3. $(1, 0)$, x-axis

5. $(5, -1)$, Quadrant IV

7. $(-4, -2)$, Quadrant III

9. $(-2, 4)$, Quadrant II

11. Distance is $\sqrt{(4 - 1)^2 + (7 - 3)^2} = \sqrt{9 + 16} = \sqrt{25} = 5$, midpoint is $(2.5, 5)$

13. Distance is $\sqrt{(-1 - 1)^2 + (-2 - 0)^2} = \sqrt{4 + 4} = 2\sqrt{2}$, midpoint is $(0, -1)$

15. Distance is $\sqrt{(12 - 5)^2 + (-11 - 13)^2} = \sqrt{49 + 576} = \sqrt{625} = 25$, and the midpoint is $\left(\dfrac{12 + 5}{2}, \dfrac{-11 + 13}{2}\right) = \left(\dfrac{17}{2}, 1\right)$

17. Distance is $\sqrt{(-1 + 3\sqrt{3} - (-1))^2 + (4 - 1)^2} = \sqrt{27 + 9} = 6$, midpoint is $\left(\dfrac{-2 + 3\sqrt{3}}{2}, \dfrac{5}{2}\right)$

19. Distance is $\sqrt{(1.2 + 3.8)^2 + (4.4 + 2.2)^2} = \sqrt{25 + 49} = \sqrt{74}$, midpoint is $(-1.3, 1.3)$

21. The distance is

$$\sqrt{(a-b)^2 + 0} = |a-b|$$

and the midpoint is

$$\left(\frac{a+b}{2}, 0\right)$$

23. The distance is

$$\frac{\sqrt{\pi^2 + 4}}{2}$$

and the midpoint is

$$\left(\frac{3\pi}{4}, \frac{1}{2}\right)$$

25. Center$(0,0)$, radius 4

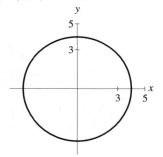

27. Center $(-6, 0)$, radius 6

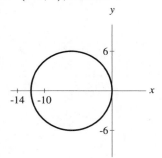

29. Center $(-1, 0)$, radius 5

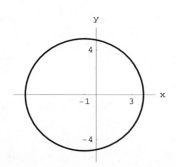

31. Center $(2, -2)$, radius $2\sqrt{2}$

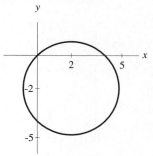

33. $x^2 + y^2 = 49$

35. $(x+2)^2 + (y-5)^2 = 1/4$

37. The distance between $(3,5)$ and the origin is $\sqrt{34}$ which is the radius. The standard equation is

$$(x-3)^2 + (y-5)^2 = 34.$$

39. The distance between $(5, -1)$ and $(1, 3)$ is $\sqrt{32}$ which is the radius. The standard equation is

$$(x-5)^2 + (y+1)^2 = 32.$$

41. Center $(0, 0)$, radius 3

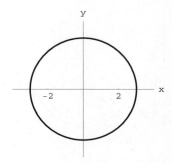

43. Completing the square, we have

$$
\begin{aligned}
x^2 + (y^2 + 6y + 9) &= 0 + 9 \\
x^2 + (y+3)^2 &= 9.
\end{aligned}
$$

The center is $(0, -3)$ and the radius is 3.

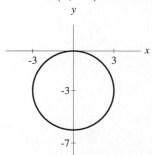

45. Completing the square, we obtain

$$
\begin{aligned}
(x^2 + 6x + 9) + (y^2 + 8y + 16) &= 9 + 16 \\
(x+3)^2 + (y+4)^2 &= 25.
\end{aligned}
$$

The center is $(-3, -4)$ and the radius is 5.

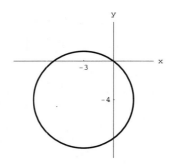

47. Completing the square, we find

$$
\begin{aligned}
\left(x^2 - 3x + \frac{9}{4}\right) + (y^2 + 2y + 1) &= \frac{3}{4} + \frac{9}{4} + 1 \\
\left(x - \frac{3}{2}\right)^2 + (y+1)^2 &= 4.
\end{aligned}
$$

The center is $\left(\frac{3}{2}, -1\right)$ and the radius is 2.

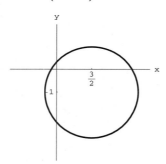

49. Completing the square, we obtain

$$
\begin{aligned}
(x^2 - 6x + 9) + (y^2 - 8y + 16) &= 9 + 16 \\
(x-3)^2 + (y-4)^2 &= 25.
\end{aligned}
$$

The center is $(3, 4)$ and the radius is 5.

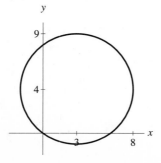

51. Completing the square, we obtain

$$
\begin{aligned}
(x^2 - 4x + 4) + \left(y^2 - 3y + \frac{9}{4}\right) &= 4 + \frac{9}{4} \\
(x-2)^2 + \left(y - \frac{3}{2}\right)^2 &= \frac{25}{4}.
\end{aligned}
$$

The center is $\left(2, \frac{3}{2}\right)$ and the radius is $\frac{5}{2}$.

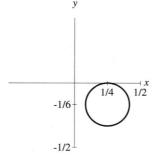

53. Completing the square, we obtain

$$
\begin{aligned}
\left(x^2 - \frac{1}{2}x + \frac{1}{16}\right) + \left(y^2 + \frac{1}{3}y + \frac{1}{36}\right) &= \frac{1}{36} \\
\left(x - \frac{1}{4}\right)^2 + \left(y + \frac{1}{6}\right)^2 &= \frac{1}{36}.
\end{aligned}
$$

The center is $\left(\frac{1}{4}, -\frac{1}{6}\right)$ and the radius is $\frac{1}{6}$.

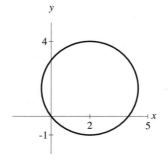

55. a. Since the center is $(0, 0)$ and the radius is 7, the standard equation is $x^2 + y^2 = 49$.

b. The radius, which is the distance between $(1, 0)$ and $(3, 4)$, is given by

$$
\sqrt{(3-1)^2 + (4-0)^2} = \sqrt{20}.
$$

Together with the center $(1, 0)$, it follows that the standard equation is

$$
(x-1)^2 + y^2 = 20.
$$

c. Using the midpoint formula, the center is

$$\left(\frac{3-1}{2}, \frac{5-1}{2}\right) = (1, 2).$$

The diameter is

$$\sqrt{(3-(-1))^2 + (5-(-1))^2} = \sqrt{52}.$$

Since the square of the radius is

$$\left(\frac{1}{2}\sqrt{52}\right)^2 = 13,$$

the standard equation is

$$(x-1)^2 + (y-2)^2 = 13.$$

57. a. Since the center is $(2, -3)$ and the radius is 2, the standard equation is

$$(x-2)^2 + (y+3)^2 = 4.$$

b. The center is $(-2, 1)$, the radius is 1, and the standard equation is

$$(x+2)^2 + (y-1)^2 = 1.$$

c. The center is $(3, -1)$, the radius is 3, and the standard equation is

$$(x-3)^2 + (y+1)^2 = 9.$$

d. The center is $(0, 0)$, the radius is 1, and the standard equation is

$$x^2 + y^2 = 1.$$

59. $y = 3x - 4$ goes through $(0, -4)$, $\left(\frac{4}{3}, 0\right)$.

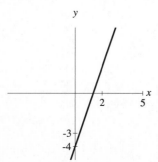

61. $3x - y = 6$ goes through $(0, -6)$, $(2, 0)$.

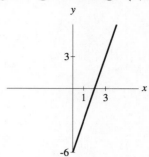

63. $x = 3y - 90$ goes through $(0, 30), (-90, 0)$.

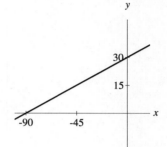

65. $\frac{2}{3}y - \frac{1}{2}x = 400$ goes through $(0, 600), (-800, 0)$.

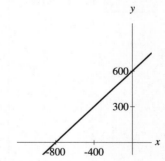

67. Intercepts are $(0, 0.0025), (0.005, 0)$.

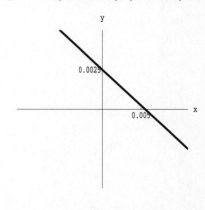

69. Intercepts are $(0, 2500), (5000, 0)$.

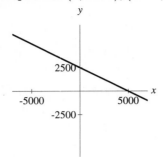

71. $x = 5$

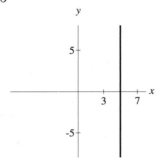

73. $y = 4$

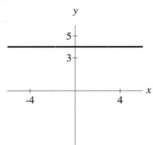

75. $x = -4$

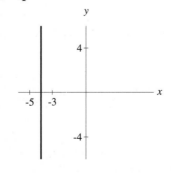

77. Solving for y, we have $y = 1$.

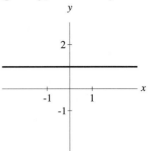

79. Since the x-intercept of
$$y = 2.4x - 8.64$$
is $(3.6, 0)$, the solution set of
$$2.4x - 8.64 = 0$$
is $\{3.6\}$.

81. Since the x-intercept of
$$y = -\frac{3}{7}x + 6$$
is $(14, 0)$, the solution set of
$$-\frac{3}{7}x + 6 = 0$$
is $\{14\}$.

83. The solution is $x = -\dfrac{3.4}{12} \approx -2.83$.

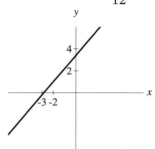

85. The solution is $\dfrac{687}{1.23} \approx 558.54$

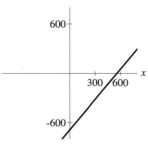

87. The solution is

$$\frac{3497}{0.03} \approx 116,566.67$$

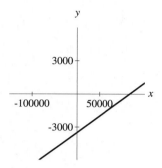

89. Note,

$$4.3 - 3.1(2.3x) + 3.1(9.9) = 0$$
$$4.3 - 7.13x + 30.69 = 0$$
$$34.99 - 7.13x = 0$$
$$x = \frac{3499}{713}$$
$$x \approx 4.91.$$

The solution set is $\{4.91\}$.

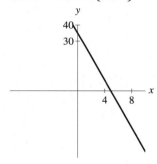

91. a) The midpoint is

$$\left(\frac{0+30}{2}, \frac{20.8+25.1}{2}\right) = (15, 22.95).$$

The median of age at first marriage in 1985 was 22.95 years.

b) The distance is

$$\sqrt{(2000-1970)^2 + (25.1-20.8)^2} \approx 30.3$$

Because of the units, the distance is meaningless.

93. Given $D = 22,800$ lbs, the graph of

$$C = \frac{4B}{\sqrt[3]{22,800}}$$

is given below.

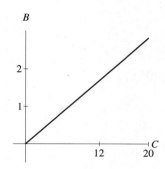

For Island Packet 40, we get

$$C = \frac{4(12 + \frac{11}{12})}{\sqrt[3]{22,800}} \approx 1.8.$$

95. From the inequalities

$$a = \frac{a+a}{2} < \frac{a+b}{2} < \frac{b+b}{2} = b$$

we conclude $a < \frac{a+b}{2} < b$.

97. Note, $AB = \sqrt{(1+4)^2 + (1+5)^2} = \sqrt{25+36} = \sqrt{61}$. Similarly, $BC = \sqrt{61}$ and $AC = \sqrt{244} = 2\sqrt{61}$. Since $AB + BC = \sqrt{61} + \sqrt{61} = 2\sqrt{61} = AC$, we conclude A, B, and C are collinear.

99. The distance between $(10, 0)$ and $(0, 0)$ is 10. The distance between $(1, 3)$ and the origin is $\sqrt{10}$.

If two points have integer coordinates, then the distance between them is of the form $\sqrt{s^2 + t^2}$ where $s^2, t^2 \in \{0, 1, 2^2, 3^2, 4^2, ...\} = \{0, 1, 4, 9, 16, ...\}$.

Note, there exists no pair s^2 and t^2 in $\{0, 1, 4, 9, 16, ...\}$ satisfying $s^2 + t^2 = 19$. Thus, one cannot find two points with integer coordinates whose distance between them is $\sqrt{19}$.

For Thought

1. False, the slope is $\dfrac{3-2}{3-2} = 1$.

2. False, the slope is $\dfrac{5-1}{-3-(-3)} = \dfrac{4}{0}$

which is undefined.

3. False, slopes of vertical lines are undefined.

4. False, it is a vertical line.   **5.**   True

6. False, $x = 1$ cannot be written in the slope-intercept form.

7. False, the slope is -2.

8. True   **9.**   False   **10.**   True

1.4 Exercises

1. $\dfrac{5-3}{4+2} = \dfrac{1}{3}$

3. $\dfrac{3+5}{1-3} = -4$

5. $\dfrac{2-2}{5+3} = 0$

7. $\dfrac{1/2 - 1/4}{1/4 - 1/8} = \dfrac{1/4}{1/8} = 2$

9. $\dfrac{3-(-1)}{5-5} = \dfrac{4}{0}$, no slope

11. The slope is $m = \dfrac{4-(-1)}{3-(-1)} = \dfrac{5}{4}$. Since $y+1 = \dfrac{5}{4}(x+1)$, we get $y = \dfrac{5}{4}x + \dfrac{5}{4} - 1$ or $y = \dfrac{5}{4}x + \dfrac{1}{4}$.

13. The slope is $m = \dfrac{-1-6}{4-(-2)} = -\dfrac{7}{6}$. Since $y + 1 = -\dfrac{7}{6}(x-4)$, we obtain $y = -\dfrac{7}{6}x + \dfrac{14}{3} - 1$ or $y = -\dfrac{7}{6}x + \dfrac{11}{3}$.

15. The slope is $m = \dfrac{5-5}{-3-3} = 0$. Since $y - 5 = 0(x-3)$, we get $y = 5$.

17. Since $m = \dfrac{12-(-3)}{4-4} = \dfrac{15}{0}$ is undefined, the equation of the vertical line is $x = 4$.

19. The slope of the line through $(0, -1)$ and $(3, 1)$ is $m = \dfrac{2}{3}$. Since the y-intercept is $(0, -1)$, the line is given by $y = \dfrac{2}{3}x - 1$.

21. The slope of the line through $(1, 4)$ and $(-1, 1)$ is $m = \dfrac{5}{2}$. Solving for y in

$$y - 1 = \dfrac{5}{2}(x+1)$$

we get $y = \dfrac{5}{2}x + \dfrac{3}{2}$.

23. The slope of the line through $(0, 4)$ and $(2, 0)$ is $m = -2$. Since the y-intercept is $(0, 4)$, the line is given by $y = -2x + 4$.

25. The slope of the line through $(1, 4)$ and $(-3, -2)$ is $m = \dfrac{3}{2}$. Solving for y in

$$y - 4 = \dfrac{3}{2}(x-1)$$

we get $y = \dfrac{3}{2}x + \dfrac{5}{2}$.

27. $y = \dfrac{3}{5}x - 2$, slope is $\dfrac{3}{5}$, y-intercept is $(0, -2)$

29. Since $y - 3 = 2x - 8$, $y = 2x - 5$. The slope is 2 and y-intercept is $(0, -5)$.

31. Since $y + 1 = \dfrac{1}{2}x + \dfrac{3}{2}$, $y = \dfrac{1}{2}x + \dfrac{1}{2}$. The slope is $\dfrac{1}{2}$ and y-intercept is $\left(0, \dfrac{1}{2}\right)$.

33. Since $y = 4$, the slope is $m = 0$ and the y-intercept is $(0, 4)$.

35. Since $y - 0.4 = 0.03x - 3$, $y = 0.03x - 2.6$. The slope is 0.03 and y-intercept is $(0, -2.6)$

37. $y = \dfrac{1}{2}x - 2$ goes through the points $(0, -2), (2, -1)$, and $(4, 0)$.

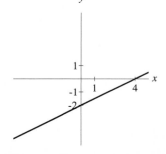

39. $y = -3x + 1$ goes through $(0, 1), (1, -2)$

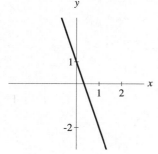

41. $y = -\dfrac{3}{4}x - 1$ goes through $(0, -1), (-4/3, 0)$

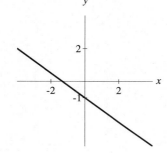

43. $x - y = 3$ goes through $(0, -3), (3, 0)$

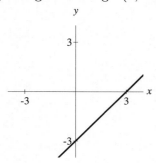

45. $y = 5$ is a horizontal line

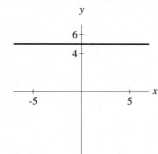

47. Since $m = \dfrac{4}{3}$ and

$$y - 0 = \frac{4}{3}(x - 3)$$

we have $4x - 3y = 12$.

49. Since $m = \dfrac{4}{5}$ and

$$y - 3 = \frac{4}{5}(x - 2)$$

we obtain $5y - 15 = 4x - 8$ and

$$4x - 5y = -7.$$

51. $x = -4$ is a vertical line.

53. Note, the slope is

$$m = \frac{\dfrac{2}{3} + 2}{2 + \dfrac{1}{2}} = \frac{8/3}{5/2} = \frac{16}{15}.$$

Using the point-slope form, we obtain a standard equation of the line using only integers.

$$
\begin{aligned}
y - \frac{2}{3} &= \frac{16}{15}(x - 2) \\
15y - 10 &= 16(x - 2) \\
15y - 10 &= 16x - 32 \\
-16x + 15y &= -22 \\
16x - 15y &= 22
\end{aligned}
$$

55. The slope is

$$m = \frac{\dfrac{1}{4} - \dfrac{1}{5}}{\dfrac{1}{2} + \dfrac{1}{3}} = \frac{1/20}{5/6} = \frac{3}{50}.$$

Using the point-slope form, we get a standard equation of the line using only integers.

$$
\begin{aligned}
y - \frac{1}{4} &= \frac{3}{50}\left(x - \frac{1}{2}\right) \\
100y - 25 &= 6\left(x - \frac{1}{2}\right) \\
100y - 25 &= 6x - 3 \\
-22 &= 6x - 100y \\
3x - 50y &= -11
\end{aligned}
$$

57. 0.5   **59.** -1 **61.**   0

63. Since $y + 2 = 2(x - 1)$, $2x - y = 4$

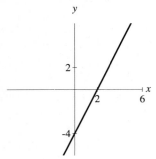

65. Since slope of $y = -3x$ is -3 and $y - 4 = -3(x - 1)$, we obtain

$$3x + y = 7.$$

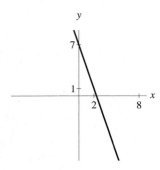

67. Since the slope of $5x - 7y = 35$ is $\dfrac{5}{7}$, we obtain

$$y - 1 = \frac{5}{7}(x - 6).$$

Multiplying by 7, we get $7y - 7 = 5x - 3$ or equivalently

$$5x - 7y = 23.$$

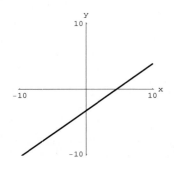

69. Since the slope of $y = \dfrac{2}{3}x + 5$ is $\dfrac{2}{3}$, we obtain $y + 3 = -\dfrac{3}{2}(x - 2)$. Multiplying by 2, we find $2y + 6 = -3x + 6$ or equivalently $3x + 2y = 0$.

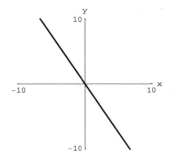

71. Since slope of $y = \dfrac{1}{2}x - \dfrac{3}{2}$ is $\dfrac{1}{2}$ and

$$y - 1 = -2(x + 3)$$

we find $2x + y = -5$.

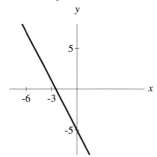

73. Since $x = 4$ is a vertical line, the horizontal line through $(2, 5)$ is $y = 5$.

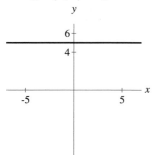

75. Since $\dfrac{5 - 3}{8 + 2} = \dfrac{1}{5} = -\dfrac{1}{a}$, we find $a = -5$.

77. Since $\dfrac{a - 3}{-2 - a} = -\dfrac{1}{2}$, we obtain $2a - 6 = 2 + a$ and $a = 8$.

79. Plot the points $A(-1,2)$, $B(2,-1)$, $C(3,3)$, and $D(-2,-2)$, respectively. The slopes of the opposite sides are $m_{AC} = m_{BD} = 1/4$ and $m_{AD} = m_{BC} = 4$. Since the opposite sides are parallel, it is a parallelogram.

81. Plot the points $A(-5,-1)$, $B(-3,-4)$, $C(3,0)$, and $D(1,3)$, respectively. The slopes of the opposite sides are $m_{AB} = m_{CD} = -3/2$ and $m_{AD} = m_{BC} = 2/3$. Since the adjacent sides are perpendicular, it is a rectangle.

83. Plot the points $A(-5,1)$, $B(-2,-3)$, and $C(4,2)$, respectively. The slopes of the sides are $m_{AB} = -4/3$, $m_{BC} = 5/6$ and $m_{AC} = 1/9$. It is not a right triangle since no two sides are perpendicular.

85. Yes, they appear to be parallel. However, they are not parallel since their slopes are not equal, i.e., $\dfrac{1}{3} \neq 0.33$.

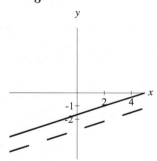

87. Since $x^3 - 8 = (x-2)(x^2 + 2x + 4)$, we obtain $\dfrac{x^3 - 8}{x^2 + 2x + 4} = x - 2$. A linear function for the graph is $y = x - 2$.

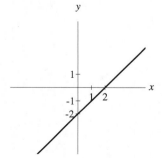

89. The slope is $\dfrac{212 - 32}{100 - 0} = \dfrac{9}{5}$.

Since $F - 32 = \dfrac{9}{5}(C - 0)$, $F = \dfrac{9}{5}C + 32$.

When $C = 150$, $F = \dfrac{9}{5}(150) + 32 = 302^o\text{F}$.

91. Thee linear function through $(1,49)$ and $(2,48)$ is $c = 50 - n$. With $n = 40$ people in a tour she would charge $10 each and make $400.

93. The slope is $\dfrac{75 - 95}{4000} = -0.005$. Since $S - 95 = -0.005(D - 0)$, we obtain $S = -0.005D + 95$.

95. Let c and p be the number of computers and printers, respectively. Since $60000 = 2000c + 1500p$, we have

$$2000c = -1500p + 60000$$
$$c = -\frac{3}{4}p + 30.$$

The slope is $-\dfrac{3}{4}$, i.e., if 4 more printers are purchased then 3 fewer computers must be bought.

97. Let $b_1 \neq b_2$. If $y = mx + b_1$ and $y = mx + b_2$ have a point (s,t) in common, then $ms + b_1 = ms + b_2$. After subtracting ms from both sides, we get $b_1 = b_2$; a contradiction. Thus, $y = mx + b_1$ and $y = mx + b_2$ have no points in common if $b_1 \neq b_2$.

99. Consider the isosceles right triangle with vertices $S(0,0)$, $T(1,2)$, and $U(3,1)$. Then the angle $\angle TUS = 45°$.

Let V be the point $(0,1)$. Note, $\angle SUV = A$ and $\angle TUV = B$. Then $A + B = 45°$.

Since C is an angle of the isosceles right triangle with vertices at $(2,0)$, $(3,0)$, and $(3,1)$, then $C = 45°$. Thus, $C = A + B$.

For Thought

1. True, a scatter diagram is a graph consisting of ordered pairs.

2. True

3. False, it is possible for the variables to have no relationship.

4. True

5. True, in fact, if $r = 1$, the data is perfectly in line.

6. True. In addition, if $r = -1$ then the data is perfectly in line.

7. False, since $r = 0.002$ is close to zero, we say that there is no positive correlation.

8. False, since $r = -0.001$ is approximatley zero, we say that there is no negative correlation.

9. False, interpolating is making a prediction within the range of the data.

10. False, exterpolating is making a prediction outside the range of the data.

1.5 Exercises

1. Linear relationship

3. No relationship

5. Nonlinear relationship

7. Linear relationship

9. Linear relationship

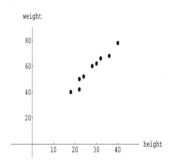

11. Linear relationship

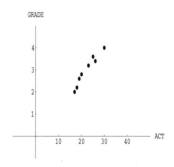

13. No relationship

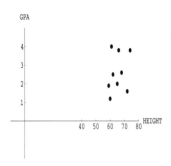

15. The missing entries are $(3.3, 160)$ and $(4.0, 193)$.

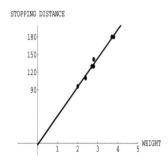

17. Missing entries are $(132, 34)$ and $(148, 25)$.

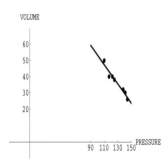

19. a) With a graphing calculator, we find
$$y = 331.5x + 4956.6$$

b) If $x = 8$, then
$$y = 331.5(8) + 4956.6 = 7608.6$$

In 2008, the number produced is $7,708,600$ cars.

c) Let $y = 0$. The solution to
$$0 = 333.5x + 4956.6$$
is $x = -\dfrac{4956.6}{3331.5} \approx -15$. Thus, no cars were produced in 1985.

21. a) With a graphing calculator, we find

$$y = 50.94285714x + 1083.809524$$

Approximately, $y = 50.9x + 1083.8$.

b) Let $y = 2000$. The solution to

$$2000 = 50.94285714x + 1083.809524$$

is $x = \dfrac{916.190476}{50.94285714} \approx 18$.

In the year 2018, the money supply will be \$2 trillion.

c) If $x = -5$, then

$$50.94285714(-5) + 1083.809524 \approx 829.$$

In 1995, the money supply was \$829 billion.

23. With a calculator, we get

$$p = -0.069A + 0.403.$$

If $A = 4$, then

$$p = -0.069(4) + 0.403 = 0.127.$$

Thus, 12.7% of the runners in age group 4 are predicted to be injured.

For Thought

1. False, since $x = 1$ is a solution of the first equation and not of the second equation.

2. False, since $x^2 + 1 = 0$ cannot be factored with real coefficients.

3. False, since

$$\left(x + \frac{2}{3}\right)^2 = x^2 + \frac{4}{3}x + \frac{9}{4}.$$

4. False, the solutions to

$$(x - 3)(2x + 5) = 0$$

are $x = 3$ and $x = -\dfrac{5}{2}$.

5. False, $x^2 = 0$ has only $x = 0$ as its solution.

6. True, since $a = 1, b = -3$, and $c = 1$, then by the quadratic formula we obtain

$$x = \frac{3 \pm \sqrt{9 - 4}}{2} = \frac{3 \pm \sqrt{5}}{2}.$$

7. False, the quadratic formula can be used to solve any quadratic equation.

8. False, $x^2 + 1 = 0$ has only imaginary zeros.

9. True, for $b^2 - 4ac = 12^2 - 4(4)(9) = 0$.

10. True, $x^2 - 6x + 9 = (x - 3)^2 = 0$ has only one real solution, namely, $x = 3$.

1.6 Exercises

1. Since $x^2 = 5$, the solution set is $\{\pm\sqrt{5}\}$.

3. Since $x^2 = -\dfrac{2}{3}$, we find $x = \pm i\dfrac{\sqrt{2}}{\sqrt{3}}$.

The solution set is $\left\{\pm i\dfrac{\sqrt{6}}{3}\right\}$.

5. Since $x - 3 = \pm 3$, we get $x = 3 \pm 3$.

The solution set is $\{0, 6\}$.

7. By the square root property, we get $3x - 1 = \pm 0 = 0$. Solving for x, we obtain $x = \dfrac{1}{3}$. The solution set is $\left\{\dfrac{1}{3}\right\}$.

9. Since $x - \dfrac{1}{2} = \pm\dfrac{5}{2}$, it follows that $x = \dfrac{1}{2} \pm \dfrac{5}{2}$. The solution set is $\{-2, 3\}$.

11. Since $x + 2 = \pm 2i$, the solution set is $\{-2 \pm 2i\}$.

13. Since $x - \dfrac{2}{3} = \pm\dfrac{2}{3}$, we get

$$x = \frac{2}{3} \pm \frac{2}{3} = \frac{4}{3}, 0.$$

The solution set is $\left\{\dfrac{4}{3}, 0\right\}$.

15. Using the factorization

$$(x - 5)(x + 4) = 0$$

we find that the solution set is $\{5, -4\}$.

17. Since we have
$$a^2 + 3a + 2 = (a+2)(a+1) = 0$$
the solution set is $\{-2, -1\}$.

19. Since we find
$$(2x+1)(x-3) = 0$$
the solution set is $\left\{-\dfrac{1}{2}, 3\right\}$.

21. Since $(2x-1)(3x-2) = 0$, the solution set is
$\left\{\dfrac{1}{2}, \dfrac{2}{3}\right\}$.

23. Note, $y^2 + y - 12 = 30$. Subtracting 30 from both sides, one obtains $y^2 + y - 42 = 0$ or $(y+7)(y-6) = 0$. The solution set is $\{-7, 6\}$.

25. $x^2 - 12x + \left(\dfrac{12}{2}\right)^2 = x^2 - 12x + 6^2 = x^2 - 12x + 36$

27. $r^2 + 3r + \left(\dfrac{3}{2}\right)^2 = r^2 + 3r + \dfrac{9}{4}$

29. $w^2 + \dfrac{1}{2}w + \left(\dfrac{1}{4}\right)^2 = w^2 + \dfrac{1}{2}w + \dfrac{1}{16}$

31. By completing the square, we derive
$$\begin{aligned}
x^2 + 6x &= -1 \\
x^2 + 6x + 9 &= -1 + 9 \\
(x+3)^2 &= 8 \\
x + 3 &= \pm 2\sqrt{2}.
\end{aligned}$$
The solution set is $\{-3 \pm 2\sqrt{2}\}$.

33. By completing the square, we find
$$\begin{aligned}
n^2 - 2n &= 1 \\
n^2 - 2n + 1 &= 1 + 1 \\
(n-1)^2 &= 2 \\
n - 1 &= \pm\sqrt{2}.
\end{aligned}$$
The solution set is $\{1 \pm \sqrt{2}\}$.

35.
$$\begin{aligned}
h^2 + 3h &= 1 \\
h^2 + 3h + \dfrac{9}{4} &= 1 + \dfrac{9}{4} \\
\left(h + \dfrac{3}{2}\right)^2 &= \dfrac{13}{4} \\
h + \dfrac{3}{2} &= \pm\dfrac{\sqrt{13}}{2}
\end{aligned}$$

The solution set is $\left\{\dfrac{-3 \pm \sqrt{13}}{2}\right\}$.

37.
$$\begin{aligned}
x^2 + \dfrac{5}{2}x &= 6 \\
x^2 + \dfrac{5}{2}x + \dfrac{25}{16} &= 6 + \dfrac{25}{16} \\
\left(x + \dfrac{5}{4}\right)^2 &= \dfrac{121}{16} \\
x &= -\dfrac{5}{4} \pm \dfrac{11}{4}
\end{aligned}$$
The solution set is $\left\{-4, \dfrac{3}{2}\right\}$.

39.
$$\begin{aligned}
x^2 + \dfrac{2}{3}x &= -\dfrac{1}{3} \\
x^2 + \dfrac{2}{3}x + \dfrac{1}{9} &= -\dfrac{3}{9} + \dfrac{1}{9} \\
\left(x + \dfrac{1}{3}\right)^2 &= -\dfrac{2}{9} \\
x &= -\dfrac{1}{3} \pm i\dfrac{\sqrt{2}}{3}
\end{aligned}$$

The solution set is $\left\{\dfrac{-1 \pm i\sqrt{2}}{3}\right\}$.

41. Since $a = 1, b = 3, c = -4$ and
$$x = \dfrac{-3 \pm \sqrt{3^2 - 4(1)(-4)}}{2(1)} = \dfrac{-3 \pm \sqrt{25}}{2} =$$
$\dfrac{-3 \pm 5}{2}$, the solution set is $\{-4, 1\}$.

43. Since $a = 2, b = -5, c = -3$ and
$$x = \dfrac{5 \pm \sqrt{(-5)^2 - 4(2)(-3)}}{2(2)} = \dfrac{5 \pm \sqrt{49}}{4} =$$
$\dfrac{5 \pm 7}{4}$, the solution set is $\left\{-\dfrac{1}{2}, 3\right\}$.

45. Since $a = 9, b = 6, c = 1$ and
$$x = \dfrac{-6 \pm \sqrt{6^2 - 4(9)(1)}}{2(9)} = \dfrac{-6 \pm 0}{18},$$
the solution set is $\left\{-\dfrac{1}{3}\right\}$.

47. Since $a = 2, b = 0, c = -3$ and
$$x = \dfrac{0 \pm \sqrt{0^2 - 4(2)(-3)}}{2(2)} = \dfrac{\pm\sqrt{24}}{4} =$$

$\pm \dfrac{2\sqrt{6}}{4}$, the solution set is $\left\{\pm \dfrac{\sqrt{6}}{2}\right\}$.

49. In $x^2 - 4x + 5 = 0$, $a = 1, b = -4, c = 5$.

Then $x = \dfrac{4 \pm \sqrt{(-4)^2 - 4(1)(5)}}{2(1)} =$

$\dfrac{4 \pm \sqrt{-4}}{2} = \dfrac{4 \pm 2i}{2}$.

The solution set is $\{2 \pm i\}$.

51. Note, $a = 1, b = -2$, and $c = 4$. Then

$$x = \dfrac{2 \pm \sqrt{4 - 16}}{2} = \dfrac{2 \pm 2i\sqrt{3}}{2}.$$

The solution set is $\left\{1 \pm i\sqrt{3}\right\}$.

53. Since $2x^2 - 2x + 5 = 0$, we find $a = 2, b = -2$,

and $c = 5$. Then

$$x = \dfrac{2 \pm \sqrt{4 - 40}}{4} = \dfrac{2 \pm 6i}{4}.$$

The solution set is

$$\left\{\dfrac{1}{2} \pm \dfrac{3}{2}i\right\}.$$

55. Since $a = 4, b = -8, c = 7$ and

$x = \dfrac{8 \pm \sqrt{64 - 112}}{8} = \dfrac{8 \pm \sqrt{-48}}{8} =$

$\dfrac{8 \pm 4i\sqrt{3}}{8}$, the solution set is $\left\{1 \pm \dfrac{\sqrt{3}}{2}i\right\}$.

57. Since $a = 3.2, b = 7.6$, and $c = -9$,

$x = \dfrac{-7.6 \pm \sqrt{(7.6)^2 - 4(3.2)(-9)}}{2(3.2)} \approx$

$\dfrac{-7.6 \pm \sqrt{172.96}}{6.4} \approx \dfrac{-7.6 \pm 13.151}{6.4}$.

The solution set is $\{-3.24, 0.87\}$.

59. Note, $a = 3.25, b = -4.6$, and $c = -22$.

Then $x = \dfrac{4.6 \pm \sqrt{(-4.6)^2 - 4(3.25)(-22)}}{2(3.25)}$

$= \dfrac{4.6 \pm \sqrt{307.16}}{6.5}$. The solution set

is $\{-1.99, 3.40\}$.

61. The discriminant is $(-30)^2 - 4(9)(25) = 900 - 900 = 0$. Only one solution and it is real.

63. The discriminant is $(-6)^2 - 4(5)(2) = 36 - 40 = -4$. There are no real solutions.

65. The discriminant is $12^2 - 4(7)(-1) = 144 + 28 = 172$. There are two distinct real solutions.

67. Note, x-intercepts are $\left(-\dfrac{2}{3}, 0\right)$ and $\left(\dfrac{1}{2}, 0\right)$.

The solution set is $\left\{-\dfrac{2}{3}, \dfrac{1}{2}\right\}$.

69. Since the x-intercepts are $(-3, 0)$ and $(5, 0)$,

the solution set is $\{-3, 5\}$.

71. Note, the graph of $y = 1.44x^2 - 8.4x + 12.25$ has exactly one x-intercept.

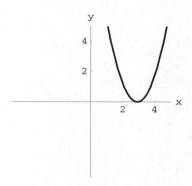

73. Note, the graph of $y = x^2 + 3x + 15$ has no x-intercept.

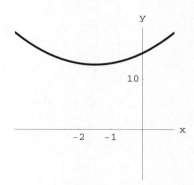

Thus, $x^2 + 3x + 15 = 0$ has no real solution.

75. The graph of $y = x^2 + 3x - 160$ has two x-intercepts.

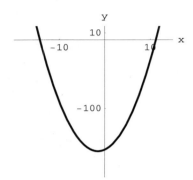

Then $x^2 + 3x - 160 = 0$ has two real solutions.

77. Set the right-hand side to 0.

$$
\begin{aligned}
x^2 - \frac{4}{3}x - \frac{5}{9} &= 0 \\
9x^2 - 12x - 5 &= 0 \\
(3x + 1)(3x - 5) &= 0
\end{aligned}
$$

The solution set is $\left\{ -\dfrac{1}{3}, \dfrac{5}{3} \right\}$.

79. Since $x^2 = \sqrt{2}$, $x = \pm\sqrt{\sqrt{2}} = \pm\sqrt[4]{2}$.

The solution set is $\left\{ \pm\sqrt[4]{2} \right\}$.

81. By the quadratic formula,

$$
x = \frac{-\sqrt{6} \pm \sqrt{(-\sqrt{6})^2 - 4(12)(-1)}}{2(12)} =
$$

$$
\frac{-\sqrt{6} \pm \sqrt{54}}{24} = \frac{-\sqrt{6} \pm 3\sqrt{6}}{24} = \frac{2\sqrt{6}}{24}, \frac{-4\sqrt{6}}{24}.
$$

The solution set is $\left\{ -\dfrac{\sqrt{6}}{6}, \dfrac{\sqrt{6}}{12} \right\}$.

83. Since $x^2 + 6x - 72 = (x + 12)(x - 6) = 0$, the solution set is $\{-12, 6\}$.

85. Multiply by x to get $x^2 = x + 1$.
So $x^2 - x - 1 = 0$ and by the quadratic formula,

$$
x = \frac{1 \pm \sqrt{1 - 4(1)(-1)}}{2} = \frac{1 \pm \sqrt{5}}{2}.
$$

The solution set is $\left\{ \dfrac{1 \pm \sqrt{5}}{2} \right\}$.

87. Multiply by x^2 to get $28x - 7 = 7x^2$.
Applying the quadratic formula to

$7x^2 - 28x + 7 = 0$, we obtain

$$
x = \frac{28 \pm \sqrt{588}}{14} = \frac{28 \pm 14\sqrt{3}}{14}.
$$

The solution set is $\left\{ 2 \pm \sqrt{3} \right\}$.

89. Multiplying by $(3 - x)(x + 7)$,

$$
\begin{aligned}
(x - 12)(x + 7) &= (x + 4)(3 - x) \\
x^2 - 5x - 84 &= -x^2 - x + 12 \\
2x^2 - 4x - 96 &= 0 \\
2(x - 8)(x + 6) &= 0
\end{aligned}
$$

the solution set is $\{8, -6\}$.

91. Multiplying by $(x + 2)(x + 3)$, we find

$$
\begin{aligned}
(x - 8)(x + 3) &= (x + 2)(2x - 1) \\
x^2 - 5x - 24 &= 2x^2 + 3x - 2 \\
0 &= 2x^2 - 8x + 22 \\
0 &= x^2 - 4x + 11 \\
-11 + 4 &= x^2 - 4x + 4 \\
-7 &= (x - 2)^2.
\end{aligned}
$$

Since the left side is not a negative number, the solution set is the empty set $\emptyset$.

93. Multiplying by $(2x + 1)(2x + 3)$, we find

$$
\begin{aligned}
(2x + 3)^2 &= 8(2x + 1) \\
4x^2 + 12x + 9 &= 16x + 8 \\
4x^2 - 4x + 1 &= 0 \\
(2x - 1)^2 &= 0.
\end{aligned}
$$

Thus, the solution set is $\left\{ \dfrac{1}{2} \right\}$.

95. Since $r^2 = \dfrac{A}{\pi}$, $r = \pm\sqrt{\dfrac{A}{\pi}}$.

97. We use the quadratic formula to solve

$$
x^2 + (2k)x + 3 = 0.
$$

Since $a = 1, b = 2k$, and $c = 3$, we obtain

$$
x = \frac{-2k \pm \sqrt{(2k)^2 - 4(1)(3)}}{2(1)}
$$

$$
x = \frac{-2k \pm \sqrt{4k^2 - 12}}{2}
$$

$$x = \frac{-2k \pm \sqrt{4\left(k^2 - 3\right)}}{2}$$

$$x = \frac{-2k \pm 2\sqrt{k^2 - 3}}{2}$$

$$x = -k \pm \sqrt{k^2 - 3}.$$

99. We use the quadratic formula to solve

$$2y^2 + (4x)y - x^2 = 0.$$

Since $a = 2, b = 4x$, and $c = -x^2$, we obtain

$$y = \frac{-4x \pm \sqrt{(4x)^2 - 4(2)(-x^2)}}{2(2)} =$$

$$\frac{-4x \pm \sqrt{16x^2 + 8x^2}}{4} = \frac{-4x \pm \sqrt{24x^2}}{4}$$

$$\frac{-4x \pm 2|x|\sqrt{6}}{4} = x\left(-1 \pm \frac{\sqrt{6}}{2}\right)$$

and note that we used $|x| = \sqrt{x^2}$.

101. From the revenue function,

$$x(40 - 0.001x) = 175,000$$
$$40x - 0.001x^2 = 175,000$$

By applying the quadratic formula to $0.001x^2 - 40x + 175,000 = 0$, we get

$$x = \frac{40 \pm \sqrt{(40)^2 - 4(0.001)(175,000)}}{0.002}$$

$$x = \frac{40 \pm \sqrt{900}}{0.002} = \frac{40 \pm 30}{0.002}$$

$$x = 5000 \text{ or } 35,000$$

Then 5000 units or 35,000 units must be produced weekly.

103. The height S (in feet) of the ball from the ground t seconds after it was tossed is given by $S = -16t^2 + 40t + 4$. When the height is 4 feet,

$$-16t^2 + 40t + 4 = 4$$
$$-16t^2 + 40t = 0$$
$$-8t(2t - 5) = 0$$
$$t = 0, \frac{5}{2}.$$

The ball returns to a height of 4 ft in 2.5 sec.

105. Let d be the diagonal distance across the field from one goal to the other. By the Pythagorean Theorem, we obtain $d = \sqrt{300^2 + 160^2} = 340$ ft.

107. Let w and $2w + 2$ be the length and width. From the given area of the court, we obtain

$$(2w + 2)w = 312$$
$$2w^2 + 2w - 312 = 0$$
$$w^2 + w - 156 = 0$$
$$(w - 12)(w + 13) = 0$$

Then $w = 12$ yd and the length is 26 yd. The distance between two opposite corners (by the Pythagorean Theorem) is $\sqrt{12^2 + 26^2} = 2\sqrt{205} \approx 28.6$ yd.

109. Substituting the values of S and A, we find that the displacement is

$$\frac{1}{2^{12}}d^2(18.8)^3 - 822^3 = 0$$

$$\frac{1}{2^{12}}d^2(18.8)^3 = 822^3$$

$$d = \sqrt{\frac{822^3(2^{12})}{18.8^3}}$$

$$d \approx 18,503.4 \text{ lbs.}$$

111. By choosing an appropriate coordinate system, we can assume the circle is given by $(x+r)^2 + (y-r)^2 = r^2$ where $r > 0$ is the radius of the circle and $(-5, 1)$ is the common point between the block and the circle. Note, the radius is less than 5 feet. Substitute $x = -5$ and $y = 1$. Then we obtain

$$(-5 + r)^2 + (1 - r)^2 = r^2$$
$$r^2 - 10r + 25 + 1 - 2r + r^2 = r^2$$
$$r^2 - 12r + 26 = 0.$$

The solutions of the last quadratic equation are $r = 6 \pm \sqrt{10}$. Since $r < 5$, the radius of the circle is $r = 6 - \sqrt{10}$ ft.

113. Let x be the normal speed of the tortoise in ft/hr.

	distance	rate	time
hwy	24	$x + 2$	$24/(x + 2)$
off hwy	24	x	$24/x$

Since 24 minutes is 2/5 of an hour, we get

$$
\begin{aligned}
\frac{2}{5} + \frac{24}{x + 2} &= \frac{24}{x} \\
2x(x + 2) + 24(5)x &= 24(5)(x + 2) \\
2x^2 + 4x + 120x &= 120x + 240 \\
x^2 + 2x - 120 &= 0 \\
(x + 12)(x - 10) &= 0 \\
x &= -12, 10.
\end{aligned}
$$

The normal speed of the tortoise is 10 ft/hr.

115. Using $v_1^2 = v_0^2 + 2gS$ with $S = 1.07$ and $v_1 = 0$,
we find that

$$
\begin{aligned}
v_0^2 + 2(-9.8)(1.07) &= 0 \\
v_0^2 - 20.972 &= 0 \\
v_0 = \pm\sqrt{20.972} &\approx \pm 4.58.
\end{aligned}
$$

His initial upward velocity is 4.58 m/sec.

Using $S = \frac{1}{2}gt^2 + v_0 t$ with $S = 0$ and $v_o = 4.58$, we find that his time t in the air satisfies

$$
\begin{aligned}
\frac{1}{2}(-9.8)t^2 + 4.58t &= 0 \\
t(4.58 - 4.9t) &= 0 \\
t &\approx 0, 0.93.
\end{aligned}
$$

Carter is in the air for 0.93 seconds.

117. a) Let x be the number of years since 1980. With the aid of a graphing calculator, the quadratic regression curve is approximately

$$ y = -0.067x^2 + 1.26x + 51.14 $$

b) Using the regression curve in part a) and the quadratic formula, we find that the positive solution to

$$ 0 = ax^2 + bx + c $$

is

$$ x = \frac{-b - \sqrt{b^2 - 4ac}}{2a} \approx 39. $$

In the year 2019, the extrapolated birth rate will be zero.

119. Let x and $x - 2$ be the number of days it takes to design a direct mail package using traditional methods and a computer, respectively.

	rate
together	$2/7$
computer	$1/(x - 2)$
traditional	$1/x$

$$
\begin{aligned}
\frac{1}{x - 2} + \frac{1}{x} &= \frac{2}{7} \\
7x + (7x - 14) &= 2(x^2 - 2x) \\
0 &= 2x^2 - 18x + 14 \\
0 &= x^2 - 9x + 7 \\
x &= \frac{9 \pm \sqrt{81 - 28}}{2} \\
x &= \frac{9 \pm \sqrt{53}}{2} \\
x &\approx 8.14, 0.86
\end{aligned}
$$

Curt using traditional methods can do the job in 8.14 days. Note, $x \approx 0.86$ days has to be excluded since $x - 2$ is negative when $x \approx 0.86$.

121. Let x and $x - 10$ be the number of pounds of white meat in a Party Size bucket and a Big Family Size bucket, respectively. From the ratios, we obtain

$$
\begin{aligned}
\frac{8}{x} &= \frac{3}{x - 10} + 0.10 \\
8(x - 10) &= 3x + 0.10x(x - 10) \\
0 &= 0.10x^2 - 6x + 80 \\
0 &= x^2 - 60x + 800 \\
0 &= (x - 40)(x - 20) \\
x &= 40, 20
\end{aligned}
$$

A Party Size bucket weighs 20 or 40 lbs.

For Thought

1. True

2. False, since $-2x < -6$ is equivalent to

$$\frac{-2x}{-2} > \frac{-6}{-2}.$$

3. False, since there is a number between any two distinct real numbers.

4. True, since $|-6-6| = |-12| = 12 > -1$.

5. False, $(-\infty, -3) \cap (-\infty, -2) = (-\infty, -3)$.

6. False, $(5, \infty) \cap (-\infty, -3) = \phi$.

7. False, no real number satisfies $|x - 2| < 0$.

8. False, it is equivalent to $|x| > 3$.

9. False, $|x| + 2 < 5$ is equivalent to $-3 < x < 3$.

10. True

1.7 Exercises

1. $x < 12$

3. $x \geq -7$

5. $[-8, \infty)$

7. $(-\infty, \pi/2)$

9. Since $3x > 15$ implies $x > 5$, the solution set is $(5, \infty)$ and the graph is

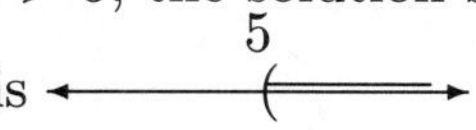

11. Since $10 \leq 5x$ implies $2 \leq x$, the solution set is $[2, \infty)$ and the graph is

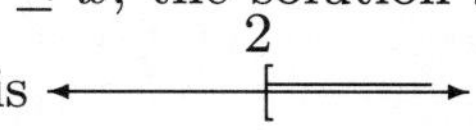

13. Multiply 6 to both sides of the inequality.

$$\begin{aligned} 3x - 24 &< 2x + 30 \\ x &< 54 \end{aligned}$$

The solution is the interval $(-\infty, 54)$ and the graph is

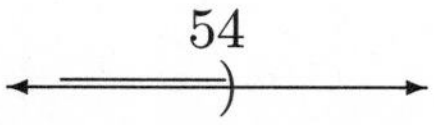

15. Multiplying the inequality by 2, we find

$$\begin{aligned} 7 - 3x &\geq -6 \\ 13 &\geq 3x \\ 13/3 &\geq x. \end{aligned}$$

The solution is the interval $(-\infty, 13/3]$ and the graph is

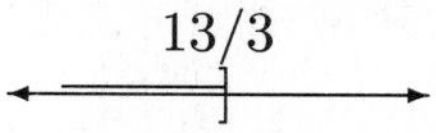

17. Multiply the inequality by -5 and reverse the direction of the inequality.

$$\begin{aligned} 2x - 3 &\leq 0 \\ 2x &\leq 3 \\ x &\leq \frac{3}{2} \end{aligned}$$

The solution is the interval $(-\infty, 3/2]$ and the graph is

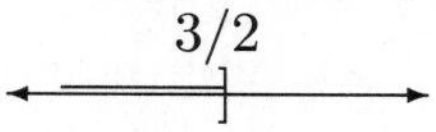

19. Multiply the left-hand side.

$$\begin{aligned} -6x + 4 &\geq 4 - x \\ 0 &\geq 5x \\ 0 &\geq x. \end{aligned}$$

The solution is the interval $(-\infty, 0]$ and the graph is

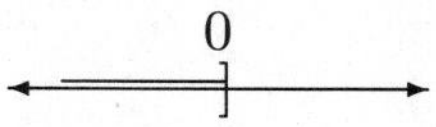

21. Using the portion of the graph below the x-axis, the solution set is $(-\infty, -3.5)$.

23. Using the part of the graph on or above the x-axis, the solution set is $(-\infty, 1.4]$.

25. By taking the part of the line $y = 2x - 3$ above the horizontal line $y = 5$ and by using $(4, 5)$, the solution set is $(4, \infty)$.

27. Note, the graph of $y = -3x - 7$ is above or on the graph of $y = x + 1$ for $x \leq -2$. Thus, the solution set is $(-\infty, -2]$.

29. $(-3, \infty)$

31. $(-3, \infty)$

33. $(-5, -2)$

35. ϕ

37. $(-\infty, 5]$

39. Solve each simple inequality and find the intersection of their solution sets.

$$x > 3 \quad \text{and} \quad 0.5x < 3$$
$$x > 3 \quad \text{and} \quad x < 6$$

The intersection of these values of x is the interval $(3, 6)$ and whose graph is

3   6

41. Solve each simple inequality and find the intersection of their solution sets.

$$2x - 5 > -4 \quad \text{and} \quad 2x + 1 > 0$$
$$x > \frac{1}{2} \quad \text{and} \quad x > -\frac{1}{2}$$

The intersection of these values of x is the interval $(1/2, \infty)$ and the graph is

1/2

43. Solve each simple inequality and find the union of their solution sets.

$$-6 < 2x \quad \text{or} \quad 3x > -3$$
$$-3 < x \quad \text{or} \quad x > -1$$

The union of these values of x is $(-3, \infty)$ and the graph is

-3

45. Solve each simple inequality and find the union of their solution sets.

$$x + 1 > 6 \quad \text{or} \quad x < 7$$
$$x > 5 \quad \text{or} \quad x < 7$$

The union of these values of x is $(-\infty, \infty)$ and

the graph is

47. Solve each simple inequality and find the intersection of their solution sets.

$$2 - 3x < 8 \quad \text{and} \quad x - 8 \le -12$$
$$-6 < 3x \quad \text{and} \quad x \le -4$$
$$-2 < x \quad \text{and} \quad x \le -4$$

The intersection is empty and there is no solution.

49.

$$6 \;<\; 3x \;<\; 12$$
$$2 \;<\; x \;<\; 4$$

The solution set is the interval $(2, 4)$ and the graph is

2   4

51.

$$-6 \;\le\; -6x \;<\; 18$$
$$1 \;\ge\; x \;>\; -3$$

The solution set is the interval $(-3, 1]$ and the graph is

-3   1

53. Solve an equivalent compound inequality.

$$-2 <\; 3x - 1 \;< 2$$
$$-1 <\; 3x \;< 3$$
$$-\frac{1}{3} <\; x \;< 1$$

The solution set is the interval $(-1/3, 1)$ and

the graph is   -1/3   1

55. Solve an equivalent compound inequality.

$$-1 \le\; 5 - 4x \;\le 1$$
$$-6 \le\; -4x \;\le -4$$
$$\frac{3}{2} \ge\; x \;\ge 1$$

The solution set is the interval $[1, 3/2]$ and

the graph is   1   3/2

57. Solve an equivalent compound inequality.

$$x - 1 \ge 1 \quad \text{or} \quad x - 1 \le -1$$
$$x \ge 2 \quad \text{or} \quad x \le 0$$

The solution set is $(-\infty, 0] \cup [2, \infty)$ and the

graph is   0   2

59. Solve an equivalent compound inequality.

$$5 - x > 3 \quad \text{or} \quad 5 - x < -3$$
$$2 > x \quad \text{or} \quad 8 < x$$

The solution set is $(-\infty, 2) \cup (8, \infty)$ and the

graph is   2   8

61. Solve an equivalent compound inequality.

$$\begin{array}{rcl} -5 \le & 4-x & \le 5 \\ -9 \le & -x & \le 1 \\ 9 \ge & x & \ge -1 \end{array}$$

The solution set is the interval $[-1, 9]$ and the graph is

$$\xleftarrow{\quad\underset{-1}{[}\!=\!=\!=\!\underset{9}{]}\quad}\rightarrow$$

63. No solution since an absolute value is never negative.

65. No solution since an absolute value is never negative.

67. Note, $3|x-2| > 3$ or $|x-2| > 1$.
We solve an equivalent compound inequality.

$$\begin{array}{ccc} x-2 > 1 & \text{or} & x-2 < -1 \\ x > 3 & \text{or} & x < 1 \end{array}$$

The solution set is $(-\infty, 1) \cup (3, \infty)$ and the graph is

$$\xleftarrow{=\!=\!\underset{1}{)}\quad\underset{3}{(}\!=\!=}\rightarrow$$

69. Solve an equivalent compound inequality.

$$\begin{array}{ccc} \dfrac{x-3}{2} > 1 & \text{or} & \dfrac{x-3}{2} < -1 \\ x-3 > 2 & \text{or} & x-3 < -2 \\ x > 5 & \text{or} & x < 1 \end{array}$$

The solution set is the interval $(-\infty, 1) \cup (5, \infty)$

and the graph is

$$\xleftarrow{=\!=\!\underset{1}{)}\quad\underset{5}{(}\!=\!=}\rightarrow$$

71. $|x| < 5$

73. $|x| > 3$

75. Since 6 is the midpoint of 4 and 8, the inequality is $|x-6| < 2$.

77. Since 4 is the midpoint of 3 and 5, the inequality is $|x-4| > 1$.

79. $|x| \ge 9$

81. Since 7 is the midpoint, the inequality is $|x-7| \le 4$.

83. Since 5 is the midpoint, the inequality is $|x-5| > 2$.

85. Since $x - 2 \ge 0$, the solution set is $[2, \infty)$.

87. Since $2 - x > 0$ is equivalent to $2 > x$, the solution set is $(-\infty, 2)$.

89. Since $|x| \ge 3$ is equivalent to $x \ge 3$ or $x \le -3$, the solution set is $(-\infty, -3] \cup [3, \infty)$.

91. If x is the price of a car excluding sales tax then it must satisfy $0 \le 1.1x + 300 \le 8000$.

This is equivalent to $0 \le x \le \dfrac{7700}{1.1} = 7000$.

The price range of Yolanda's car is the interval $[\$0, \$7000]$.

93. Let x be Lucky's score on the final exam.

$$\begin{array}{rcl} 79 < & \dfrac{65+x}{2} & < 90 \\ 158 < & 65+x & < 180 \\ 93 < & x & < 115. \end{array}$$

The final exam score must lie in $(93, 115)$.

95. Let x be Ingrid's final exam score. Since $\dfrac{2x+65}{3}$ is her weighted average, we obtain

$$\begin{array}{rcl} 79 < & \dfrac{2x+65}{3} & < 90 \\ 237 < & 2x+65 & < 270 \\ 172 < & 2x & < 205 \\ 86 < & x & < 102.5 \end{array}$$

Ingrid's final exam score must lie in $(86, 102.5)$.

97. If h is the height of the box, then

$$\begin{array}{rcl} 40 + 2(30) + 2h & \le & 130 \\ 100 + 2h & \le & 130 \\ 2h & \le & 30. \end{array}$$

The range of the height is $(0 \text{ in.}, 15 \text{ in.}]$.

99. By substituting $N = 50$ and $w = 27$ into $r = \dfrac{Nw}{n}$ we find $r = \dfrac{1350}{n}$. Moreover if $n = 14$, then $r = \dfrac{1350}{14} = 96.4 \approx 96$.

Similarly, the other gear ratios are the following.

n	14	17	20	24	29
r	96	79	68	56	47

Yes, the bicycle has a gear ratio for each of the four types.

101. If x is the price of a BMW 760 Li, then $|x - 74,595| > 25,000$. An equivalent inequality is

$$x - 74,595 > 25,000 \quad \text{or} \quad x - 74,595 < -25,000$$
$$x > 99,595 \quad \text{or} \quad x < 49,595.$$

The price of a BMW 760 Li is either under $\$49,595$ or over $\$99,595$.

103. If x is the actual temperature, then

$$\left| \frac{x - 35}{35} \right| < .01$$

$$-.35 < x - 35 < .35$$
$$34.65 < x < 35.35.$$

The actual temperature must lie in the interval $(34.65°, 35.35°)$.

105. If c is the actual circumference, then $c = \pi d$ and

$$|\pi d - 7.2| \leq 0.1$$
$$-0.1 \leq \pi d - 7.2 \leq 0.1$$
$$7.1 \leq \pi d \leq 7.3$$
$$2.26 \leq d \leq 2.32.$$

The actual diameter must lie in the interval [2.26 cm, 2.32 cm].

107. a) The inequality $|a - 31,632| < 2000$ is equivalent to

$$-2000 < a - 31,632 < 2000$$
$$29,632 < a < 33,632.$$

The states within this range are Alaska, Florida, and Hawaii.

b) The inequality $|a - 31,632| > 3000$ is equivalent to

$$x - 31,632 > 3000 \quad \text{or} \quad x - 31,632 < -3000$$
$$x > 34,632 \quad \text{or} \quad x < 28,632.$$

The states satisfying the inequality are Alabama, Maryland, New Jersey, and Connecticut.

Chapter 1 Review Exercises

1. Since $3x = 2$, the solution set is $\{2/3\}$.

3. Multiply by 60 to get $30y - 20 = 15y + 12$, or $15y = 32$. The solution set is

$$\{32/15\}.$$

5. Multiply by $x(x - 1)$ to get $2x - 2 = 3x$. The solution set is

$$\{-2\}.$$

7. Multiply by $(x + 1)(x - 3)$ and get $-2x - 3 = x - 2$. Then $-1 = 3x$. The solution set is

$$\{-1/3\}.$$

9. The distance is $\sqrt{(-3 - 2)^2 + (5 - (-6))^2} = \sqrt{(-5)^2 + 11^2} = \sqrt{25 + 121} = \sqrt{146}$. The midpoint is $\left(\dfrac{-3 + 2}{2}, \dfrac{5 - 6}{2} \right) = \left(-\dfrac{1}{2}, -\dfrac{1}{2} \right)$.

11. Distance is $\sqrt{\left(\dfrac{1}{2} - \dfrac{1}{4} \right)^2 + \left(\dfrac{1}{3} - 1 \right)^2} = \sqrt{\left(\dfrac{1}{4} \right)^2 + \left(-\dfrac{2}{3} \right)^2} = \sqrt{\dfrac{1}{16} + \dfrac{4}{9}} = \sqrt{\dfrac{73}{144}} = \dfrac{\sqrt{73}}{12}$. Midpoint is $\left(\dfrac{1/2 + 1/4}{2}, \dfrac{1/3 + 1}{2} \right) = \left(\dfrac{3/4}{2}, \dfrac{4/3}{2} \right) = \left(\dfrac{3}{8}, \dfrac{2}{3} \right)$.

13. Circle with radius 5 and center at the origin.

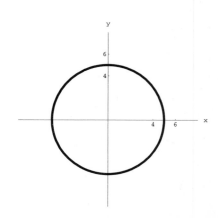

15. Equivalently, by using the method of completing the square, the circle is given by $(x+2)^2 + y^2 = 4$. It has radius 2 and center $(-2, 0)$.

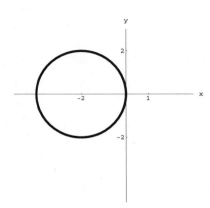

17. The line $y = -x + 25$ has intercepts $(0, 25)$, $(25, 0)$.

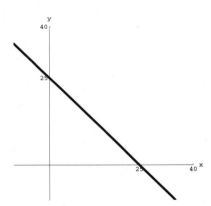

19. The line $y = 3x - 4$ has intercepts $(0, -4)$, $(4/3, 0)$.

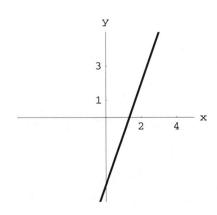

21. Vertical line $x = 5$ has intercept $(5, 0)$.

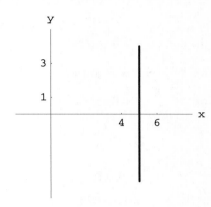

23. Simplify $(x - (-3))^2 + (y - 5)^2 = \left(\sqrt{3}\right)^2$. The standard equation is $(x+3)^2 + (y-5)^2 = 3$.

25. Substitute $y = 0$ in $3x - 4y = 12$. Then $3x = 12$ or $x = 4$. The x-intercept is $(4, 0)$. Substitute $x = 0$ in $3x - 4y = 12$ to get $-4y = 12$ or $y = -3$. The y-intercept is $(0, -3)$.

27. $\dfrac{2 - (-6)}{-1 - 3} = \dfrac{8}{-4} = -2$

29. Note, $m = \dfrac{-1 - 3}{5 - (-2)} = -\dfrac{4}{7}$. Solving for y in $y - 3 = -\dfrac{4}{7}(x + 2)$, we obtain $y = -\dfrac{4}{7}x + \dfrac{13}{7}$.

31. Note, the slope of $3x + y = -5$ is -3. The standard form for the line through $(2, -4)$ with slope $\dfrac{1}{3}$ is derived below.

$$
\begin{aligned}
y + 4 &= \frac{1}{3}(x - 2) \\
3y + 12 &= x - 2 \\
-x + 3y &= -14 \\
x - 3y &= 14
\end{aligned}
$$

33. Since $2x - 6 = 3y$, $y = \dfrac{2}{3}x - 2$.

35. Note, $y(x - 3) = 1$. Then $y = \dfrac{1}{x - 3}$.

37. Note, $by = -ax + c$. Then $y = -\dfrac{a}{b}x + \dfrac{c}{b}$ provided $b \neq 0$.

39. The discriminant of $x^2 - 4x + 2$ is $(-4)^2 - 4(2) = 8$. There are two distinct real solutions.

41. The discriminant is $(-20)^2 - 4(4)(25) = 0$. Only one real solution.

43. Since $x^2 = 5$, the solution set is $\{\pm\sqrt{5}\}$.

45. Since $x^2 = -8$, the solution set is $\{\pm 2i\sqrt{2}\}$.

47. Since $x^2 = -\dfrac{2}{4}$, the solution set is $\left\{\pm i\dfrac{\sqrt{2}}{2}\right\}$.

49. Since $x - 2 = \pm\sqrt{17}$, we get $x = 2 \pm \sqrt{17}$. The solution set is $\left\{2 \pm \sqrt{17}\right\}$.

51. Since $(x + 3)(x - 4) = 0$, the solution set is $\{-3, 4\}$.

53. We apply the method of completing the square.
$$\begin{aligned} b^2 - 6b + 10 &= 0 \\ b^2 - 6b + 9 &= -1 \\ (b - 3)^2 &= -1 \\ b - 3 &= \pm i \end{aligned}$$
The solution set is $\{3 \pm i\}$.

55. We apply the method of completing the square.
$$\begin{aligned} s^2 - 4s &= -1 \\ s^2 - 4s + 4 &= -1 + 4 \\ (s - 2)^2 &= 3 \\ s - 2 &= \pm\sqrt{3} \end{aligned}$$
The solution set is $\left\{2 \pm \sqrt{3}\right\}$.

57. Use the quadratic formula to solve $4x^2 - 4x - 5 = 0$.
$$\begin{aligned} x &= \frac{4 \pm \sqrt{(-4)^2 - 4(4)(-5)}}{2(4)} \\ &= \frac{4 \pm \sqrt{96}}{8} \\ &= \frac{4 \pm 4\sqrt{6}}{8} \\ &= \frac{1 \pm \sqrt{6}}{2} \end{aligned}$$
The solution set is $\left\{\dfrac{1 \pm \sqrt{6}}{2}\right\}$.

59. Subtracting 1 from both sides, we find
$$\begin{aligned} x^2 - 2x + 1 &= -1 \\ (x - 1)^2 &= -1 \\ x - 1 &= \pm i. \end{aligned}$$
The solution set is $\{1 \pm i\}$.

61. Multiplying by $2x(x - 1)$, we obtain
$$\begin{aligned} 2(x - 1) + 2x &= 3x(x - 1) \\ 0 &= 3x^2 - 7x + 2 \\ 0 &= (x - 2)(3x - 1). \end{aligned}$$
The solution set is $\left\{\dfrac{1}{3}, 2\right\}$.

63. Solve an equivalent statement
$$\begin{aligned} 3q - 4 = 2 \quad &\text{or} \quad 3q - 4 = -2 \\ 3q = 6 \quad &\text{or} \quad 3q = 2. \end{aligned}$$
The solution set is $\{2/3, 2\}$.

65. We obtain
$$\begin{aligned} |2h - 3| &= 0 \\ 2h - 3 &= 0 \\ h = \frac{3}{2}. \end{aligned}$$
The solution set is $\left\{\dfrac{3}{2}\right\}$.

67. No solution since absolute values are nonnegative.

69. The solution set of $x > 3$ is the interval $(3, \infty)$ and the graph is

71. The solution set of $8 > 2x$ is the interval $(-\infty, 4)$ and the graph is

73. Since $-\dfrac{7}{3} > \dfrac{1}{2}x$, the solution set is $(-\infty, -14/3)$ and the graph is

75. After multiplying the inequality by 2 we have
$$\begin{aligned} -4 < x - 3 &\leq 10 \\ -1 < x &\leq 13. \end{aligned}$$
The solution set is the interval $(-1, 13]$ and the graph is

77. The solution set of $\dfrac{1}{2} < x$ and $x < 1$ is the interval $(1/2, 1)$ and the graph is

79. The solution set of $x > -4$ or $x > -1$ is the interval $(-4, \infty)$ and the graph is

81. Solving an equivalent statement, we get

$$x - 3 > 2 \quad \text{or} \quad x - 3 < -2$$
$$x > 5 \quad \text{or} \quad x < 1.$$

The solution set is $(-\infty, 1) \cup (5, \infty)$ and the graph is

83. Since an absolute value is nonnegative, $2x - 7 = 0$. The solution set is $\{7/2\}$ and the graph is

85. Since absolute values are nonnegative, the solution set is $(-\infty, \infty)$ and

the graph is

87. The solution set is $\{10\}$ since the x-intercept is $(10, 0)$.

89. Since the x-intercept is $(8, 0)$ and the y-values are negative in quadrants 3 and 4, the solution set is $(-\infty, 8)$.

91. Let x be the length of one side of the square. Since dimensions of the base are $8 - 2x$ and $11 - 2x$, we obtain

$$
\begin{aligned}
(11 - 2x)(8 - 2x) &= 50 \\
4x^2 - 38x + 38 &= 0 \\
2x^2 - 19x + 19 &= 0 \\
x = \frac{19 \pm \sqrt{209}}{4} &\approx 8.36, 1.14.
\end{aligned}
$$

But $x = 8.36$ is too big and so $x = 1.14$ inch.

93. Let x be the number of hours it takes Lisa or Taro to drive to the restaurant. Since the sum of the driving distances is 300, we obtain

$300 = 50x + 60x$. Thus, $x = \dfrac{300}{110} \approx 2.7272$ and Lisa drove $50(2.7272) \approx 136.4$ miles.

95. Let x and $8000 - x$ be the number of fish in Homer Lake and Mirror lake, respectively. Then

$$
\begin{aligned}
0.2x + 0.3(8000 - x) &= 0.28(8000) \\
-0.1x + 2400 &= 2240 \\
1600 &= x.
\end{aligned}
$$

There were originally 1600 fish in Homer Lake.

97. Let x be the distance she hiked in the northern direction. Then she hiked $32 - x$ miles in the eastern direction. By the Pythagorean Theorem, we obtain

$$
\begin{aligned}
x^2 + (32 - x)^2 &= (4\sqrt{34})^2 \\
2x^2 - 64x + 480 &= 0 \\
2 \cdot (x - 20)(x - 12) &= 0 \\
x &= 20, 12.
\end{aligned}
$$

Since the eastern direction was the shorter leg of the journey, the northern direction was 20 miles.

99. Let x and $x + 50$ be the cost of a haircut at Joe's and Renee's, respectively. Since 5 haircuts at Joe's is less than one haircut at Renee's, we have

$$5x < x + 50.$$

Thus, the price range of a haircut at Joe's is $x < \$12.50$ or $(0, \$12.50)$.

101. Let x and $x + 2$ be the length and width of a picture frame in inches, respectively. Since there are between 32 and 50 inches of molding, we get

$$
\begin{aligned}
32 &< 2x + 2(x + 2) < 50 \\
32 &< 4x + 4 < 50 \\
28 &< 4x < 46 \\
7 \text{ in.} &< x < 11.5 \text{ in.}
\end{aligned}
$$

The set of possible widths is $(7 \text{ in}, 11.5 \text{ in})$.

103. If the average gas mileage is increased from 29.5 mpg to 31.5 mpg, then the amount of gas saved is

$$\frac{10^{12}}{29.5} - \frac{10^{12}}{31.5} \approx 2.15 \times 10^9 \text{ gallons.}$$

Suppose the mileage is increased to x from 29.5 mpg. Then x must satisfy

$$\frac{10^{12}}{29.5} - \frac{10^{12}}{x} = \frac{10^{12}}{27.5} - \frac{10^{12}}{29.5}$$

$$\frac{1}{29.5} - \frac{1}{x} = \frac{1}{27.5} - \frac{1}{29.5}$$

$$-\frac{1}{x} \approx -0.031433$$

$$x \approx 31.8.$$

The mileage must be increased to 31.8 mpg.

105. Let x be the year. The equation of the line passing through $(1985, 21.5)$ and $(2005, 240)$ is

$$y = 10.925x - 21,664.625.$$

In 1993, the number of computers in use was $10.925(1993) - 21,664.625 = 108.9$ computers.

107. Let a be the age in years and p be the percentage. The equation of the line passing through $(20, 0.23)$ and $(50, 0.47)$ is

$$p = 0.008a + 0.07.$$

If $a = 65$, then $p = 0.008(65) + 0.07 \approx 0.59$. Thus, the percentage of body fat in a 65-year old woman is 59%.

109. Let C be the cost and let y be the year.

a) The regression line is $C = ay + b$ where

$$a \approx 263.35714285714$$

and

$$b \approx -521,811.42857142.$$

Approximately, the regression line is

$$y = 263.4y - 521,811.4.$$

b) If $y = 2010$, we obtain

$$C = a(201) + b \approx \$7536.$$

Chapter 1 Test

1. Since $2x - x = -6 - 1$, the solution set is $\{-7\}$.

2. Multiplying the original equation by 6, we get $3x - 2x = 1$. The solution set is $\{1\}$.

3. Since $x^2 = \frac{2}{3}$, one obtains $x = \pm\frac{\sqrt{2}}{\sqrt{3}} = \pm\frac{\sqrt{6}}{3}$.

The solution set is $\left\{\pm\frac{\sqrt{6}}{3}\right\}$.

4. By completing the square, we obtain

$$\begin{aligned} x^2 - 6x &= -1 \\ (x-3)^2 &= -1 + 9 \\ (x-3)^2 &= 8 \\ x - 3 &= \pm\sqrt{8}. \end{aligned}$$

The solution set is $\{3 \pm 2\sqrt{2}\}$.

5. Since $x^2 - 9x + 14 = (x-2)(x-7) = 0$, the solution set is $\{2, 7\}$.

6. After cross-multiplying, we get

$$\begin{aligned} (x-1)(x-6) &= (x+3)(x+2) \\ x^2 - 7x + 6 &= x^2 + 5x + 6 \\ -7x &= 5x \\ 0 &= 12x \end{aligned}$$

The solution set is $\{0\}$.

7. We use the method of completing the square.

$$\begin{aligned} x^2 - 2x &= -5 \\ (x-1)^2 &= -5 + 1 \\ (x-1)^2 &= -4 \\ x - 1 &= \pm 2i \end{aligned}$$

The solution set is $\{1 \pm 2i\}$.

8. Since $x^2 = -1$, the solution set is $\{\pm i\}$.

9. The line $3x - 4y = 120$ passes through $(0, -30)$ and $(40, 0)$.

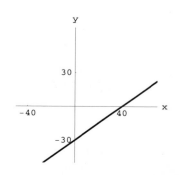

10. Circle with center $(0,0)$ and radius 20.

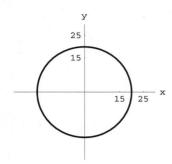

11. By using the method of completing the square, we obtain $x^2+(y+2)^2 = 4$. A circle with center $(0,-2)$ and radius 2.

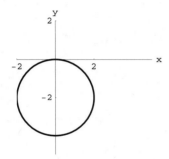

12. The line $y = -\dfrac{2}{3}x + 4$ passes through $(0,4)$ and $(6,0)$.

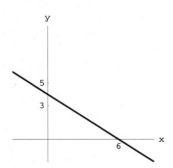

13. The horizontal line $y = 4$.

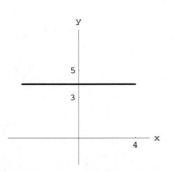

14. The vertical line $x = -2$.

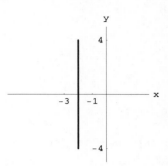

15. Since $y = \dfrac{3}{5}x - \dfrac{8}{5}$, the slope is $\dfrac{3}{5}$.

16. $\dfrac{-4 - 6}{5 - (-3)} = \dfrac{-10}{8} = -\dfrac{5}{4}$

17. We rewrite $2x - 3y = 6$ as

$$y = \frac{2}{3}x - 2.$$

Note, the slope of the above linear equation is $2/3$. Then we use $m = -3/2$ and the point $(1,-2)$.

$$y + 2 = -\frac{3}{2}(x - 1)$$
$$y = -\frac{3}{2}x + \frac{3}{2} - 2.$$

The perpendicular line is $y = -\dfrac{3}{2}x - \dfrac{1}{2}$.

18. The slope is

$$\frac{-1 - 2}{3} = -1.$$

Using the point $(3,-4)$, we find

$$y + 4 = -(x - 3)$$
$$y = -x + 3 - 4.$$

The parallel line is $y = -x - 1$.

19. $\sqrt{(-3 - 2)^2 + (1 - 4)^2} = \sqrt{25 + 9} = \sqrt{34}$

20. $\left(\dfrac{-1 + 1}{2}, \dfrac{1 + 0}{2} \right) = \left(0, \dfrac{1}{2} \right)$

21. Since the discriminant is negative, namely,

$$(-5)^2 - 4(1)(9) = -11$$

then there are no real solutions.

22. We solve for y.

$$
\begin{aligned}
5 - 4 &= 3xy + 2y \\
1 &= y(3x + 2) \\
\frac{1}{3x + 2} &= y
\end{aligned}
$$

23. Since $-4 > 2x$, the solution set is $(-\infty, -2)$ and the graph is

24. The solution set to $x > 6$ and $x > 5$ is the interval $(6, \infty)$ and the graph is

25. Solving an equivalent statement, we obtain

$$
\begin{aligned}
-3 &\leq 2x - 1 \leq 3 \\
-2 &\leq 2x \leq 4 \\
-1 &\leq x \leq 2.
\end{aligned}
$$

The solution set is the interval $[-1, 2]$ and the graph is

26. We rewrite $|x - 3| > 2$ without any absolute values. Then

$$
\begin{aligned}
x - 3 &> 2 \quad \text{or} \quad x - 3 < -2 \\
x &> 5 \quad \text{or} \quad x < 1.
\end{aligned}
$$

The solution set is $(-\infty, 1) \cup (5, \infty)$ and the graph is

27. If x is the original length of one side of the square, then

$$
\begin{aligned}
(x + 20)(x + 10) &= 999 \\
x^2 + 30x + 200 &= 999 \\
x^2 + 30x - 799 &= 0 \\
\frac{-30 \pm \sqrt{900 + 4(799)}}{2} &= x \\
\frac{-30 \pm 64}{2} &= x \\
17, -47 &= x.
\end{aligned}
$$

Thus, $x = 17$ and the original area is $17^2 = 289$ ft^2.

28. Let x be the number of gallons of the 20% solution. From the concentrations,

$$
\begin{aligned}
0.3(10 + x) &= 0.5(10) + 0.2x \\
3 + 0.3x &= 5 + 0.2x \\
0.1x &= 2 \\
x &= 20.
\end{aligned}
$$

Then 20 gallons of the 20% solution are needed.

29. Let x be the number of years since 1994 and let y be the median price of a new home in Springfield. Since the slope is given by

$$
m = \frac{122,000 - 88,000}{10 - 0} = 3400
$$

and the y-intercept is $(0, 88000)$, an equation of the line is

$$
y = 3400x + 88,000.
$$

In 2015 (i.e., $x = 21$), the median price is

$$
3400(21) + 88,000 = \$159,400.
$$

30. a) Let y be the number of farms, and let x be the number of years since 1999. The linear regression line is

$$
y = -14.4x + 2189.8.
$$

The quadratic regression curve is

$$
y = ax^2 + bx + c
$$

where

$$
a = -0.28571428571429,
$$

$$
b = -13.257142857143,
$$

and

$$
c = 2189.2285714285.
$$

b) Let $x = 10$. Using the linear regression line, the number of farms in the year 2009 is

$$
-14.4(10) + 2189.8 = 2045.8 \text{ thousands}
$$

or about $2,046,000$ farms.

Using the quadratic regression curve, the number of farms in the year 2009 is

$$a(10)^2 + b(10) + c \approx 2028.086 \text{ thousands}$$

or approximately $2,028,000$ farms.
The difference of the two extrapolations is $18,000$ farms.

Tying It All Together

1. $7x$ **2.** $30x^2$ **3.** $\dfrac{2}{2x} + \dfrac{1}{2x} = \dfrac{3}{2x}$

4. $x^2 + 6x + 9$ **5.** $6x^2 + x - 2$

6. $\dfrac{x^2 + 2xh + h^2 - x^2}{h} = \dfrac{2xh + h^2}{h} = 2x + h$

7. $\dfrac{x+1}{(x-1)(x+1)} + \dfrac{x-1}{(x+1)(x-1)} = \dfrac{2x}{x^2-1}$

8. $x^2 + 3x + \dfrac{9}{4}$

9. An identity and the solution set is $(-\infty, \infty)$.

10. Since $30x^2 - 11x = x(30x - 11) = 0$, the solution set is $\left\{0, \dfrac{11}{30}\right\}$.

11. Since $\dfrac{3}{2x} = \dfrac{3}{2x}$, the solution set is $(-\infty, 0) \cup (0, \infty)$.

12. Subtract $x^2 + 9$ from $x^2 + 6x + 9 = x^2 + 9$ and get $6x = 0$. The solution set is $\{0\}$.

13. Since $(2x - 1)(3x + 2) = 0$, the solution set is $\left\{-\dfrac{2}{3}, \dfrac{1}{2}\right\}$.

14. Multiply the equation by $8(x + 1)(x - 1)$.

$$
\begin{aligned}
8x + 8 + 8x - 8 &= 5(x^2 - 1) \\
0 &= 5x^2 - 16x - 5 \\
x &= \frac{16 \pm \sqrt{356}}{10} \\
x &= \frac{16 \pm 2\sqrt{89}}{10}
\end{aligned}
$$

The solution set is $\left\{\dfrac{8 \pm \sqrt{89}}{5}\right\}$.

15. Since $7x - 7x^2 = 7x(1 - x) = 0$, the solution set is $\{0, 1\}$.

16. Since $7x - 7 = 7(x - 1) = 0$, the solution set is $\{1\}$.

17. 0 **18.** $-1 - 3 + 2 = -2$

19. $-\dfrac{1}{8} - 3\left(\dfrac{2}{8}\right) + \dfrac{16}{8} = \dfrac{9}{8}$

20. $-\dfrac{1}{27} - 3\left(\dfrac{3}{27}\right) + \dfrac{54}{27} = \dfrac{44}{27}$

21. $-1 - 3 - 4 = -8$

22. $-4 + 6 - 4 = -2$ **23.** -4

24. $-0.25 + 1.5 - 4 = -2.75$

For Thought

1. False, since $\{(1,2),(1,3)\}$ is not a function.

2. False, since $f(5)$ is not defined.   **3.**   True

4. False, since a student's exam grade is a function of the student's preparation. If two classmates had the same IQ and only one prepared then the one who prepared will most likely achieve a higher grade.

5. False, since $(x+h)^2 = x^2 + 2xh + h^2$

6. False, since the domain is all real numbers.

7. True   **8.**   True   **9.**   True

10. False, since $\left(\dfrac{3}{8},8\right)$ and $\left(\dfrac{3}{8},5\right)$ are two ordered pairs with the same first coordinate and different second coordinates.

2.1 Exercises

1. Note, $b = 2\pi a$ is equivalent to $a = \dfrac{b}{2\pi}$.

Thus, a is a function of b, and b is a function of a.

3. a is a function of b since a given denomination has a unique length. Since a dollar bill and a five-dollar bill have the same length, then b is not a function of a.

5. Since an item has only one price, b is a function of a. Since two items may have the same price, a is not a function of b.

7. a is not a function of b since it is possible that two different students can obtain the same final exam score but the times spent on studying are different.

b is not a function of a since it is possible that two different students can spend the same time studying but obtain different final exam scores.

9. Since 1 in ≈ 2.54 cm, a is a function of b and b is a function of a.

11. No   **13.**   Yes   **15.**   Yes   **17.**   Yes

19. Not a function since 25 has two different second coordinates.

21. Not a function since 3 has two different second coordinates.

23. Yes   **24.**   Yes

25. Since the ordered pairs in the graph of $y = 3x - 8$ are $(x, 3x - 8)$, there are no two ordered pairs with the same first coordinate and different second coordinates. We have a function.

27. Since $y = (x + 9)/3$, the ordered pairs are $(x, (x+9)/3)$. Thus, there are no two ordered pairs with the same first coordinate and different second coordinates. We have a function.

29. Since $y = \pm x$, the ordered pairs are $(x, \pm x)$. Thus, there are two ordered pairs with the same first coordinate and different second coordinates. We do not have a function.

31. Since $y = x^2$, the ordered pairs are (x, x^2). Thus, there are no two ordered pairs with the same first coordinate and different second coordinates. We have a function.

33. Since $y = |x| - 2$, the ordered pairs are $(x, |x| - 2)$. Thus, there are no two ordered pairs with the same first coordinate and different second coordinates. We have a function.

35. Since $(2, 1)$ and $(2, -1)$ are two ordered pairs with the same first coordinate and different second coordinates, the equation does not define a function.

37. Domain $\{-3, 4, 5\}$, range $\{1, 2, 6\}$

39. Domain $(-\infty, \infty)$, range $\{4\}$

41. Domain $(-\infty, \infty)$;
since $|x| \geq 0$, the range of $y = |x| + 5$ is $[5, \infty)$

43. Since $x = |y| - 3 \geq -3$, the domain of $x = |y| - 3$ is $[-3, \infty)$; range $(-\infty, \infty)$

45. Since $\sqrt{x - 4}$ is a real number whenever $x \geq 4$, the domain of $y = \sqrt{x - 4}$ is $[4, \infty)$.
Since $y = \sqrt{x - 4} \geq 0$ for $x \geq 4$, the range is $[0, \infty)$.

47. Since $x = -y^2 \leq 0$, the domain of $x = -y^2$ is $(-\infty, 0]$; range is $(-\infty, \infty)$;

49. 6

51. $g(2) = 3(2) + 5 = 11$

53. Since $(3, 8)$ is the ordered pair, one obtains $f(3) = 8$. The answer is $x = 3$.

55. Solving $3x + 5 = 26$, we find $x = 7$.

57. $f(4) + g(4) = 5 + 17 = 22$

59. $3a^2 - a$

61. $4(a + 2) - 2 = 4a + 6$

63. $3(x^2 + 2x + 1) - (x + 1) = 3x^2 + 5x + 2$

65. $4(x + h) - 2 = 4x + 4h - 2$

67. $(3x^2 + 6xh + 3h^2 - x - h) - 3x^2 + x = 3h^2 + 6xh - h$

69. The average rate of change is
$$\frac{4,000 - 16,000}{5} = -\$2,400 \text{ per year.}$$

71. The average rate of change on $[0, 2]$ is
$$\frac{h(2) - h(0)}{2 - 0} = \frac{0 - 64}{2 - 0} = -32 \text{ ft/sec.}$$
The average rate of change on $[1, 2]$ is
$$\frac{h(2) - h(1)}{2 - 1} = \frac{0 - 48}{2 - 1} = -48 \text{ ft/sec.}$$
The average rate of change on $[1.9, 2]$ is
$$\frac{h(2) - h(1.9)}{2 - 1.9} = \frac{0 - 6.24}{0.1} = -62.4 \text{ ft/sec.}$$
The average rate of change on $[1.99, 2]$ is
$$\frac{h(2) - h(1.99)}{2 - 1.99} = \frac{0 - 0.6384}{0.01} = -63.84 \text{ ft/sec.}$$
The average rate of change on $[1.999, 2]$ is
$$\frac{h(2) - h(1.999)}{2 - 1.999} = \frac{0 - 0.063984}{0.001} = -63.984$$
ft/sec.

73. The average rate of change is $\dfrac{896 - 1056}{2004 - 1988} = -10$ million hectares per year.

75.
$$\frac{f(x + h) - f(x)}{h} = \frac{4(x + h) - 4x}{h}$$
$$= \frac{4h}{h}$$
$$= 4$$

77.
$$\frac{f(x + h) - f(x)}{h} = \frac{3(x + h) + 5 - 3x - 5}{h}$$
$$= \frac{3h}{h}$$
$$= 3$$

79. Let $g(x) = x^2 + x$. Then we obtain
$$\frac{g(x + h) - g(x)}{h} =$$
$$\frac{(x + h)^2 + (x + h) - x^2 - x}{h} =$$
$$\frac{2xh + h^2 + h}{h} =$$
$$2x + h + 1.$$

81. Difference quotient is
$$= \frac{-(x + h)^2 + (x + h) - 2 + x^2 - x + 2}{h}$$
$$= \frac{-2xh - h^2 + h}{h}$$
$$= -2x - h + 1$$

83. Difference quotient is
$$= \frac{3\sqrt{x + h} - 3\sqrt{x}}{h} \cdot \frac{3\sqrt{x + h} + 3\sqrt{x}}{3\sqrt{x + h} + 3\sqrt{x}}$$
$$= \frac{9(x + h) - 9x}{h(3\sqrt{x + h} + 3\sqrt{x})}$$
$$= \frac{9h}{h(3\sqrt{x + h} + 3\sqrt{x})}$$
$$= \frac{3}{\sqrt{x + h} + \sqrt{x}}$$

85. Difference quotient is
$$= \frac{\sqrt{x + h + 2} - \sqrt{x + 2}}{h} \cdot \frac{\sqrt{x + h + 2} + \sqrt{x + 2}}{\sqrt{x + h + 2} + \sqrt{x + 2}}$$
$$= \frac{(x + h + 2) - (x + 2)}{h(\sqrt{x + h + 2} + \sqrt{x + 2})}$$
$$= \frac{h}{h(\sqrt{x + h + 2} + \sqrt{x + 2})}$$
$$= \frac{1}{\sqrt{x + h + 2} + \sqrt{x + 2}}$$

87. Difference quotient is

$$= \frac{\dfrac{1}{x+h} - \dfrac{1}{x}}{h} \cdot \frac{x(x+h)}{x(x+h)}$$

$$= \frac{x - (x+h)}{xh(x+h)}$$

$$= \frac{-h}{xh(x+h)}$$

$$= \frac{-1}{x(x+h)}$$

89. Difference quotient is

$$= \frac{\dfrac{3}{x+h+2} - \dfrac{3}{x+2}}{h} \cdot \frac{(x+h+2)(x+2)}{(x+h+2)(x+2)}$$

$$= \frac{3(x+2) - 3(x+h+2)}{h(x+h+2)(x+2)}$$

$$= \frac{-3h}{h(x+h+2)(x+2)}$$

$$= \frac{-3}{(x+h+2)(x+2)}$$

91. a) $A = s^2$ **b)** $s = \sqrt{A}$ **c)** $s = \dfrac{d\sqrt{2}}{2}$

d) $d = s\sqrt{2}$ **e)** $P = 4s$ **f)** $s = P/4$

g) $A = P^2/16$ **h)** $d = \sqrt{2A}$

93. $C = 50 + 35n$

95.

(a) The quantity

$$C(4) = (0.95)(4) + 5.8 = \$9.6 \text{ billion}$$

represents the amount spent on computers in the year 2004.

(b) By solving $0.95n + 5.8 = 15$, we obtain

$$n = \frac{9.2}{0.95} \approx 10.$$

Thus, spending for computers will be $15 billion in 2010.

97. Let a be the radius of each circle. Note, triangle $\triangle ABC$ is an equilateral triangle with side $2a$ and height $\sqrt{3}a$.

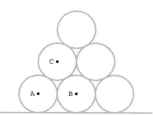

Thus, the height of the circle centered at C from the horizontal line is $\sqrt{3}a + 2a$. Hence, by using a similar reasoning, we obtain that height of the highest circle from the line is

$$2\sqrt{3}a + 2a$$

or equivalently $(2\sqrt{3} + 2)a$.

99. When $x = 18$ and $h = 0.1$, we have

$$\frac{R(18.1) - R(18)}{0.1} = 1,950.$$

The revenue from the concert will increase by approximately $1,950 if the price of a ticket is raised from $18 to $19.

If $x = 22$ and $h = 0.1$, then

$$\frac{R(22.1) - R(22)}{0.1} = -2,050.$$

The revenue from the concert will decrease by approximately $2,050 if the price of a ticket is raised from $22 to $23.

For Thought

1. True, since the graph is a parabola opening down with vertex at the origin.

2. False, the graph is decreasing.

3. True

4. True, since $f(-4.5) = [-1.5] = -2$.

5. False, since the range is $\{\pm 1\}$.

6. True   **7.**   True   **8.**   True

9. False, since the range is the interval $[0, 4]$.

10. True

2.2 Exercises

1. Function $y = 2x$ includes the points $(0,0), (1,2)$, domain and range are both $(-\infty, \infty)$

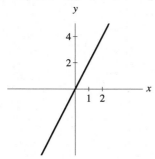

3. Function $x - y = 0$ includes the points $(-1,-1)$, $(0,0), (1,1)$, domain and range are both $(-\infty, \infty)$

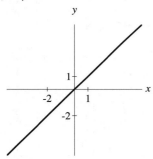

5. Function $y = 5$ includes the points $(0,5)$, $(\pm 2, 5)$, domain is $(-\infty, \infty)$, range is $\{5\}$

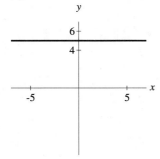

7. Function $y = 2x^2$ includes the points $(0,0)$, $(\pm 1, 2)$, domain is $(-\infty, \infty)$, range is $[0, \infty)$

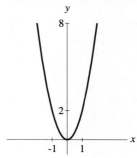

9. Function $y = 1 - x^2$ includes the points $(0,1)$, $(\pm 1, 0)$, domain is $(-\infty, \infty)$, range is $(-\infty, 1]$

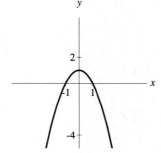

11. Function $y = 1 + \sqrt{x}$ includes the points $(0,1)$, $(1,2), (4,3)$, domain is $[0, \infty)$, range is $[1, \infty)$

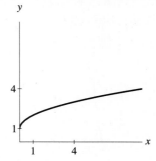

13. $x = y^2 + 1$ is not a function and includes the points $(1,0), (2, \pm 1)$, domain is $[1, \infty)$, range is $(-\infty, \infty)$

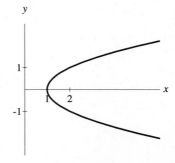

15. Function $x = \sqrt{y}$ goes through $(0,0), (2,4), (3,9)$, domain and range is $[0, \infty)$

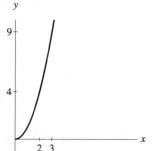

17. Function $y = \sqrt[3]{x} + 1$ goes through $(-1,0), (1,2), (8,3)$, domain $(-\infty, \infty)$, and range $(-\infty, \infty)$

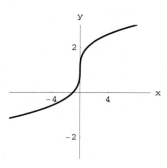

19. Function, $x = \sqrt[3]{y}$ goes through $(0,0), (1,1), (2,8)$, domain $(-\infty, \infty)$, and range $(-\infty, \infty)$

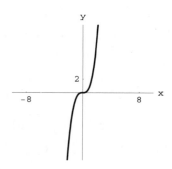

21. Not a function, $y^2 = 1 - x^2$ goes through $(1,0), (0,1), (-1,0)$, domain $[-1,1]$, and range $[-1,1]$

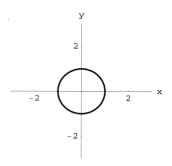

23. Function, $y = \sqrt{1 - x^2}$ goes through $(\pm 1, 0), (0,1)$, domain $[-1,1]$, and range $[0,1]$

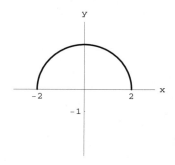

25. Function $y = x^3$ includes the points $(0,0)$, $(1,1), (2,8)$, domain and range are both $(-\infty, \infty)$

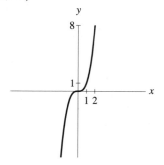

27. Function $y = 2|x|$ includes the points $(0,0)$, $(\pm 1, 2)$, domain is $(-\infty, \infty)$, range is $[0, \infty)$

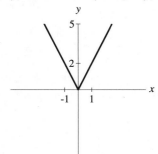

29. Function $y = -|x|$ includes the points $(0,0)$, $(\pm 1, -1)$, domain is $(-\infty, \infty)$, range is $(-\infty, 0]$

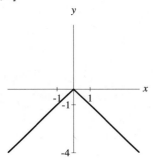

31. Not a function, graph of $x = |y|$ includes the points $(0,0), (2,2), (2,-2)$, domain is $[0, \infty)$, range is $(-\infty, \infty)$

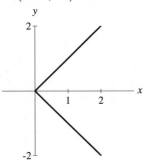

33. Domain is $(-\infty, \infty)$, range is $\{\pm 2\}$, some points are $(-3, -2)$, $(1, -2)$

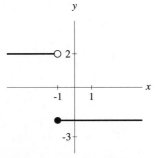

35. Domain is $(-\infty, \infty)$, range is $(-\infty, -2] \cup (2, \infty)$, some points are $(2,3)$, $(1, -2)$

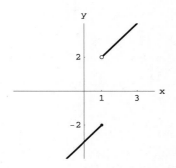

37. Domain is $[-2, \infty)$, range is $(-\infty, 2]$, some points are $(2, 2)$, $(-2, 0)$, $(3, 1)$

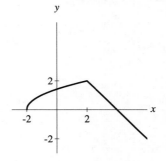

39. Domain is $(-\infty, \infty)$, range is $[0, \infty)$, some points are $(-1, 1)$, $(-4, 2)$, $(4, 2)$

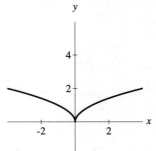

41. Domain is $(-\infty, \infty)$, range is $(-\infty, \infty)$, some points are $(-2, 4)$, $(1, -1)$

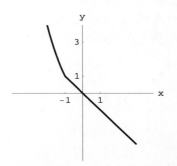

43. Domain is $(-\infty, \infty)$, range is the set of integers, some points are $(0, 1)$, $(1, 2)$, $(1.5, 2)$

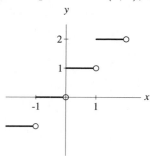

45. Domain $[0, 4)$, range is $\{2, 3, 4, 5\}$, some points are $(0, 2)$, $(1, 3)$, $(1.5, 3)$

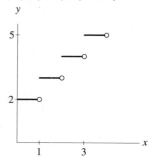

47. a. Domain and range are both $(-\infty, \infty)$, decreasing on $(-\infty, \infty)$

　　b. Domain is $(-\infty, \infty)$, range is $(-\infty, 4]$ increasing on $(-\infty, 0)$, decreasing on $(0, \infty)$

49. a. Domain is $[-2, 6]$, range is $[3, 7]$ increasing on $(-2, 2)$, decreasing on $(2, 6)$

　　b. Domain $(-\infty, 2]$, range $(-\infty, 3]$, increasing on $(-\infty, -2)$, constant on $(-2, 2)$

51. a. Domain is $(-\infty, \infty)$, range is $[0, \infty)$ increasing on $(0, \infty)$, decreasing on $(-\infty, 0)$

　　b. Domain and range are both $(-\infty, \infty)$ increasing on $(-2, -2/3)$, decreasing on $(-\infty, -2)$ and $(-2/3, \infty)$

53. a. Domain and range are both $(-\infty, \infty)$, increasing on $(-\infty, \infty)$

　　b. Domain is $[-2, 5]$, range is $[1, 4]$ increasing on $(1, 2)$, decreasing on $(-2, 1)$, constant on $(2, 5)$

55. Domain and range are both $(-\infty, \infty)$ increasing on $(-\infty, \infty)$, some points are $(0, 1)$, $(1, 3)$

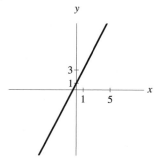

57. Domain is $(-\infty, \infty)$, range is $[0, \infty)$, increasing on $(1, \infty)$, decreasing on $(-\infty, 1)$, some points are $(0, 1)$, $(1, 0)$

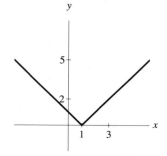

59. Domain is $(-\infty, 0) \cup (0, \infty)$, range is $\{\pm 1\}$, constant on $(-\infty, 0)$ and $(0, \infty)$, some points are $(1, 1)$, $(-1, -1)$

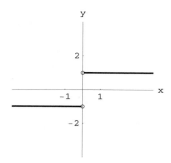

61. Domain is $[-3, 3]$, range is $[0, 3]$, increasing on $(-3, 0)$, decreasing on $(0, 3)$, some points are $(\pm 3, 0)$, $(0, 3)$

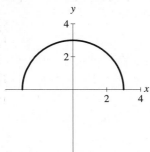

63. Domain and range are both $(-\infty, \infty)$, increasing on $(-\infty, 3)$ and $(3, \infty)$, some points are $(4, 5)$, $(0, 2)$

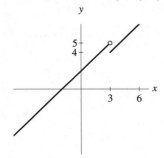

65. Domain is $(-\infty, \infty)$, range is $(-\infty, 2]$, increasing on $(-\infty, -2)$ and $(-2, 0)$, decreasing on $(0, 2)$ and $(2, \infty)$, some points are $(-3, 0)$, $(0, 2)$, $(4, -1)$

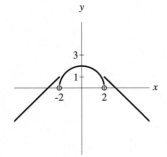

67. $f(x) = \begin{cases} 2 & \text{for} \quad x > -1 \\ -1 & \text{for} \quad x \le -1 \end{cases}$

69. The line joining $(-1, 1)$ and $(-3, 3)$ is $y = -x$, and the line joining $(-1, -2)$ and $(3, 2)$ is $y = x - 1$. The piecewise function is

$$f(x) = \begin{cases} x - 1 & \text{for} \quad x \ge -1 \\ -x & \text{for} \quad x < -1. \end{cases}$$

71. The line joining $(0, -2)$ and $(2, 2)$ is $y = 2x - 2$, and the line joining $(0, -2)$ and $(-3, 1)$ is $y = -x - 2$. The piecewise function is

$$f(x) = \begin{cases} 2x - 2 & \text{for} \quad x \ge 0 \\ -x - 2 & \text{for} \quad x < 0. \end{cases}$$

73. increasing on the interval $(0.83, \infty)$, decreasing on $(-\infty, 0.83)$

75. increasing on $(-\infty, -1)$ and $(1, \infty)$, decreasing on $(-1, 1)$

77. increasing on $(-1.73, 0)$ and $(1.73, \infty)$, decreasing on $(-\infty, -1.73)$ and $(0, 1.73)$

79. increasing on $(30, 50)$, and $(70, \infty)$, decreasing on $(-\infty, 30)$ and $(50, 70)$

81. c, graph was increasing at first, then suddenly dropped and became constant, then increased slightly

83. d, graph was decreasing at first, then fluctuated between increases and decreases, then the market increased

85. The independent variable is time t where t is the number of minutes after 7:45 and the dependent variable is distance D from the holodeck.

D is increasing on the intervals $(0, 3)$ and $(6, 15)$, decreasing on $(3, 6)$ and $(30, 39)$, and constant on $(15, 30)$.

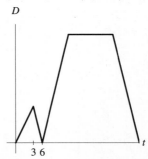

87. Independent variable is time t in years, dependent variable is savings s in dollars

s is increasing on the interval $(0, 2)$; s is constant on $(2, 2.5)$; s is decreasing on $(2.5, 4.5)$.

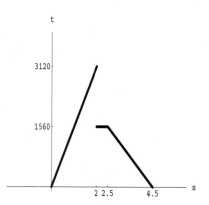

89. In 1988, there were $M(18) = 565$ million cars. In 2010, it is projected that there will be $M(40) = 800$ million cars.
The average rate of change from 1984 to 1994 is $\dfrac{M(24) - M(14)}{10} = 14.5$ million cars per year.

91. Constant on $(0, 10^4)$, increasing on $(10^4, \infty)$

93. The cost is over \$235 for t in $[5, \infty)$.

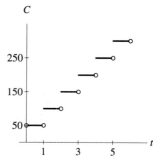

95. $f(x) = \begin{cases} 150 & \text{if} \quad 0 < x < 3 \\ 50x & \text{if} \quad 3 \le x \le 10 \end{cases}$

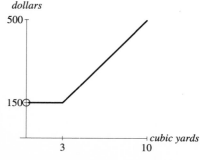

97. A mechanic's fee is \$20 for each half-hour of work with any fraction of a half-hour charged as a half-hour. If y is the fee in dollars and x is the number of hours, then $y = -20[-2x]$.

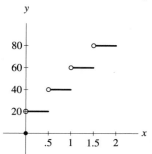

For Thought

1. False, it is a reflection in the y-axis.

2. True   **3.**   False, rather it is a left translation.

4. True   **5.**   True

6. False, the down shift should come after the reflection.   **7.**   True

8. False, since their domains are different.

9. True   **10.**   True

2.3 Exercises

1. $f(x) = |x|, g(x) = |x| - 4$

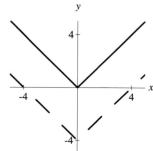

3. $f(x) = x, g(x) = x + 3$

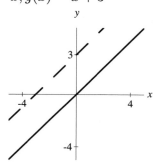

5. $y = x^2, y = (x - 3)^2$

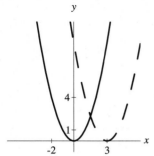

7. $y = \sqrt{x}, y = \sqrt{x + 9}$

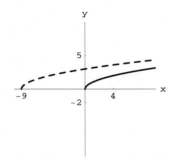

9. $f(x) = \sqrt{x}, g(x) = -\sqrt{x}$

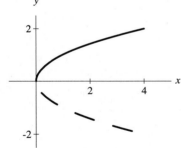

11. $y = \sqrt{x}, y = 3\sqrt{x}$

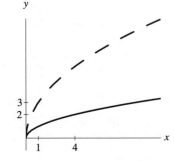

13. $y = x^2, y = \dfrac{1}{4}x^2$

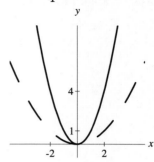

15. $y = \sqrt{4 - x^2}, y = -\sqrt{4 - x^2}$

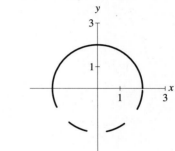

17. g   **19.**   b

21. c   **23.**   f

25. $y = \sqrt{x} + 2$

27. $y = (x - 5)^2$

29. $y = (x - 10)^2 + 4$

31. $y = -(3\sqrt{x} + 5)$ or $y = -3\sqrt{x} - 5$

33. $y = -3|x - 7| + 9$

35. $y = (x - 1)^2 + 2$; right by 1, up by 2, domain $(-\infty, \infty)$, range $[2, \infty)$

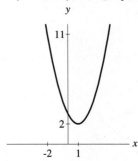

37. $y = |x - 1| + 3$; right by 1, up by 3
domain $(-\infty, \infty)$, range $[3, \infty)$

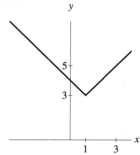

39. $y = 3x - 40$,
domain and range are both $(-\infty, \infty)$

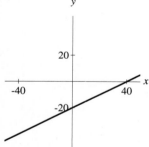

41. $y = \frac{1}{2}x - 20$,
domain and range are both $(-\infty, \infty)$

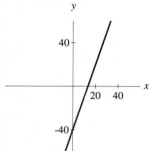

43. $y = -\frac{1}{2}|x| + 40$, shrink by 1/2,
reflect about x-axis, up by 40,
domain $(-\infty, \infty)$, range $(-\infty, 40]$

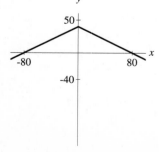

45. $y = -\frac{1}{2}|x + 4|$, left by 4,

reflect about x-axis, shrink by 1/2,

domain $(-\infty, \infty)$, range $(-\infty, 0]$

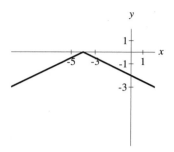

47. $y = -\sqrt{x - 3} + 1$, right by 3,
reflect about x-axis, up by 1,
domain $[3, \infty)$, range $(-\infty, 1]$

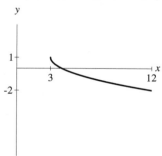

49. $y = -2\sqrt{x + 3} + 2$, left by 3, stretch by 2,
reflect about x-axis, up by 2,
domain $[-3, \infty)$, range $(-\infty, 2]$

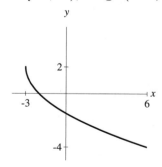

51. Symmetric about y-axis, even function
since $f(-x) = f(x)$

53. No symmetry, neither even nor odd
since $f(-x) \neq f(x)$ and $f(-x) \neq -f(x)$

55. Symmetric about $x = -3$, neither even nor
odd since $f(-x) \neq f(x)$ and $f(-x) \neq -f(x)$

57. Symmetry about $x = 2$, not an even or odd function since $f(-x) \neq f(x)$ and $f(-x) \neq -f(x)$

59. Symmetric about the origin, odd function since $f(-x) = -f(x)$

61. No symmetry, not an even or odd function since $f(-x) \neq f(x)$ and $f(-x) \neq -f(x)$

63. No symmetry, not an even or odd function since $f(-x) \neq f(x)$ and $f(-x) \neq -f(x)$

65. Symmetric about the y-axis, even function since $f(-x) = f(x)$

67. No symmetry, not an even or odd function since $f(-x) = -f(x)$ and $f(-x) \neq -f(x)$

69. Symmetric about the y-axis, even function since $f(-x) = f(x)$

71. e   **73.**   g

75. b   **77.**   c

79. $(-\infty, -1] \cup [1, \infty)$

81. $(-\infty, -1) \cup (5, \infty)$

83. Using the graph of $y = (x - 1)^2 - 9$, we find that the solution is $(-2, 4)$.

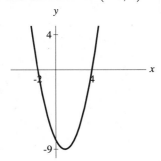

85. From the graph of $y = 5 - \sqrt{x}$, we find that the solution is $[0, 25]$.

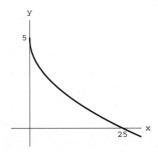

87. Note, the points of intersection of $y = 3$ and $y = (x - 2)^2$ are $(2 \pm \sqrt{3}, 3)$. The solution set of $(x-2)^2 > 3$ is $\left(-\infty, 2 - \sqrt{3}\right) \cup \left(2 + \sqrt{3}, \infty\right)$.

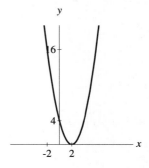

89. From the graph of $y = \sqrt{25 - x^2}$, we conclude that the solution is $(-5, 5)$.

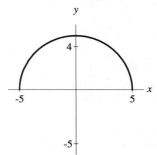

91. From the graph of $y = \sqrt{3}x^2 + \pi x - 9$,

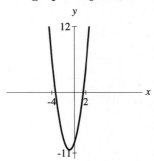

we observe that the solution set of $\sqrt{3}x^2 + \pi x - 9 < 0$ is $(-3.36, 1.55)$.

93. a. Stretch the graph of f by a factor of 2.

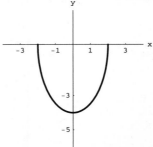

b. Reflect the graph of f about the x-axis.

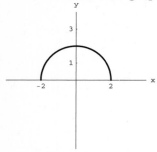

c. Translate the graph of f to the left by 1-unit.

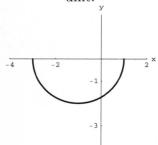

d. Translate the graph of f to the right by 3-units.

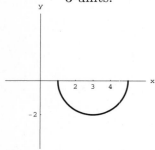

e. Stretch the graph of f by a factor of 3 and reflect about the x-axis.

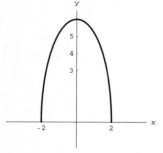

f. Translate the graph of f to the left by 2-units and down by 1-unit.

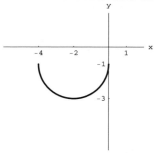

g. Translate the graph of f to the right by 1-unit and up by 3-units.

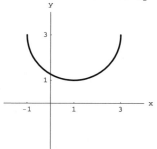

h. Translate the graph of f to the right by 2-units, stretch by a factor of 3, and up by 1-unit.

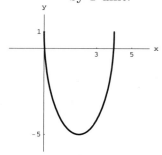

95. $N(x) = x + 2000$

97. If inflation rate is less than 50%, then $1 - \sqrt{x} < \dfrac{1}{2}$. This simplifies to $\dfrac{1}{2} < \sqrt{x}$. After squaring we have $\dfrac{1}{4} < x$ and so $x > 25\%$.

99.

(a) Both functions are even functions and the graphs are identical

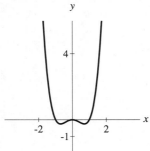

(b) One graph is a reflection of the other about the y-axis. Both functions are odd functions.

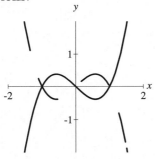

(c) The second graph is obtained by shifting the first one to the left by 1 unit.

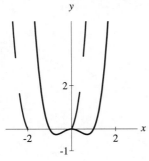

(d) The second graph is obtained by translating the first one to the right by 2 units and 3 units up.

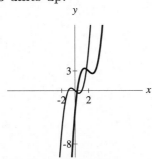

For Thought

1. False, since $f + g$ has an empty domain.

2. True **3.** True **4.** True

5. True, since $A = P^2/16$. **6.** True

7. False, since $(f \circ g)(x) = \sqrt{x-2}$ **8.** True

9. False, since $(h \circ g)(x) = x^2 - 9$.

10. True, since x belongs to the domain if $\sqrt{x-2}$ is a real number, i.e., if $x \geq 2$.

2.4 Exercises

1. $-1 + 2 = 1$

3. $-5 - 6 = -11$

5. $(-4) \cdot 2 = -8$

7. $1/12$

9. $(a - 3) + (a^2 - a) = a^2 - 3$

11. $(a - 3)(a^2 - a) = a^3 - 4a^2 + 3a$

13. $f + g = \{(-3, 1 + 2), (2, 0 + 6)\} = \{(-3, 3), (2, 6)\}$, domain $\{-3, 2\}$

15. $f - g = \{(-3, 1 - 2), (2, 0 - 6)\} = \{(-3, -1), (2, -6)\}$, domain $\{-3, 2\}$

17. $f \cdot g = \{(-3, 1 \cdot 2), (2, 0 \cdot 6) = \{(-3, 2), (2, 0)\}$, domain $\{-3, 2\}$

19. $g/f = \{(-3, 2/1)\} = \{(-3, 2)\}$, domain $\{-3\}$

21. $(f + g)(x) = \sqrt{x} + x - 4$, domain is $[0, \infty)$

23. $(f - h)(x) = \sqrt{x} - \dfrac{1}{x - 2}$, domain is $[0, 2) \cup (2, \infty)$

25. $(g \cdot h)(x) = \dfrac{x - 4}{x - 2}$, domain is $(-\infty, 2) \cup (2, \infty)$

27. $\left(\dfrac{g}{f}\right)(x) = \dfrac{x - 4}{\sqrt{x}}$, domain is $(0, \infty)$

29. $\{(-3, 0), (1, 0), (4, 4)\}$

31. $\{(1, 4)\}$

33. $\{(-3, 4), (1, 4)\}$

35. $f(2) = 5$

37. $f(2) = 5$

39. $f(20.2721) = 59.8163$

41. $(g \circ h \circ f)(2) = (g \circ h)(5) = g(2) = 5$

43. $(f \circ g \circ h)(2) = (f \circ g)(1) = f(2) = 5$

45. $(f \circ h)(a) = f\left(\dfrac{a+1}{3}\right) =$

$$3\left(\dfrac{a+1}{3}\right) - 1 = (a+1) - 1 = a$$

47. $(f \circ g)(t) = f\left(t^2 + 1\right) =$
$3(t^2 + 1) - 1 = 3t^2 + 2$

49. $(f \circ g)(x) = \sqrt{x} - 2$, domain $[0, \infty)$

51. $(f \circ h)(x) = \dfrac{1}{x} - 2$, domain $(-\infty, 0) \cup (0, \infty)$

53. $(h \circ g)(x) = \dfrac{1}{\sqrt{x}}$, domain $(0, \infty)$

55. $(f \circ f)(x) = (x - 2) - 2 = x - 4$,
domain $(-\infty, \infty)$

57. $(h \circ g \circ f)(x) = h(\sqrt{x} - 2) = \dfrac{1}{\sqrt{x} - 2}$,

domain $(2, \infty)$

59. $(h \circ f \circ g)(x) = h\left(\sqrt{x} - 2\right) = \dfrac{1}{\sqrt{x} - 2}$,

domain $(0, 4) \cup (4, \infty)$

61. $F = g \circ h$

63. $H = h \circ g$

65. $N = h \circ g \circ f$

67. $P = g \circ f \circ g$

69. $S = g \circ g$

71. If $g(x) = x^3$ and $h(x) = x - 2$, then

$$(h \circ g)(x) = g(x) - 2 = x^3 - 2 = f(x).$$

73. If $g(x) = x + 5$ and $h(x) = \sqrt{x}$, then

$$(h \circ g)(x) = \sqrt{g(x)} = \sqrt{x + 5} = f(x).$$

75. If $g(x) = 3x - 1$ and $h(x) = \sqrt{x}$, then

$$(h \circ g)(x) = \sqrt{g(x)} = \sqrt{3x - 1} = f(x).$$

77. If $g(x) = |x|$ and $h(x) = 4x + 5$, then

$$(h \circ g)(x) = 4g(x) + 5 = 4|x| + 5 = f(x).$$

79. $y = 2(3x + 1) - 3 = 6x - 1$

81. $y = (x^2 + 6x + 9) - 2 = x^2 + 6x + 7$

83. $y = 3 \cdot \dfrac{x+1}{3} - 1 = x + 1 - 1 = x$

85. Since $m = n - 4$ and $y = m^2$, $y = (n - 4)^2$.

87. Since $w = x + 16$, $z = \sqrt{w}$, and $y = \dfrac{z}{8}$,

we obtain $y = \dfrac{\sqrt{x + 16}}{8}$.

89. After multiplying y by $\dfrac{x+1}{x+1}$ we have

$$y = \dfrac{\dfrac{x-1}{x+1} + 1}{\dfrac{x-1}{x+1} - 1} = \dfrac{(x-1) + (x+1)}{(x-1) - (x+1)} = -x$$

The domain of the original function is
$(-\infty, -1) \cup (-1, \infty)$ while the domain of
the simplified function is $(-\infty, \infty)$.
The two functions are not the same.

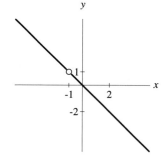

91. Domain $[-1, \infty)$, range $[-7, \infty)$

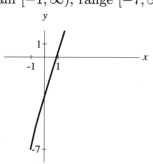

93. Domain $[1, \infty)$, range $[0, \infty)$

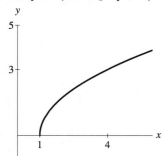

95. Domain $[0, \infty)$, range $[4, \infty)$

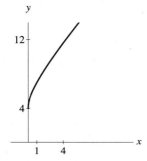

97. $P(x) = 68x - (40x + 200) = 28x - 200$.

Since $200/28 \approx 7.1$, the profit is positive when the number of trimmers satisfies $x \geq 8$.

99. $A = d^2/2$

101. $(f \circ f)(x) = 0.899x$ and $(f \circ f \circ f)(x) = 0.852x$ are the amounts of forest land at the start of 2002 and 2003, respectively.

103. Total cost is $(T \circ C)(x) = 1.05(1.20x) = 1.26x$.

105. Note, $D = \dfrac{d/2240}{x} = \dfrac{d/2240}{L^3/100^3} = \dfrac{100^3 d}{2240 L^3}$

$$= \frac{100^3(26000)}{2240 L^3} = \frac{100^4(26)}{224 L^3} = \frac{100^4(13)}{112 L^3}.$$

Expressing D as a function of L, we write $D = \dfrac{(13)100^4}{112 L^3}$ or $D = \dfrac{1.16 \times 10^7}{L^3}$.

107. The area of a semicircle with radius $s/2$ is $(1/2)\pi(s/2)^2 = \pi s^2/8$. The area of the square is s^2. The area of the window is

$$W = s^2 + \frac{\pi s^2}{8} = \frac{(8+\pi)s^2}{8}.$$

109. Form a right triangle with two sides of length s and a hypotenuse of length d. By the Pythagorean Theorem, we obtain

$$d^2 = s^2 + s^2.$$

Solving for s, we have $s = \dfrac{d\sqrt{2}}{2}$.

111. If a coat is on sale at 25% off and there is an additional 10% off, then the coat will cost $0.90(.75x) = 0.675x$ where x is the regular price. Thus, the discount sale is 32.5% off and not 35% off.

For Thought

1. False, since the inverse function is $\{(3,2), (5,5)\}$.

2. False, since it is not one-to-one.

3. False, $g^{-1}(x)$ does not exist since g is not one-to-one.

4. True

5. False, a function that fails the horizontal line test has no inverse.

6. False, since it fails the horizontal line test.

7. False, since $f^{-1}(x) = \left(\dfrac{x}{3}\right)^2 + 2$ where $x \geq 0$.

8. False, $f^{-1}(x)$ does not exist since f is not one-to-one.

9. False, since $y = |x|$ is V-shaped and the horizontal line test fails.

10. True

2.5 Exercises

1. Not invertible, there can be two different items with the same price.

3. Invertible, since the playing time is a function of the length of the VCR tape.

5. Invertible, assuming that cost is simply a multiple of the number of days. If cost includes extra charges, then the function may not be invertible.

7. Invertible, $\{(3,9),(2,2)\}$

9. Not invertible

11. Invertible, $\{(3,3),(2,2),(4,4),(7,7)\}$

13. Not invertible

15. $f^{-1} = \{(1,2),(5,3)\}$, $f^{-1}(5) = 3$,
$(f^{-1} \circ f)(2) = 2$

17. $f^{-1} = \{(-3,-3),(5,0),(-7,2)\}$, $f^{-1}(5) = 0$,
$(f^{-1} \circ f)(2) = 2$

19. Not one-to-one

21. One-to-one

23. Not one-to-one

25. One-to-one; since the graph of $y = 2x-3$ shows
$y = 2x - 3$ is an increasing function, the
Horizontal Line Test implies $y = 2x - 3$ is
one-to-one.

27. One-to-one; for if $q(x_1) = q(x_2)$ then

$$
\begin{aligned}
\frac{1 - x_1}{x_1 - 5} &= \frac{1 - x_2}{x_2 - 5} \\
(1 - x_1)(x_2 - 5) &= (1 - x_2)(x_1 - 5) \\
x_2 - 5 - x_1 x_2 + 5x_1 &= x_1 - 5 - x_2 x_1 + 5x_2 \\
x_2 + 5x_1 &= x_1 + 5x_2 \\
4(x_1 - x_2) &= 0 \\
x_1 - x_2 &= 0.
\end{aligned}
$$

Thus, if $q(x_1) = q(x_2)$ then $x_1 = x_2$.
Hence, q is one-to-one.

29. Not one-to-one for $p(-2) = p(0) = 1$.

31. Not one-to-one for $w(1) = w(-1) = 4$.

33. One-to-one; for if $k(x_1) = k(x_2)$ then

$$
\begin{aligned}
\sqrt[3]{x_1 + 9} &= \sqrt[3]{x_2 + 9} \\
(\sqrt[3]{x_1 + 9})^3 &= (\sqrt[3]{x_2 + 9})^3 \\
x_1 + 9 &= x_2 + 9 \\
x_1 &= x_2.
\end{aligned}
$$

Thus, if $k(x_1) = k(x_2)$ then $x_1 = x_2$.
Hence, k is one-to-one.

35. Not invertible since it fails the Horizontal Line
Test.

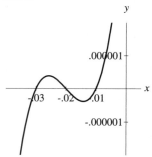

37. Not invertible since it fails the Horizontal Line
Test.

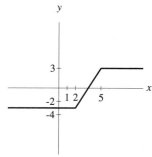

39. Interchange x and y then solve for y.

$$
\begin{aligned}
x &= 3y - 7 \\
\frac{x + 7}{3} &= y \\
\frac{x + 7}{3} &= f^{-1}(x)
\end{aligned}
$$

41. Interchange x and y then solve for y.

$$
\begin{aligned}
x &= 2 + \sqrt{y - 3} \quad \text{for } x \geq 2 \\
(x - 2)^2 &= y - 3 \quad \text{for } x \geq 2 \\
f^{-1}(x) &= (x - 2)^2 + 3 \quad \text{for } x \geq 2
\end{aligned}
$$

43. Interchange x and y then solve for y.

$$
\begin{aligned}
x &= -y - 9 \\
y &= -x - 9 \\
f^{-1}(x) &= -x - 9
\end{aligned}
$$

45. Interchange x and y then solve for y.

$$
\begin{aligned}
x &= \frac{y + 3}{y - 5} \\
xy - 5x &= y + 3
\end{aligned}
$$

$$xy - y = 5x + 3$$
$$y(x - 1) = 5x + 3$$
$$f^{-1}(x) = \frac{5x + 3}{x - 1}$$

47. Interchange x and y then solve for y.

$$x = -\frac{1}{y}$$
$$xy = -1$$
$$f^{-1}(x) = -\frac{1}{x}$$

49. Interchange x and y then solve for y.

$$x = \sqrt[3]{y - 9} + 5$$
$$x - 5 = \sqrt[3]{y - 9}$$
$$(x - 5)^3 = y - 9$$
$$f^{-1}(x) = (x - 5)^3 + 9$$

51. Interchange x and y then solve for y.

$$x = (y - 2)^2 \quad x \ge 0$$
$$\sqrt{x} = y - 2$$
$$f^{-1}(x) = \sqrt{x} + 2$$

53. Since $(g \circ f)(x) = 0.25(4x + 4) - 1 = x$ and $(f \circ g)(x) = 4(0.25x - 1) + 4 = x$, g and f are inverse functions of each other.

55. Since $(f \circ g)(x) = \left(\sqrt{x - 1}\right)^2 + 1 = x$ and and $(g \circ f)(x) = \sqrt{x^2 + 1 - 1} = \sqrt{x^2} = |x|$, g and f are not inverse functions of each other.

57. We find

$$(f \circ g)(x) = \frac{1}{1/(x - 3)} + 3$$
$$= x - 3 + 3$$
$$(f \circ g)(x) = x$$

and

$$(g \circ f)(x) = \frac{1}{\left(\frac{1}{x} + 3\right) - 3}$$
$$= \frac{1}{1/x}$$
$$(g \circ f)(x) = x.$$

Then g and f are inverse functions of each other.

59. We obtain

$$(f \circ g)(x) = \sqrt[3]{\frac{5x^3 + 2 - 2}{5}}$$
$$= \sqrt[3]{\frac{5x^3}{5}}$$
$$= \sqrt[3]{x^3}$$
$$(f \circ g)(x) = x$$

and

$$(g \circ f)(x) = 5\left(\sqrt[3]{\frac{x - 2}{5}}\right)^3 + 2$$
$$= 5\left(\frac{x - 2}{5}\right) + 2$$
$$= (x - 2) + 2$$
$$(g \circ f)(x) = x.$$

Thus, g and f are inverse functions of each other.

61. y_1 and y_2 are inverse functions of each other and $y_3 = y_2 \circ y_1$.

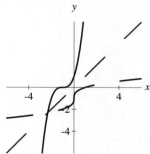

63. No, since they fail the Horizontal Line Test.

65. Yes, since the graphs are symmetric about the line $y = x$.

67. Graph of f^{-1}

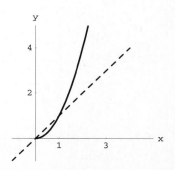

69. Graph of f^{-1}

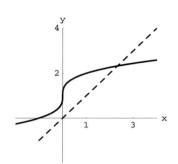

71. $f^{-1}(x) = \dfrac{x-2}{3}$

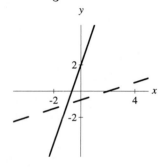

73. $f^{-1}(x) = \sqrt{x+4}$

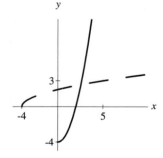

75. $f^{-1}(x) = \sqrt[3]{x}$

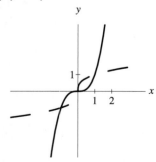

77. $f^{-1}(x) = (x+3)^2$ for $x \geq -3$

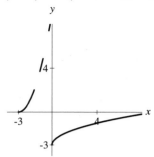

79. a) $f^{-1}(x) = x/5$

    b) $f^{-1}(x) = x + 88$

    c) $f^{-1}(x) = (x+7)/3$

    d) $f^{-1}(x) = \dfrac{x-4}{-3}$

    e) $f^{-1}(x) = 2(x+9) = 2x + 18$

    f) $f^{-1}(x) = -x$

    g) $f^{-1}(x) = (x+9)^3$

    h) $f^{-1}(x) = \sqrt[3]{\dfrac{x+7}{3}}$

81. $C = 1.08P$ expresses the total cost as a function of the purchase price; and $P = C/1.08$ is the purchase price as a function of the total cost.

83. The graph of t as a function of r satisfies the Horizontal Line Test and is invertible. Solving for r we find,

$$\begin{aligned} t - 7.89 &= -0.39r \\ r &= \frac{t - 7.89}{-0.39} \end{aligned}$$

and the inverse function is $r = \dfrac{7.89 - t}{0.39}$.

If $t = 5.55$ min., then $r = \dfrac{7.89 - 5.55}{0.39} = 6$ rowers.

85. Solving for w, we obtain

$$\begin{aligned} 1.496w &= V^2 \\ w &= \frac{V^2}{1.496} \end{aligned}$$

and the inverse function is $w = \dfrac{V^2}{1.496}$. If $V = 115$ ft./sec., then $w = \dfrac{115^2}{1.496} \approx 8,840$ lb.

87. a) Let $V = \$28,000$. The depreciation rate is

$$r = 1 - \left(\frac{28,000}{50,000}\right)^{1/5} \approx 0.109$$

or $r \approx 10.9\%$.

b) Writing V as a function of r we find

$$1 - r = \left(\frac{V}{50,000}\right)^{1/5}$$
$$(1-r)^5 = \frac{V}{50,000}$$

and $V = 50,000(1-r)^5$.

89. Since $g^{-1}(x) = \dfrac{x+5}{3}$ and $f^{-1}(x) = \dfrac{x-1}{2}$, we have

$$g^{-1} \circ f^{-1}(x) = \frac{\frac{x-1}{2}+5}{3} = \frac{x+9}{6}.$$

Likewise, since $(f \circ g)(x) = 6x - 9$, we get

$$(f \circ g)^{-1}(x) = \frac{x+9}{6}.$$

Hence, $(f \circ g)^{-1} = g^{-1} \circ f^{-1}$.

91. One can easily see that the slope of the line joining (a, b) to (b, a) is -1, and that their midpoint is $\left(\dfrac{a+b}{2}, \dfrac{a+b}{2}\right)$. This midpoint lies on the line $y = x$ whose slope is 1. Then $y = x$ is the perpendicular bisector of the line segment joining the points (a, b) and (b, a)

93. Dividing we get $\dfrac{x-3}{x+2} = 1 - \dfrac{5}{x+2}$

For Thought

1. False

2. False, since cost varies directly with the number of pounds purchased.

3. True **4.** True

5. True, since the area of a circle varies directly with the square of its radius.

6. False, since $y = k/x$ is undefined when $x = 0$.

7. True **8.** True **9.** True

10. False, the surface area is not equal to

$$k \cdot length \cdot width \cdot height$$

for any constant k.

2.6 Exercises

1. $G = kn$ **3.** $V = k/P$ **5.** $C = khr$

7. $Y = \dfrac{kx}{\sqrt{z}}$

9. A varies directly as the square of r

11. y varies inversely as x

13. Not a variation expression

15. a varies jointly as z and w

17. H varies directly as the square root of t and inversely as s

19. D varies jointly as L and J and inversely as W

21. Since $y = kx$ and $5 = k \cdot 9$, $k = 5/9$. Then $y = 5x/9$.

23. Since $T = k/y$ and $-30 = k/5$, $k = -150$. Thus, $T = -150/y$.

25. Since $m = kt^2$ and $54 = k \cdot 18$, $k = 3$. Thus, $m = 3t^2$.

27. Since $y = kx/\sqrt{z}$ and $2.192 = k(2.4)/\sqrt{2.25}$, we obtain $k = 1.37$. Hence, $y = 1.37x/\sqrt{z}$.

29. Since $y = kx$ and $9 = k(2)$, we obtain

$$y = \frac{9}{2} \cdot (-3) = -27/2.$$

31. Since $P = k/w$ and $2/3 = \dfrac{k}{1/4}$, we find

$$k = \frac{2}{3} \cdot \frac{1}{4} = \frac{1}{6}. \text{ Thus, } P = \frac{1/6}{1/6} = 1.$$

33. Since $A = kLW$ and $30 = k(3)(5\sqrt{2})$,

we obtain $A = \sqrt{2}(2\sqrt{3})\dfrac{1}{2} = \sqrt{6}$.

35. Since $y = ku/v^2$ and $7 = k \cdot 9/36$,

we find $y = 28 \cdot 4/64 = 7/4$.

37. Let L_i and L_f be the length in inches and feet, respectively. Then $L_i = 12L_f$ is a direct variation.

39. Let P and n be the cost per person and the number of persons, respectively. Then $P = 20/n$ is an inverse variation.

41. Let S_m and S_k be the speeds of the car in mph and kph, respectively. Then $S_m \approx S_k/1.6 \approx 0.6S_k$ is a direct variation.

43. Not a variation

45. Let A and W be the area and width, respectively. Then $A = 30W$ is a direct variation.

47. Let n and p be the number of gallons and price per gallon, respectively. Since $np = 5$, we obtain that $n = \dfrac{5}{p}$ is an inverse variation.

49. If p is the pressure at depth d, then $p = kd$. Since $4.34 = k(10)$, $k = 0.434$. At $d = 6000$ ft, the pressure is $p = 0.434(6000) = 2604$ lb per square inch.

51. If h is the number of hours, p is the number of pounds, and w is the number of workers then $h = kp/w$. Since $8 = k(3000)/6$, $k = 0.016$. Five workers can process 4000 pounds in $h = (0.016)(4000)/5 = 12.8$ hours.

53. Since $I = kPt$ and $20.80 = k(4000)(16)$, we find $k = 0.000325$. The interest from a deposit of \$6500 for 24 days is

$$I = (0.000325)(6500)(24) \approx \$50.70.$$

55. Since $C = kDL$ and $18.60 = k(6)(20)$, we obtain $k = 0.155$. The cost of a 16 ft pipe with a diameter of 8 inches is

$$C = 0.155(8)(16) = \$19.84.$$

57. Since $w = khd^2$ and $14.5 = k(4)(6^2)$, we find $k = \dfrac{14.5}{144}$. Then a 5-inch high can with a

diameter of 6 inches has weight

$$w = \frac{14.5}{144}(5)(6^2) = 18.125 \text{ oz.}$$

59. Since $V = kh/l$ and $10 = k(50)/(200)$, we get $k = 40$. The velocity, if the head is 60 ft and the length is 300 ft, is $V = (40)(60)/(300) = 8$ ft/year.

61. No, it is not directly proportional otherwise the following ratios $\dfrac{42,506}{1.34} \approx 31,720$,

$\dfrac{59,085}{0.295} \approx 200,288$, and

$\dfrac{738,781}{0.958} \approx 771,170$ would be

the same but they are not.

63. Since $g = ks/p$ and $76 = k(12)/(10)$, $k = \dfrac{190}{3}$.

If Calvin studies for 9 hours and plays for 15 hours, then his score is

$$g = \frac{190}{3} \cdot \frac{s}{p} = \frac{190}{3} \cdot \frac{9}{15} = 38.$$

65. Since $h = kv^2$ and $16 = k(32)^2$, we get $k = \dfrac{1}{64}$.

To reach a height of $20'2.5''$, the velocity v must satisfy

$$20 + \frac{2.5}{12} = \frac{1}{64}v^2.$$

Solving for v, we find $v \approx 35.96$ ft/sec.

Review Exercises

1. Function, domain and range are both $\{-2, 0, 1\}$

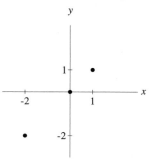

3. $y = 3 - x$ is a function, domain and range are both $(-\infty, \infty)$

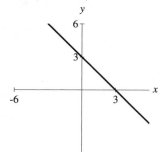

5. Not a function, domain is $\{2\}$, range is $(-\infty, \infty)$

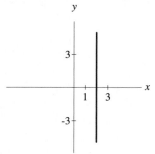

7. $x^2 + y^2 = 0.01$ is not a function, domain and range are both $[-0.1, 0.1]$

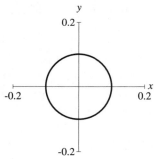

9. $x = y^2 + 1$ is not a function, domain is $[1, \infty)$, range is $(-\infty, \infty)$

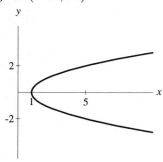

11. $y = \sqrt{x} - 3$ is a function, domain is $[0, \infty)$, range is $[-3, \infty)$

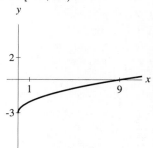

13. $9 + 3 = 12$

15. $24 - 7 = 17$

17. If $x^2 + 3 = 19$, then $x^2 = 16$ or $x = \pm 4$.

19. $g(12) = 17$

21. $7 + (-3) = 4$

23. $(4)(-9) = -36$

25. $f(-3) = 12$

27.

$$
\begin{aligned}
f(g(x)) &= f(2x - 7) \\
&= (2x - 7)^2 + 3 \\
&= 4x^2 - 28x + 52
\end{aligned}
$$

29. $(x^2 + 3)^2 + 3 = x^4 + 6x^2 + 12$

31. $(a + 1)^2 + 3 = a^2 + 2a + 4$

33.

$$
\begin{aligned}
\frac{f(3 + h) - f(3)}{h} &= \frac{(9 + 6h + h^2) + 3 - 12}{h} \\
&= \frac{6h + h^2}{h} \\
&= 6 + h
\end{aligned}
$$

35.

$$
\begin{aligned}
\frac{f(x + h) - f(x)}{h} &= \\
\frac{(x^2 + 2xh + h^2) + 3 - x^2 - 3}{h} &= \\
\frac{2xh + h^2}{h} &= \\
2x + h &=
\end{aligned}
$$

37. $g\left(\dfrac{x+7}{2}\right) = (x+7) - 7 = x$

39. $g^{-1}(x) = \dfrac{x+7}{2}$

41. $f(x) = \sqrt{x}, g(x) = 2\sqrt{x+3}$; left by 3, stretch by 2

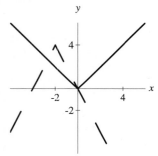

43. $f(x) = |x|, g(x) = -2|x+2| + 4$; left by 2, stretch by 2, reflect about x-axis, up by 4

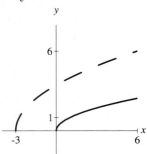

45. $f(x) = x^2, g(x) = \dfrac{1}{2}(x-2)^2 + 1$; right by 2, stretch by $\dfrac{1}{2}$, up by 1

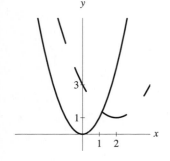

47. Translate the graph of f to the right by 2-units, stretch by a factor of 2, shift up by 1-unit.

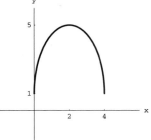

49. Translate the graph of f to the left by 1-unit, reflect about the x-axis, shift down by 3-units.

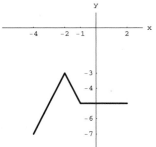

51. Translate the graph of f to the left by 2-units, stretch by a factor of 2, reflect about the x-axis.

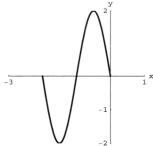

53. Stretch the graph of f by a factor of 2, reflect about the x-axis, shift up by 3-units.

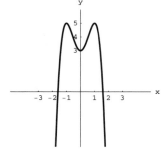

55. $F = f \circ g$

57. $H = f \circ h \circ g \circ j$

59. $N = h \circ f \circ j$

61. $R = g \circ h \circ j$

63.

$$\frac{f(x+h) - f(x)}{h} = \frac{-5(x+h) + 9 + 5x - 9}{h}$$

$$= \frac{-5h}{h}$$

$$= -5$$

65.

$$\frac{f(x+h) - f(x)}{h} =$$

$$= \frac{\dfrac{1}{2x+2h} - \dfrac{1}{2x}}{h} \cdot \frac{(2x+2h)(2x)}{(2x+2h)(2x)}$$

$$= \frac{(2x) - (2x+2h)}{h(2x+2h)(2x)}$$

$$= \frac{-2}{(2x+2h)(2x)}$$

$$= \frac{-1}{(x+h)(2x)}$$

67. Domain is $[-10, 10]$, range is $[0, 10]$, increasing on $(-10, 0)$, decreasing on $(0, 10)$

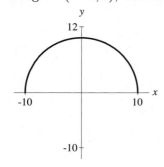

69. Domain and range are both $(-\infty, \infty)$, increasing on $(-\infty, \infty)$

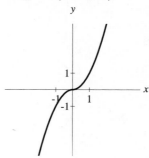

71. Domain is $(-\infty, \infty)$, range is $[-2, \infty)$, increasing on $(-2, 0)$ and $(2, \infty)$, decreasing on $(-\infty, -2)$ and $(0, 2)$

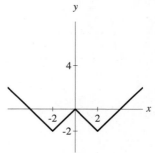

73. $y = |x| - 3$, domain is $(-\infty, \infty)$, range is $[-3, \infty)$

75. $y = -2|x| + 4$, domain is $(-\infty, \infty)$, range is $(-\infty, 4]$

77. $y = |x + 2| + 1$, domain is $(-\infty, \infty)$, range is $[1, \infty)$

79. Symmetry: y-axis

81. Symmetric about the origin

83. Neither symmetry

85. Symmetric about the y-axis

87. From the graph of $y = |x - 3| - 1$, the solution set is $(-\infty, 2] \cup [4, \infty)$

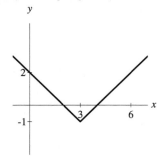

89. From the graph of $y = -2x^2 + 4$, the solution set is $(-\sqrt{2}, \sqrt{2})$

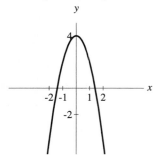

91. No solution since $-\sqrt{x + 1} - 2 \leq -2$

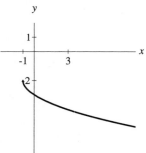

93. Inverse functions,
$$f(x) = \sqrt{x + 3}, g(x) = x^2 - 3 \text{ for } x \geq 0$$

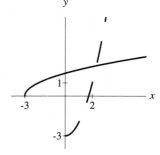

95. Inverse functions,
$$f(x) = 2x - 4, g(x) = \frac{1}{2}x + 2$$

97. Not invertible

99. Inverse is $f^{-1}(x) = \dfrac{x + 21}{3}$ with domain and range both $(-\infty, \infty)$

101. Not invertible

103. Inverse is $f^{-1}(x) = x^2 + 9$ for $x \geq 0$ with domain $[0, \infty)$ and range $[9, \infty)$

105. Inverse is $f^{-1}(x) = \dfrac{5x + 7}{1 - x}$ with domain $(-\infty, 1) \cup (1, \infty)$, and range $(-\infty, -5) \cup (-5, \infty)$

107. Inverse is $f^{-1}(x) = -\sqrt{x - 1}$ with domain $[1, \infty)$ and range $(-\infty, 0]$

109. Let x be the number of roses. The cost function is $C(x) = 1.20x + 40$, the revenue function is $R(x) = 2x$, and the profit function is $P(x) = R(x) - C(x)$ or $P(x) = 0.80x - 40$.

Since $P(50) = 0$, to make a profit she must sell at least 51 roses.

111. Since $h(0) = 64$ and $h(2) = 0$, the range of $h = -16t^2 + 64$ is the interval $[0, 64]$.
Then the domain of the inverse function is $[0, 64]$. Solving for t, we obtain

$$\begin{aligned} 16t^2 &= 64 - h \\ t^2 &= \frac{64 - h}{16} \\ t &= \frac{\sqrt{64 - h}}{4}. \end{aligned}$$

The inverse function is $t = \dfrac{\sqrt{64 - h}}{4}$.

113. Since $A = \pi \left(\dfrac{d}{2}\right)^2$, $d = 2\sqrt{\dfrac{A}{\pi}}$.

115. The average rate of change is
$$\frac{8 - 6}{4} = 0.5 \text{ inch/lb.}$$

117. Since $D = kW$ and $9 = k \cdot 25$, we obtain
$$D = \frac{9}{25}100 = 36.$$

119. Since $V = k\sqrt{h}$ and $45 = k\sqrt{1.5}$, the velocity of a Triceratops is $V = \dfrac{45}{\sqrt{1.5}} \cdot \sqrt{2.8} \approx 61$ kph.

121. Since $C = kd^2$ and $4.32 = k \cdot 36$, a 16-inch diameter globe costs $C = \dfrac{4.32}{36} \cdot 16^2 = \$30.72$.

Chapter 2 Test

1. No, since $(0, 5)$ and $(0, -5)$ are two ordered pairs with the same first coordinate and different second coordinates.

2. Yes, since to each x-coordinate there is exactly one y-coordinate, namely, $y = \dfrac{3x - 20}{5}$.

3. No, since $(1, -1)$ and $(1, -3)$ are two ordered pairs with the same first coordinate and different second coordinates.

4. Yes, since to each x-coordinate there is exactly one y-coordinate, namely, $y = x^3 - 3x^2 + 2x - 1$.

5. Domain is $\{2, 5\}$, range is $\{-3, -4, 7\}$

6. Domain is $[9, \infty)$, range is $[0, \infty)$

7. Domain is $[0, \infty)$, range is $(-\infty, \infty)$

8. Graph of $3x - 4y = 12$ includes the points $(4, 0), (0, -3)$

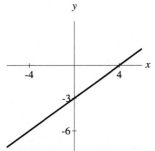

9. Graph of $y = 2x - 3$ includes the points $(3/2, 0), (0, -3)$

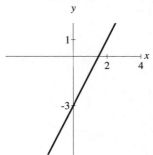

10. $y = \sqrt{25 - x^2}$ is a semicircle with radius 5

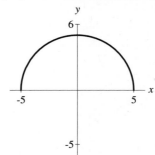

11. $y = -(x - 2)^2 + 5$ is a parabola with vertex $(2, 5)$

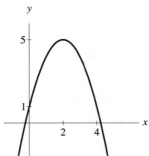

12. $y = 2|x| - 4$ includes the points $(0, -4), (\pm 3, 2)$

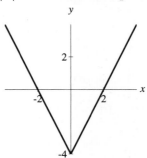

13. $y = \sqrt{x + 3} - 5$ includes the points $(1, -3), (6, -2)$

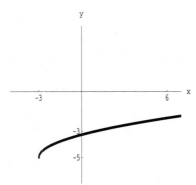

14. Graph includes the points $(-2, -2), (0, 2), (3, 2)$

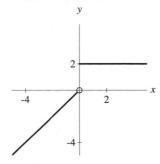

15. $\sqrt{9} = 3$ **16.** $f(5) = \sqrt{7}$

17. $f(3x - 1) = \sqrt{(3x - 1) + 2} = \sqrt{3x + 1}$.

18. $g^{-1}(x) = \dfrac{x + 1}{3}$ **19.** $\sqrt{16} + 41 = 45$

20.

$$
\begin{aligned}
\frac{g(x + h) - g(x)}{h} &= \frac{3(x + h) - 1 - 3x + 1}{h} \\
&= \frac{3h}{h} \\
&= 3
\end{aligned}
$$

21. Increasing on $(3, \infty)$, decreasing on $(-\infty, 3)$

22. Symmetric about the y-axis

23. Add 1 to $-3 < x - 1 < 3$ to obtain $-2 < x < 4$. Thus, the solution set $(-2, 4)$.

24. $g^{-1}(x) = (x - 3)^3 + 2$

25. $\dfrac{60 - 35}{200} = \$0.125$ per envelope

26. Since $I = k/d^2$ and $300 = k/4$, we get $k = 1200$. If $d = 10$, then $I = 1200/100 = 12$ candlepower.

27. Let s be the length of one side of the cube. By the Pythagorean Theorem we have $s^2 + s^2 = d^2$. Then $s = \dfrac{d}{\sqrt{2}}$ and the volume is $V = \left(\dfrac{d}{\sqrt{2}}\right)^3 = \dfrac{\sqrt{2}d^3}{4}$.

Tying It All Together

1. Add 3 to both sides of $2x - 3 = 0$ to obtain $2x = 3$. Thus, the solution set is $\left\{\dfrac{3}{2}\right\}$.

2. Add $2x$ to both sides of $-2x + 6 = 0$ to get $6 = 2x$. Then the solution set is $\{3\}$.

3. Note, $|x| = 100$ is equivalent to $x = \pm 100$. The solution set is $\{\pm 100\}$.

4. The equation is equivalent to $\dfrac{1}{2} = |x + 90|$.
Then $x + 90 = \pm\dfrac{1}{2}$ and $x = \pm\dfrac{1}{2} - 90$.
The solution set is $\{-90.5, -89.5\}$.

5. Note, $3 = 2\sqrt{x + 30}$. If we divide by 2 and square both sides, we get $x + 30 = \dfrac{9}{4}$.
Since $x = \dfrac{9}{4} - 30 = -27.75$, the solution set is $\{-27.75\}$.

6. Note, $\sqrt{x - 3}$ is not a real number if $x < 3$. Since $\sqrt{x - 3}$ is nonnegative for $x \geq 3$, it follows that $\sqrt{x - 3} + 15$ is at least 15. In particular, $\sqrt{x - 3} + 15 = 0$ has no real solution.

7. Rewriting, we obtain $(x - 2)^2 = \dfrac{1}{2}$. By the square root property, $x - 2 = \pm\dfrac{\sqrt{2}}{2}$.
Thus, $x = 2 \pm \dfrac{\sqrt{2}}{2}$. The solution set is $\left\{\dfrac{4 \pm \sqrt{2}}{2}\right\}$.

8. Rewriting, we get $(x+2)^2 = \dfrac{1}{4}$. By the square root property, $x + 2 = \pm\dfrac{1}{2}$. Thus, $x = -2 \pm \dfrac{1}{2}$. The solution set is $\left\{ -\dfrac{3}{2}, -\dfrac{5}{2} \right\}$.

9. Squaring both sides of $\sqrt{9 - x^2} = 2$, we get $9 - x^2 = 4$. Then $5 = x^2$. The solution set is $\left\{ \pm\sqrt{5} \right\}$.

10. Note, $\sqrt{49 - x^2}$ is not a real number if $49 - x^2 < 0$. Since $\sqrt{49 - x^2}$ is nonnegative for $49 - x^2 \geq 0$, we get that $\sqrt{49 - x^2} + 3$ is at least 3. In particular, $\sqrt{49 - x^2} + 3 = 0$ has no real solution.

11. Domain is $(-\infty, \infty)$, range is $(-\infty, \infty)$, x-intercept $(3/2, 0)$

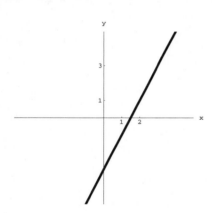

12. Domain is $(-\infty, \infty)$, range is $(-\infty, \infty)$, x-intercept $(3, 0)$

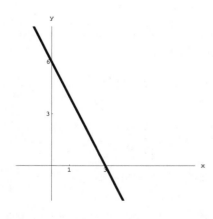

13. Domain is $(-\infty, \infty)$, range is $[-100, \infty)$, x-intercepts $(\pm 100, 0)$

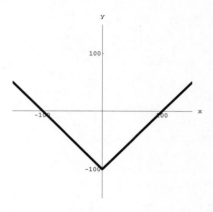

14. Domain is $(-\infty, \infty)$, range is $(-\infty, 1]$, x-intercepts $(-90.5, 0)$ and $(-89.5, 0)$.

15. Domain is $[-30, \infty)$ since we need to require $x + 30 \geq 0$, range is $(-\infty, 3]$, x-intercept is $(-27.75, 0)$ since the solution to $3 - 2\sqrt{x + 30} = 0$ is $x = -27.75$

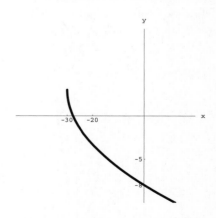

16. Domain is $[3, \infty)$ since we need to require $x - 3 \geq 0$, range is $[15, \infty)$, no x-intercept

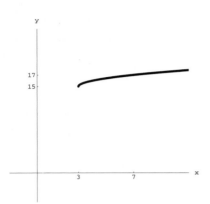

17. Domain is $(-\infty, \infty)$, range is $(-\infty, 1]$ since the vertex is $(2, 1)$, and the x-intercepts are $\left(\dfrac{4 \pm \sqrt{2}}{2}, 0\right)$ since the solutions of $0 = -2(x - 2)^2 + 1$ are $x = \dfrac{4 \pm \sqrt{2}}{2}$.

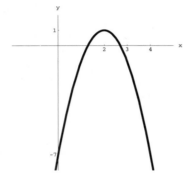

18. Domain is $(-\infty, \infty)$, range is $[-1, \infty)$ since the vertex is $(-2, -1)$, and the x-intercepts are $(-5/2, 0)$ and $(-3/2, 0)$ for the solutions to $0 = 4(x + 2)^2 - 1$ are $x = -5/2, -3/2$.

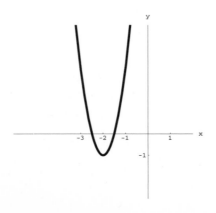

19. Domain is $[-3, 3]$, range is $[-2, 1]$ since the graph is a semi-circle with center $(0, -2)$ and radius 3. The x-intercepts are $(\pm\sqrt{5}, 0)$ for the solutions to $0 = \sqrt{9 - x^2} - 2$ are $x = \pm\sqrt{5}$. The graph is shown in the next column.

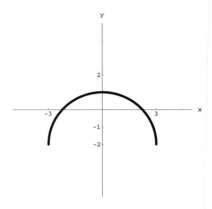

20. Domain is $[-7, 7]$, range is $[3, 10]$ since the graph is a semi-circle with center $(0, 3)$ and radius 7, there are no x-intercepts as seen from the graph

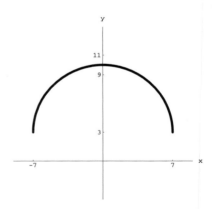

21. Since $2x > 3$ or $x > 3/2$, the solution set is $(3/2, \infty)$.

22. Since $-2x \leq -6$ or $x \geq 3$, the solution set is $[3, \infty)$.

23. Based on the portion of the graph of $y = |x| - 100$ above the x-axis, and its x-intercepts, the solution set of $|x| - 100 \geq 0$ is $(-\infty, -100] \cup [100, \infty)$.

24. Based on the part of the graph of $y = 1 - 2|x + 90|$ above the x-axis, and its x-intercepts, the solution set of $1 - 2|x + 90| > 0$ is $(-90.5, -89.5)$.

25. Based on the part of the graph of $y = 3 - 2\sqrt{x + 30}$ below the x-axis, and its x-intercepts, the solution set of $3 - 2\sqrt{x + 30} \leq 0$ is $[-27.75, \infty)$.

26. Solution set is $(-\infty, \infty)$ since the graph is entirely above the x-axis

27. Based on the portion of the graph of $y = -2(x - 2)^2 + 1$ below the x-axis, and its x-intercepts, the solution set of $-2(x - 2)^2 + 1 < 0$ is
$$\left(-\infty, \frac{4 - \sqrt{2}}{2}\right) \cup \left(\frac{4 + \sqrt{2}}{2}, \infty\right).$$

28. Based on the part of the graph of $y = 4(x + 2)^2 - 1$ above the x-axis, and its x-intercepts, the solution set of $4(x + 2)^2 - 1 < 0$ is $(-\infty, -5/2] \cup [-3/2, \infty)$.

29. Based on the part of the graph of $y = \sqrt{9 - x^2} - 2$ above the x-axis, and its x-intercepts, the solution set of $\sqrt{9 - x^2} - 2 \geq 0$ is $[-\sqrt{5}, \sqrt{5}]$.

30. Since the graph of $y = \sqrt{49 - x^2} + 3$ is entirely above the line $y = 3$, the solution to $\sqrt{49 - x^2} + 3 \leq 0$ is the empty set $\emptyset$.

31. $f(2) = 3(2 + 1)^3 - 24 = 3(27) - 24 = 57$

32. The graph of $f(x) = 3(x + 1)^3 - 24$ is given.

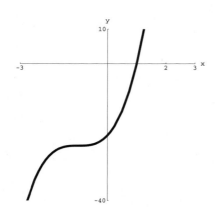

33. $f = K \circ F \circ H \circ G$

34. Solving for x, we get
$$\begin{aligned} 3(x + 1)^3 &= 24 \\ (x + 1)^3 &= 8 \\ x + 1 &= 2 \\ x &= 1. \end{aligned}$$

The solution set is $\{1\}$.

35. Based on the part of the graph of $y = 3(x + 1)^3 - 24$ above the x-axis, and its x-intercept $(1, 0)$, the solution set of $3(x + 1)^3 - 24 \geq 0$ is $[1, \infty)$.

36. Solving for x, we obtain
$$\begin{aligned} y + 24 &= 3(x + 1)^3 \\ \frac{y + 24}{3} &= (x + 1)^3 \\ \sqrt[3]{\frac{y + 24}{3}} &= x + 1 \\ \sqrt[3]{\frac{y + 24}{3}} - 1 &= x. \end{aligned}$$

37. Based on the answer from number 36, the inverse is $f^{-1}(x) = \sqrt[3]{\dfrac{x + 24}{3}} - 1$

38. The graph of f^{-1} is given below.

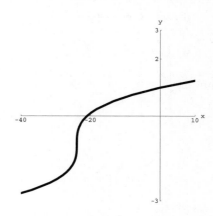

39. Based on the part of the graph of f^{-1} above the x-axis, and its x-intercept $(-21, 0)$, the solution set of $f^{-1}(x) > 0$ is $(-21, \infty)$.

40. Since $f = K \circ F \circ H \circ G$, we get
$$f^{-1} = G^{-1} \circ H^{-1} \circ F^{-1} \circ K^{-1}.$$

For Thought

1. False, the range of $y = x^2$ is $[0, \infty)$.

2. False, the vertex is the point $(3, -1)$.

3. True **4.** True **5.** True, since $\dfrac{-b}{2a} = \dfrac{6}{2 \cdot 3} = 1$.

6. True, the x-intercept of $y = (3x + 2)^2$ is the vertex $(-2/3, 0)$ and the y-intercept is $(0, 4)$.

7. True

8. True, since $(x - \sqrt{3})^2$ is always nonnegative.

9. True, since if x and $\dfrac{p - 2x}{2}$ are the length and the width, respectively, of a rectangle with perimeter p, then the area is $y = x \cdot \dfrac{p - 2x}{2}$. This is a parabola opening down with vertex $\left(\dfrac{p}{4}, \dfrac{p^2}{16} \right)$. Thus, the maximum area is $\dfrac{p^2}{16}$.

10. False

3.1 Exercises

1. Completing the square, we get

$$
\begin{aligned}
y &= \left(x^2 + 4x + \left(\frac{4}{2}\right)^2 \right) - \left(\frac{4}{2}\right)^2 \\
y &= \left(x^2 + 4x + 4 \right) - 4 \\
y &= (x + 2)^2 - 4.
\end{aligned}
$$

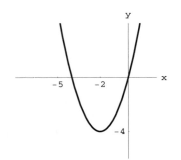

3. $y = \left(x^2 - 3x + \dfrac{9}{4} \right) - \dfrac{9}{4} = \left(x - \dfrac{3}{2} \right)^2 - \dfrac{9}{4}$

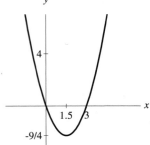

5. Completing the square, we get

$$
\begin{aligned}
y &= 2\left(x^2 - 6x + \left(\frac{6}{2}\right)^2 \right) - 2\left(\frac{6}{2}\right)^2 + 22 \\
y &= 2(x - 3)^2 + 4.
\end{aligned}
$$

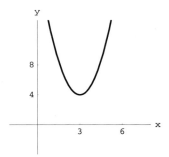

7. Completing the square, we find

$$
\begin{aligned}
y &= -3\left(x^2 - 2x + \left(\frac{2}{2}\right)^2 \right) + 3\left(\frac{2}{2}\right)^2 - 3 \\
y &= -3(x - 1)^2.
\end{aligned}
$$

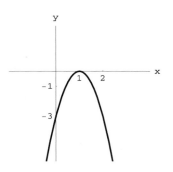

9. $y = \left(x^2 + 3x + \dfrac{9}{4}\right) - \dfrac{9}{4} + \dfrac{5}{2} = \left(x + \dfrac{3}{2}\right)^2 + \dfrac{1}{4}$

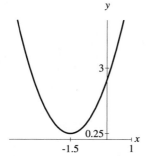

11. $y = -2\left(x^2 - \dfrac{3}{2}x + \dfrac{9}{16}\right) + \dfrac{9}{8} - 1 =$

$-2\left(x - \dfrac{3}{4}\right)^2 + \dfrac{1}{8}$

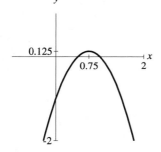

13. Since $\dfrac{-b}{2a} = \dfrac{12}{6} = 2$ and $f(2) = 12 - 24 + 1 =$

-11, the vertex is $(2, -11)$.

15. Vertex: $(4, 1)$

17. Since $\dfrac{-b}{2a} = \dfrac{1/3}{-1} = -\dfrac{1}{3}$ and $f(-1/3) =$

$-\dfrac{1}{18} + \dfrac{1}{9} = \dfrac{1}{18}$, the vertex is $\left(-\dfrac{1}{3}, \dfrac{1}{18}\right)$.

19. Up, vertex $(1, -4)$, axis of symmetry $x = 1$, range $[-4, \infty)$, minimum value -4, decreasing on $(-\infty, 1)$, inreasing on $(1, \infty)$.

21. Since it opens down with vertex $(0, 3)$, the range is $(-\infty, 3]$, maximum value is 3, decreasing on $(0, \infty)$, and increasing on $(-\infty, 0)$.

23. Since it opens up with vertex $(1, -1)$, the range is $[-1, \infty)$, minimum value is -1, decreasing on $(-\infty, 1)$, and increasing on $(1, \infty)$.

25. Since it opens up with vertex $(-4, -18)$, range is $[-18, \infty)$, minimum value is -18, decreasing on $(-\infty, -4)$, and increasing on $(-4, \infty)$.

27. Since it opens up with vertex is $(3, 4)$, the range is $[4, \infty)$, minimum value is 4, decreasing on $(-\infty, 3)$, and increasing on $(3, \infty)$.

29. Since it opens down with vertex $(3/2, 27/2)$, the range is $(-\infty, 27/2]$, maximum value is $27/2$, decreasing on $(3/2, \infty)$, and increasing on $(-\infty, 3/2)$.

31. Since it opens down with vertex is $(1/2, 9)$, the range is $(-\infty, 9]$, maximum value is 9, decreasing on $(1/2, \infty)$, and increasing on $(-\infty, 1/2)$.

33. Vertex $(0, -3)$, axis $x = 0$, y-intercept $(0, -3)$, x-intercepts $(\pm\sqrt{3}, 0)$, opening up

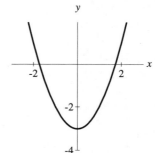

35. Vertex $(1/2, -1/4)$, axis $x = 1/2$, y-intercept $(0, 0)$, x-intercepts $(0, 0), (1, 0)$, opening up

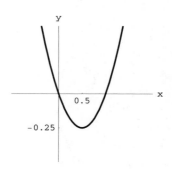

37. Vertex $(-3, 0)$, axis $x = -3$, y-intercept $(0, 9)$, x-intercept $(-3, 0)$, opening up

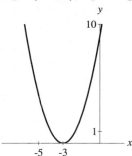

39. Vertex $(3, -4)$, axis $x = 3$, y-intercept $(0, 5)$, x-intercepts $(1, 0), (5, 0)$, opening up

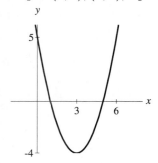

41. Vertex $(2, 12)$, axis $x = 2$, y-intercept $(0, 0)$, x-intercepts $(0, 0), (4, 0)$, opening down

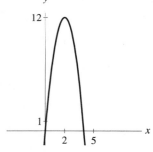

43. Vertex $(1, 3)$, axis $x = 1$, y-intercept $(0, 1)$, x-intercepts $\left(1 \pm \dfrac{\sqrt{6}}{2}, 0\right)$, opening down

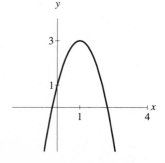

45. The sign graph of $(2x - 3)(x + 1) < 0$ is

```
- - - - - - - - - 0 + + + +
- - - - - 0 + + + + + + + +
<------------------------->
        -1        3/2
```

The solution set is the interval $(-1, 3/2)$

and the graph is
```
        -1    3/2
<------(======)------->
```

47. The sign graph of $(x + 3)(x - 5) > 0$ is

```
- - - - - - - - - 0 + + + +
- - - - - 0 + + + + + + + +
<------------------------->
        -3        5
```

The solution set is $(-\infty, -3) \cup (5, \infty)$

and the graph is
```
     -3   5
<====)   (====>
```

49. The sign graph of $(w + 2)(w - 6) \geq 0$ is

```
- - - - - - - - - 0 + + + +
- - - - - 0 + + + + + + + +
<------------------------->
        -2        6
```

The solution set is $(-\infty, -2] \cup [6, \infty)$

and the graph is
```
     -2  6
<====]   [====>
```

51. The sign graph of $(t - 4)(t + 4) \leq 0$ is

```
- - - - - - - - - 0 + + + +
- - - - - 0 + + + + + + + +
<------------------------->
        -4        4
```

The solution set is $[-4, 4]$ and the

graph is
```
     -4    4
<----[====]---->
```

53. The sign graph of $(a + 3)^2 \leq 0$ is

```
+ + + + 0 + + + + + + + + +
<------------------------->
        -3
```

The solution set is $\{-3\}$ and the

graph is
```
        -3
<-------●------->
```

55. The sign graph of $(2z - 3)^2 > 0$ is

```
+ + + + 0 + + + + + + + + +
<------------------------->
        3/2
```

The solution set is $(-\infty, 3/2) \cup (3/2, \infty)$

and the graph is 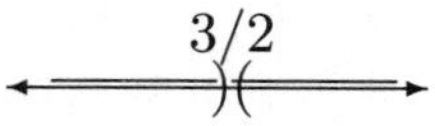

57. The roots of $x^2 - 4x + 2 = 0$ are $x_1 = 2 - \sqrt{2}$ and $x_2 = 2 + \sqrt{2}$.

If $x = -5$, then $(-5)^2 - 4(-5) + 2 > 0$.
If $x = 2$, then $(2)^2 - 4(2) + 2 < 0$.
If $x = 5$, then $(5)^2 - 4(5) + 2 > 0$.

$+ \quad 0 \quad - \quad 0 \quad +$

$-5 \quad x_1 \quad 0 \quad x_2 \quad 5$

The solution set of $x^2 - 4x + 2 < 0$

is $(2 - \sqrt{2}, 2 + \sqrt{2})$ and its graph follows.

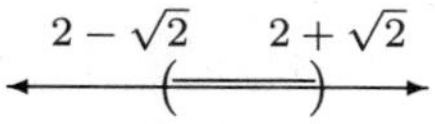

59. The roots of $x^2 - 10 = 0$ are $x_1 = -\sqrt{10}$ and $x_2 = \sqrt{10}$.

If $x = -4$, then $(-4)^2 - 10 > 0$.
If $x = 0$, then $(0)^2 - 10 < 0$.
If $x = 4$, then $(4)^2 - 10 > 0$.

$+ \quad 0 \quad - \quad 0 \quad +$

$-4 \quad x_1 \quad 0 \quad x_2 \quad 4$

The solution set of $x^2 - 9 \geq 1$

is $(-\infty, -\sqrt{10}] \cup [\sqrt{10}, \infty)$ and its

graph is 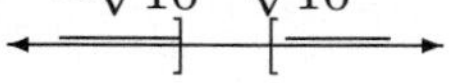

61. The roots of $y^2 - 10y + 18 = 0$ are $y_1 = 5 - \sqrt{7}$ and $y_2 = 5 + \sqrt{7}$.

If $y = 2$, then $(2)^2 - 10(2) + 18 > 0$.
If $y = 5$, then $(5)^2 - 10(5) + 18 < 0$.
If $y = 8$, then $(8)^2 - 10(8) + 18 > 0$.

$+ \quad 0 \quad - \quad 0 \quad +$

$2 \quad y_1 \quad 5 \quad y_2 \quad 8$

The solution set of $y^2 - 10y + 18 > 0$

is $(-\infty, 5 - \sqrt{7}) \cup (5 + \sqrt{7}, \infty)$ and its

graph is 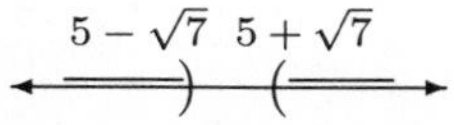

63. Note, $p^2 + 9 = 0$ has no real roots. If $p = 0$, then $(0)^2 + 9 > 0$. The signs of $p^2 + 9$ are shown below.

$+$

0

The solution set of $p^2 + 9 > 0$ is $(-\infty, \infty)$

and its graph follows.

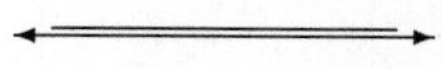

65. Note, $a^2 - 8a + 20 = 0$ has no real roots. If $a = 0$, then $(0)^2 - 8(0) + 20 > 0$. The signs of $a^2 - 8a + 20$ are shown below.

$+$

0

The solution set of $a^2 - 8a + 20 \leq 0$ is $\emptyset$.

67. Note, $2w^2 - 5w + 6 = 0$ has no real roots. If $w = 0$, then $2(0)^2 - 5(0) + 6 > 0$. The signs of $2w^2 - 5w + 6$ are shown below.

$+$

0

The solution set of $2w^2 - 5w + 6 > 0$ is $(-\infty, \infty)$

and its graph follows.

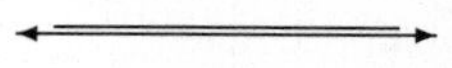

69. $(-\infty, -1] \cup [3, \infty)$

71. $(-3, 1)$

73. $[-3, 1]$

75. **a)** Since $x^2 - 3x - 10 = (x - 5)(x + 2) = 0$, the solution set is $\{-2, 5\}$.

b) Since $x^2 - 3x - 10 = -10$, we get $x^2 - 3x = 0$ or $x(x - 3) = 0$. The solution set is $\{0, 3\}$.

c) If $x = -3$, then $(-3)^2 - 3(-3) - 10 > 0$.
If $x = 0$, then $(0)^2 - 3(0) - 10 < 0$.
If $x = 6$, then $(6)^2 - 3(6) - 10 > 0$.
The signs of $x^2 - 3x - 10$ are shown below.

$+ \quad 0 \quad - \quad 0 \quad +$

$-3 \quad -2 \quad 0 \quad 5 \quad 6$

The solution set of $x^2 - 3x - 10 > 0$. is $(-\infty, -2) \cup (5, \infty)$

d) Using the sign graph of $x^2 - 3x - 10$ given in part c), the solution set of $x^2 - 3x - 10 \le 0$ is $[-2, 5]$.

e) By using the method of completing the square, one obtains

$$x^2 - 3x - 10 = \left(x - \frac{3}{2}\right)^2 - 10 - \frac{9}{4}$$

$$x^2 - 3x - 10 = \left(x - \frac{3}{2}\right)^2 - \frac{49}{4}.$$

The graph of f is obtained from the graph of $y = x^2$ by shifting to the right by $\frac{3}{2}$ unit, and down by $\frac{49}{4}$.

f) Domain is $(-\infty, \infty)$, range is $\left[-\frac{49}{4}, \infty\right)$, minimum y-value is $-\frac{49}{4}$

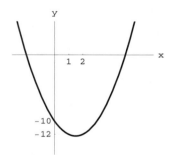

g) The solution to $f(x) > 0$ may be obtained by considering the part of the parabola that is above the x-axis, i.e., when x lies in $(-\infty, -2) \cup (5, \infty)$.

While, the solution to $f(x) \le 0$ may be obtained by considering the part of the parabola on or below the x-axis. i.e., when x is in $[-2, 5]$.

h) x-intercepts are $(5, 0)$ and $(-2, 0)$, the y-intercept is $(0, -10)$, axis of symmetry $x = \frac{3}{2}$, vertex $\left(\frac{3}{2}, -\frac{49}{4}\right)$, opens up, increasing on $\left(\frac{3}{2}, \infty\right)$, decreasing on $\left(-\infty, \frac{3}{2}\right)$

77. Since $\dfrac{-b}{2a} = \dfrac{-128}{-2(-16)} = 4$, the maximum height is $h(4) = 261$ ft.

79. a) Finding the vertex involves the number

$$\frac{-b}{2a} = \frac{-160}{-32} = 5.$$

Thus, the maximum height is

$$h(5) = -16(5)^2 + 160(5) + 8 = 408 \text{ ft.}$$

b) When the arrow reaches the ground, one has $h(t) = 0$.

$$-16t^2 + 160t + 8 = 0$$
$$t^2 - 10t = \frac{1}{2}$$
$$(t - 5)^2 = 25 + \frac{1}{2}$$
$$t = 5 \pm \sqrt{\frac{51}{2}}$$

Since $t \ge 0$, the arrow reaches the ground in

$$5 + \frac{\sqrt{102}}{2} = \frac{10 + \sqrt{102}}{2} \approx 10.05 \text{ sec.}$$

81. a) About 100 mph

b) The value of A that would maximize M is
$$A = \frac{-b}{2a} = \frac{-0.127}{-0.001306} \approx 97.24 \text{ mph.}$$

c) Lindbergh flying at 97 mph would use $\dfrac{97}{1.2} \approx 80.83$ lbs. of fuel or

$$\frac{80.83 \text{ lbs}}{6.12 \text{ lbs per gal}} \approx 13.2 \text{ gallons per hour.}$$

83. Let x and y be the length and width, respectively. Since $2x + 2y = 200$, we find $y = 100 - x$. The area as a function of x is

$$f(x) = x(100 - x) = 100x - x^2.$$

The graph of f is a parabola and its vertex is $(50, 2500)$. Thus, the maximum area is 2500 yd^2. Using $x = 50$ from the vertex, we get $y = 100 - x = 100 - 50 = 50$. The dimensions are 50 yd by 50 yd.

85. Let the length of the sides be x, x, x, y, y.

Then $3x + 2y = 120$ and $y = \dfrac{120 - 3x}{2}$.

The area of rectangular enclosure is

$$A(x) = xy = x\left(\left(\frac{120 - 3x}{2}\right)\right) = \frac{1}{2}(120x - 3x^2).$$

This is a parabola opening down. Since

$-\dfrac{b}{2a} = 20$ and $y = \dfrac{120 - 3(20)}{2} = 30$, the

optimal dimensions are 20 ft by 30 ft.

87. Let the length of the sides be x, x, and $30 - 2x$. The area of rectangular enclosure is

$$A(x) = x(30 - 2x) = 30x - 2x^2.$$

We have a parabola opening down. Since $-\dfrac{b}{2a} = 7.5$ and $y = 30 - 2(7.5) = 15$, the optimal dimensions are 15 ft by 7.5 ft.

89. Let x be the length of a folded side. The area of the cross-section is $A = x(10 - 2x)$. This is a parabola opening down with $-b/2a = 2.5$. The dimensions of the cross-section are 2.5 in. high and 5 in. wide.

91. Let n and p be the number of persons and the price of a tour per person, respectively.

a) The function expressing p as a function of n is $p = 50 - n$.

b) The revenue is $R = (50 - n)n$ or

$$R = 50n - n^2.$$

c) Since the graph of R is a parabola opening down with vertex $(25, 625)$, we find that 25 persons will give her the maximum revenue of \$625.

93. Since $v = 50p - 50p^2$ is a parabola opening down, to maximize v choose

$$p = \frac{-b}{2a} = \frac{50}{100} = 1/2.$$

95. (a) A graph of atmospheric pressure versus altitude is given below.

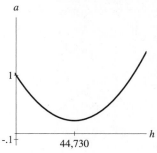

(b) From the graph, one finds the atmospheric pressure is decreasing as they went to $h = 29,029$ feet.

(c) Since we have

$$\frac{-b}{2a} = \frac{3.48 \times 10^{-5}}{2(3.89 \times 10^{-10})} \approx 44,730 \text{ feet}$$

the function is decreasing on the interval $(0, 44,730)$ and increasing on $(44,730, \infty)$.

(d) No, it does not make sense to speak of atmospheric pressure at heights that lie in $(44,730, \infty)$ since it is higher than the summit of 29,029 ft.

(e) It is valid for altitudes less than $30,000$ feet which is less than the height of the summit.

97. (a) Using a graphing calculator, we find that the equation of the regression line is

$$y = -2655x + 40,032.$$

The equation of the quadratic regression curve is $y = ax^2 + bx + c$ where

$$a = 45.04978355$$

$$b = -3105.464502$$

and

$$c = 40,857.85714$$

or approximately

$$y = 45x^2 - 3105x + 40,858.$$

(b) Judging from the graph, it is too close to tell which function seems more reasonable. [The green graph is the line, and the pink graph is the parabola.]

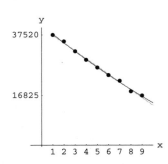

(c) We substitute $x = 11$ into the regression equations in part (a). Using the linear function, the price of an eleven year old car is $10,827.

Using $y = ax^2 + bx + c$, we obtain $12,149. The estimate using the quadratic regression curve seems more reasonable since it is closer to the value of a 9-year-old car.

For Thought

1. False   **2.**   True, by the Remainder Theorem.

3. True, by the Factor Theorem.

4. True, since $2^5 - 1 = 31$.

5. False, since $P(x) = 1$ has no zero.

6. False, rather $c^3 - c^2 + 4c - 5 = b$.

7. False, since $P(4) = -15$.   **8.**   True

9. True, since 1 is a root.

10. False, since -3 is not a root.

3.2 Exercises

1. Quotient $x - 3$, remainder 1

$$
\begin{array}{r}
x - 3 \\
x - 2 \ \overline{)x^2 - 5x + 7} \\
\underline{x^2 - 2x} \\
-3x + 7 \\
\underline{-3x + 6} \\
1
\end{array}
$$

3. Quotient $-2x^2 + 6x - 14$, remainder 33

$$
\begin{array}{r}
-2x^2 + 6x - 14 \\
x + 3 \ \overline{)-2x^3 + 0x^2 + 4x - 9} \\
\underline{-2x^3 - 6x^2} \\
6x^2 + 4x \\
\underline{6x^2 + 18x} \\
-14x - 9 \\
\underline{-14x - 42} \\
33
\end{array}
$$

5. Quotient $s^2 + 2$, remainder 16

$$
\begin{array}{r}
s^2 + 2 \\
s^2 - 5 \ \overline{)s^4 - 3s^2 + 6} \\
\underline{s^4 - 5s^2} \\
2s^2 + 6 \\
\underline{2s^2 - 10} \\
16
\end{array}
$$

7. Quotient $x + 6$, remainder 13

$$
\begin{array}{c|ccc}
2 & 1 & 4 & 1 \\
  &   & 2 & 12 \\
\hline
  & 1 & 6 & 13
\end{array}
$$

9. Quotient $-x^2 + 4x - 16$, remainder 57

$$
\begin{array}{c|cccc}
-3 & -1 & 1 & -4 & 9 \\
   &   & 3 & -12 & 48 \\
\hline
   & -1 & 4 & -16 & 57
\end{array}
$$

11. Quotient $4x^2 + 2x - 4$, remainder 0

$$
\begin{array}{c|cccc}
1/2 & 4 & 0 & -5 & 2 \\
    &   & 2 & 1 & -2 \\
\hline
    & 4 & 2 & -4 & 0
\end{array}
$$

13. Quotient $2a^2 - 4a + 6$, remainder 0

$$
\begin{array}{c|cccc}
-1/2 & 2 & -3 & 4 & 3 \\
     &   & -1 & 2 & -3 \\
\hline
     & 2 & -4 & 6 & 0
\end{array}
$$

15. Quotient $x^3 + x^2 + x + 1$, remainder -2

$$
\begin{array}{c|ccccc}
1 & 1 & 0 & 0 & 0 & -3 \\
  &   & 1 & 1 & 1 & 1 \\
\hline
  & 1 & 1 & 1 & 1 & -2
\end{array}
$$

17. Quotient $x^4 + 2x^3 - 2x^2 - 4x - 4$,
remainder -13

2	1	0	−6	0	4	−5
		2	4	−4	−8	−8
	1	2	−2	−4	−4	−13

19. $f(1) = 0$

1	1	0	0	0	0	-1
		1	1	1	1	1
	1	1	1	1	1	0

21. $f(-2) = -33$

-2	1	0	0	0	0	-1
		-2	4	-8	16	-32
	1	-2	4	-8	16	-33

23. $g(1) = 5$

1	1	-4	0	8
		1	-3	-3
	1	-3	-3	5

25. $g(-1/2) = 55/8$

-1/2	1	-4	0	8
		-1/2	9/4	-9/8
	1	-9/2	9/4	55/8

27. $h(-1) = 0$

-1	2	1	-1	3	3
		-2	1	0	-3
	2	-1	0	3	0

29. $h(1) = 8$

1	2	1	-1	3	3
		2	3	2	5
	2	3	2	5	8

31. Yes, $(x+3)(x^2+x-2) = (x+3)(x+2)(x-1)$

-3	1	4	1	-6
		-3	-3	6
	1	1	-2	0

33. Yes, $(x-4)(x^2+8x+15) = (x-4)(x+5)(x+3)$

4	1	4	-17	-60
		4	32	60
	1	8	15	0

35. Yes, since the remainder below is zero

3	2	-5	-4	3
		6	3	-3
	2	1	-1	0

37. No, since the remainder below is not zero

-2	1	2	3	1
		-2	0	-6
	1	0	3	-5

39. Yes, since the remainder below is zero

-1	1	2	4	6	3
		-1	-1	-3	-3
	1	1	3	3	0

41. No, since the remainder below is not zero

1/2	1	3	-5	7
		1/2	7/4	-13/8
	1	7/2	-13/4	43/8

43. $\pm\{1, 2, 3, 4, 6, 8, 12, 24\}$

45. $\pm\{1, 3, 5, 15\}$

47. $\pm\left\{1, 3, 5, 15, \dfrac{1}{2}, \dfrac{1}{4}, \dfrac{1}{8}, \dfrac{3}{2}, \dfrac{3}{4}, \right.$
$\left. \dfrac{3}{8}, \dfrac{5}{2}, \dfrac{5}{4}, \dfrac{5}{8}, \dfrac{15}{2}, \dfrac{15}{4}, \dfrac{15}{8}\right\}$

49. $\pm\left\{1, 2, \dfrac{1}{2}, \dfrac{1}{3}, \dfrac{1}{6}, \dfrac{1}{9}, \dfrac{1}{18}, \dfrac{2}{3}, \dfrac{2}{9}\right\}$

51. Zeros are $2, 3, 4$ since

2	1	-9	26	-24
		2	-14	24
	1	-7	12	0

and $x^2 - 7x + 12 = (x-4)(x-3)$

53. Zeros are $-3, 2 \pm i$ since

-3	1	-1	-7	15
		-3	12	-15
	1	-4	5	0

$x^2 - 4x + 5 = (x-2)^2 + 1 = 0$ or $x - 2 = \pm i$

55. Zeros are $1/2, 3/2, 5/2$ since

1/2	8	-36	46	-15
		4	-16	15
	8	-32	30	0

and $8a^2 - 32a + 30 = 2(4a^2 - 16a + 15) = 2(2a-3)(2a-5)$

57. Zeros are $\dfrac{1}{2}, \dfrac{1\pm i}{3}$ since

1/2	18	-21	10	-2
		9	-6	2
	18	-12	4	0

and the zeros of $18t^2 - 12t + 4$ are (by the quadratic formula) $\dfrac{1\pm i}{3}$

59. Zeros are $1, -2, \pm i$ since

1	1	1	-1	1	-2
		1	2	1	2
	1	2	1	2	0

-2	1	2	1	2
		-2	0	-2
	1	0	1	0

and the zeros of $w^2 + 1$ are $\pm i$

61. Zeros are $-1, \pm\sqrt{2}$ since

-1	1	2	-1	-4	-2
		-1	-1	2	2
	1	1	-2	-2	0

-1	1	1	-2	-2
		-1	0	2
	1	0	-2	0

and the zeros of $x^2 - 2$ are $\pm\sqrt{2}$

63. Zeros are $1/2, 1/3, 1/4$ since

1/2	24	-26	9	-1
		12	-7	1
	24	-14	2	0

and $2(12x^2 - 7x + 1) = 2(4x-1)(3x-1)$

65. Rational zero is $1/16$ since

1/16	16	-33	82	-5
		1	-2	5
	16	-32	80	0

and by the quadratic formula $16x^2 - 32x + 80$ has imaginary zeros $1 \pm 2i$.

67. Rational zeros are $7/3, -6/7$ since

7/3	21	-31	-21	-31	-42
		49	42	49	42
	21	18	21	18	0

-6/7	21	18	21	18
		-18	0	-18
	21	0	21	0

and $21x^2 + 21 = 0$ has imaginary zeros $\pm i$.

69. Dividing $x^3 + 6x^2 + 3x - 10$ by $x - 1$, we find

$$
\begin{array}{c|cccc}
1 & 1 & 6 & 3 & -10 \\
  &   & 1 & 7 & 10 \\
\hline
  & 1 & 7 & 10 & 0
\end{array}
$$

Moreover, $x^2 + 7x + 10 = (x + 5)(x + 2)$.
Note, the zeros of $x^2 + 9$ are $\pm 3i$.
Thus, the zeros of $f(x)$ are $x = -5, -2, 1, \pm 3i$.

71. Dividing $x^3 - 9x^2 + 23x - 15$ by $x - 1$, we find

$$
\begin{array}{c|cccc}
1 & 1 & -9 & 23 & -15 \\
  &   & 1 & -8 & 15 \\
\hline
  & 1 & -8 & 15 & 0
\end{array}
$$

For the quotient, we find

$$x^2 - 8x + 15 = (x - 5)(x - 3).$$

Then the zeros of $x^3 - 9x^2 + 23x - 15$ are $x = 1, 3, 5$.

By using the method of completing the square, we find

$$x^2 - 4x + 1 = (x - 2)^2 - 3.$$

Then the zeros of $(x - 2)^2 - 3$ are $2 \pm \sqrt{3}$.

Thus, all the zeros are $x = 1, 3, 5, 2 \pm \sqrt{3}$.

73.

$$\frac{2x + 1}{x - 2} = 2 + \frac{5}{x - 2} \quad \text{since}$$

$$
\begin{array}{c|cc}
2 & 2 & 1 \\
  &   & 4 \\
\hline
  & 2 & 5
\end{array}
$$

75.

$$\frac{a^2 - 3a + 5}{a - 3} = a + \frac{5}{a - 3} \quad \text{since}$$

$$
\begin{array}{c|ccc}
3 & 1 & -3 & 5 \\
  &   & 3 & 0 \\
\hline
  & 1 & 0 & 5
\end{array}
$$

77.

$$\frac{c^2 - 3c - 4}{c^2 - 4} = 1 + \frac{-3c}{c^2 - 4} \quad \text{since}$$

$$
\begin{array}{r}
1 \\
c^2 - 4 \overline{\smash{)}c^2 - 3c - 4} \\
\underline{c^2 + 0c - 4} \\
-3c
\end{array}
$$

79.

$$\frac{4t - 5}{2t + 1} = 2 + \frac{-7}{2t + 1} \quad \text{since}$$

$$
\begin{array}{r}
2 \\
2t + 1 \overline{\smash{)}4t - 5} \\
\underline{4t + 2} \\
-7
\end{array}
$$

81. a) Note, $\dfrac{P(t)}{t} = -t^3 + 12t^2 - 58t + 132$.

$$
\begin{array}{c|cccc}
6 & -1 & 12 & -58 & 132 \\
  &   & -6 & 36 & -132 \\
\hline
  & -1 & 6 & -22 & 0
\end{array}
$$

The drug will be eliminated in $t = 6$ hr.

b) About 120 ppm

c) About 3 hours

d) Between 1 and 5 hours approximately, the concentration is above 80 ppm. Thus, the concentration is above 80 ppm for about 4 hours.

83. If w is the width, then $w(w + 4)(w + 9) = 630$. This can be re-written as

$$w^3 + 13w^2 + 36w - 630 = 0.$$

Using synthetic division, we find

$$
\begin{array}{c|cccc}
5 & 1 & 13 & 36 & -630 \\
  &   & 5 & 90 & 630 \\
\hline
  & 1 & 18 & 126 & 0
\end{array}
$$

Since $w^2 + 18w + 126 = 0$ has non-real roots, the width of the HP box is $w = 5$ in. The dimensions are 5 in. by 9 in. by 14 in.

85. Remainder is $c = 10$ since

$$
\begin{array}{c|ccc}
3 & 1 & -2 & 7 \\
  &   & 3 & 3 \\
\hline
  & 1 & 1 & 10
\end{array}
$$

87. If $x - c$ is a factor of $P(x)$ then $P(x) = Q(x) \cdot (x-c)$ for some polynomial $Q(x)$. Thus, $P(c) = Q(c) \cdot 0 = 0$ and c is a zero of $P(x)$.

For Thought

1. False, since 1 has multiplicity 1.   **2.**   True

3. True   **4.**   False, it factors as $(x - 5)^4(x + 2)$.

5. False, rather $4+5i$ is also a solution.   **6.**   True

7. False, since they are solutions to a polynomial with real coefficients of degree at least 4.

8. False, 2 is not a solution.

9. True, since $-x^3 - 5x^2 - 6x - 1 = 0$ has no sign changes.

10. True

3.3 Exercises

1. Degree 2; 5 with multiplicity 2 since $(x-5)^2 = 0$

3. Degree 5; 0 with multiplicity 3, and ± 3 since $x^3(x - 3)(x + 3) = 0$

5. Degree 4; 0 and 1 each have multiplicity 2 since $x^2(x^2 - 2x + 1) = x^2(x - 1)^2$

7. Degree 4; 3/2 and $-4/3$ each with multiplicity 2

9. Degree 3; the roots are $0, 2 \pm \sqrt{10}$ since $x(x^2 - 4x - 6) = x((x - 2)^2 - 10) = 0$

11. $x^2 + 9$

13. $\left[(x - 1) - \sqrt{2}\right]\left[(x - 1) + \sqrt{2}\right] = (x-1)^2 - 2 = x^2 - 2x - 1$

15. $[(x - 3) - 2i][(x - 3) + 2i] = (x - 3)^2 + 4 = x^2 - 6x + 13$

17. $(x - 2)[(x - 3) - 4i][(x - 3) + 4i] = (x - 2)[(x - 3)^2 + 16] = x^3 - 8x^2 + 37x - 50$

19. $(x + 3)(x - 5) = 0$ or $x^2 - 2x - 15 = 0$

21. $(x + 4i)(x - 4i) = 0$ or $x^2 + 16 = 0$

23. $(x - (3-i))(x - (3+i)) = 0$ or $x^2 - 6x + 10 = 0$

25. $(x+2)(x-i)(x+i) = 0$ or $x^3 + 2x^2 + x + 2 = 0$

27. $x(x - i\sqrt{3})(x + i\sqrt{3}) = 0$ or $x^3 + 3x = 0$

29. $(x - 3)[x - (1 - i)][x - (1 + i)] = 0$ or $x^3 - 5x^2 + 8x - 6 = 0$

31. $(x - 1)(x - 2)(x - 3) = 0$ or $x^3 - 6x^2 + 11x - 6 = 0$

33. $(x - 1)[x - (2 - 3i)][x - (2 + 3i)] = 0$ or $x^3 - 5x^2 + 17x - 13 = 0$

35. $(2x - 1)(3x - 1)(4x - 1) = 0$ or $24x^3 - 26x^2 + 9x - 1 = 0$

37. $(x - i)(x + i)[x - (1 + i)][x - (1 - i)] = 0$ or $x^4 - 2x^3 + 3x^2 - 2x + 2 = 0$

39. $P(x) = x^3 + 5x^2 + 7x + 1$ has no sign change and $P(-x) = -x^3 + 5x^2 - 7x + 1$ has 3 sign changes. There are (a) 3 negative roots, or (b) 1 negative root & 2 imaginary roots.

41. $P(x) = -x^3 - x^2 + 7x + 6$ has 1 sign change and $P(-x) = x^3 - x^2 - 7x + 6$ has 2 sign changes. There are (a) 1 positive root and 2 negative roots, or (b) 1 positive root and 2 imaginary roots.

43. $P(y) = y^4 + 5y^2 + 7 = P(-y)$ has no sign change. There are 4 imaginary roots.

45. $P(t) = t^4 - 3t^3 + 2t^2 - 5t + 7$ has 4 sign changes and $P(-t) = t^4 + 3t^3 + 2t^2 + 5t + 7$ has no sign change. There are (a) 4 positive roots, or (b) 2 positive roots and 2 imaginary roots, or (c) 4 imaginary roots.

47. $P(x) = x^5 + x^3 + 5x$ and $P(-x) = -x^5 - x^3 - 5x$ have no sign changes; 4 imaginary roots and 0.

49. Best integral bounds are $-1 < x < 3$. One checks that $1, 2$ are not upper bounds and that 3 is a bound.

$$
\begin{array}{r|rrrr}
3 & 2 & -5 & 0 & 6 \\
  &   & 6 & 3 & 9 \\
\hline
  & 2 & 1 & 3 & 15
\end{array}
$$

-1 is a lower bound since

$$
\begin{array}{r|rrrr}
-1 & 2 & -5 & 0 & 6 \\
   &   & -2 & 7 & -7 \\
\hline
   & 2 & -7 & 7 & -1
\end{array}
$$

51. Best integral bounds are $-3 < x < 2$. One checks that 1 is not an upper bound and that 2 is a bound.

$$
\begin{array}{r|rrrr}
2 & 4 & 8 & -11 & -15 \\
  &   & 8 & 32 & 42 \\
\hline
  & 4 & 16 & 21 & 27
\end{array}
$$

$-1, -2$ are not lower bounds but -3 is a bound since

$$
\begin{array}{r|rrrr}
-3 & 4 & 8 & -11 & -15 \\
   &   & -12 & 12 & -3 \\
\hline
   & 4 & -4 & 1 & -18
\end{array}
$$

53. Best integral bounds are $-1 < x < 5$. One checks that $1, 2, 3, 4$ are not upper bounds and that 5 is a bound

$$
\begin{array}{r|rrrrr}
5 & 1 & -5 & 3 & 2 & -1 \\
  &   & 5 & 0 & 15 & 85 \\
\hline
  & 1 & 0 & 3 & 17 & 84
\end{array}
$$

-1 is a lower bound since

$$
\begin{array}{r|rrrrr}
-1 & 1 & -5 & 3 & 2 & -1 \\
   &   & -1 & 6 & -9 & 7 \\
\hline
   & 1 & -6 & 9 & -7 & 6
\end{array}
$$

55. Best integral bounds are $-1 < x < 3$. Multiply equation by -1, this makes the leading coefficient positive. One checks that $1, 2$ are not upper bounds and that 3 is a bound.

$$
\begin{array}{r|rrrr}
3 & 2 & -5 & 3 & -9 \\
  &   & 6 & 3 & 18 \\
\hline
  & 2 & 1 & 6 & 9
\end{array}
$$

-1 is a lower bound since

$$
\begin{array}{r|rrrr}
-1 & 2 & -5 & 3 & -9 \\
   &   & -2 & 7 & -10 \\
\hline
   & 2 & -7 & 10 & -19
\end{array}
$$

57. Roots are $1, 5, -2$ since

$$
\begin{array}{r|rrrr}
1 & 1 & -4 & -7 & 10 \\
  &   & 1 & -3 & -10 \\
\hline
  & 1 & -3 & -10 & 0
\end{array}
$$

and $x^2 - 3x - 10 = (x - 5)(x + 2)$

59. Roots are $-3, \dfrac{3 \pm \sqrt{13}}{2}$ since

$$
\begin{array}{r|rrrr}
-3 & 1 & 0 & -10 & -3 \\
   &   & -3 & 9 & 3 \\
\hline
   & 1 & -3 & -1 & 0
\end{array}
$$

and by the quadratic formula the roots of $x^2 - 3x - 1 = 0$ are $\dfrac{3 \pm \sqrt{13}}{2}$.

61. Roots are $2, -4, \pm i$ since

$$
\begin{array}{r|rrrrr}
2 & 1 & 2 & -7 & 2 & -8 \\
  &   & 2 & 8 & 2 & 8 \\
\hline
  & 1 & 4 & 1 & 4 & 0
\end{array}
$$

$$
\begin{array}{r|rrrr}
-4 & 1 & 4 & 1 & 4 \\
   &   & -4 & 0 & -4 \\
\hline
   & 1 & 0 & 1 & 0
\end{array}
$$

and the root of $x^2 + 1 = 0$ are $\pm i$.

63. Roots are $1/3, 1/2, -5$ since

$$
\begin{array}{r|rrrr}
1/3 & 6 & 25 & -24 & 5 \\
    &   & 2 & 9 & -5 \\
\hline
    & 6 & 27 & -15 & 0
\end{array}
$$

and $6x^2 + 27x - 15 = 3(2x - 1)(x + 5)$.

65. $1, -2$ each have multiplicity 2 since

$$
\begin{array}{r|rrrrr}
1 & 1 & 2 & -3 & -4 & 4 \\
  &   & 1 & 3 & 0 & -4 \\
\hline
  & 1 & 3 & 0 & -4 & 0
\end{array}
$$

$$
\begin{array}{r|rrrr}
-2 & 1 & 3 & 0 & -4 \\
   &   & -2 & -2 & 4 \\
\hline
   & 1 & 1 & -2 & 0
\end{array}
$$

and $x^2 + x - 2 = (x+2)(x-1)$.

67. Use synthetic division on the cubic factor in $x(x^3 - 6x^2 + 12x - 8) = 0$.

$$
\begin{array}{r|rrrr}
2 & 1 & -6 & 12 & -8 \\
  &   & 2 & -8 & 8 \\
\hline
  & 1 & -4 & 4 & 0
\end{array}
$$

Since $x^2 - 4x + 4 = (x-2)^2$, the roots are 2 (with multiplicity 3) and 0.

69. Use synthetic division on the 5th degree factor in $x(x^5 - x^4 - x^3 + x^2 - 12x + 12) = 0$.

$$
\begin{array}{r|rrrrrr}
1 & 1 & -1 & -1 & 1 & -12 & 12 \\
  &   & 1 & 0 & -1 & 0 & -12 \\
\hline
  & 1 & 0 & -1 & 0 & -12 & 0
\end{array}
$$

$$
\begin{array}{r|rrrrr}
2 & 1 & 0 & -1 & 0 & -12 \\
  &   & 2 & 4 & 6 & 12 \\
\hline
  & 1 & 2 & 3 & 6 & 0
\end{array}
$$

$$
\begin{array}{r|rrrr}
-2 & 1 & 2 & 3 & 6 \\
   &   & -2 & 0 & -6 \\
\hline
   & 1 & 0 & 3 & 0
\end{array}
$$

Note, the roots of $x^2 + 3 = 0$ are $\pm i\sqrt{3}$.
Thus, the roots are $x = 0, 1, \pm 2, \pm i\sqrt{3}$.

71. The roots are $\pm 1, -2, 1/4, 3/2$ since

$$
\begin{array}{r|rrrrrr}
-1 & 8 & 2 & -33 & 4 & 25 & -6 \\
   &   & -8 & 6 & 27 & -31 & 6 \\
\hline
   & 8 & -6 & -27 & 31 & -6 & 0
\end{array}
$$

$$
\begin{array}{r|rrrrr}
-2 & 8 & -6 & -27 & 31 & -6 \\
   &   & -16 & 44 & -34 & 6 \\
\hline
   & 8 & -22 & 17 & -3 & 0
\end{array}
$$

$$
\begin{array}{r|rrrr}
1 & 8 & -22 & 17 & -3 \\
  &   & 8 & -14 & 3 \\
\hline
  & 8 & -14 & 3 & 0
\end{array}
$$

and the last quotient is

$$8x^2 - 14x + 4 = (4x - 1)(2x - 3).$$

73. By the Theorem of Bounds and the graph below

$$
\begin{array}{r|rrrr}
6 & 2 & -3 & -50 & 18 \\
  &   & 12 & 54 & 24 \\
\hline
  & 2 & 9 & 4 & 42
\end{array}
$$

$$
\begin{array}{r|rrrr}
-5 & 2 & -3 & -50 & 18 \\
   &   & -10 & 65 & -75 \\
\hline
   & 2 & -13 & 15 & -57
\end{array}
$$

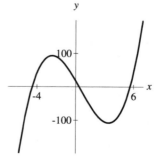

the intervals are $-5 < x < 6, -5 < x < 6$

75. By the Theorem of Bounds and the graph below

$$
\begin{array}{r|rrrrr}
6 & 1 & 0 & -26 & 0 & 153 \\
  &   & 6 & 36 & 60 & 360 \\
\hline
  & 1 & 6 & 10 & 60 & 513
\end{array}
$$

$$
\begin{array}{r|rrrrr}
-6 & 1 & 0 & -26 & 0 & 153 \\
   &   & -6 & 36 & -60 & 360 \\
\hline
   & 1 & -6 & 10 & -60 & 513
\end{array}
$$

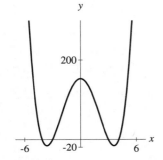

the intervals are $-6 < x < 6, -5 < x < 5$.

77. By the Theorem of Bounds and the graph below

$$
\begin{array}{r|rrrr}
23 & 4 & -90 & -2 & 45 \\
   &   & 92 & 46 & 1012 \\
\hline
   & 4 & 2 & 44 & 1057
\end{array}
$$

$$
\begin{array}{r|rrrr}
-1 & 4 & -90 & -2 & 45 \\
   &   & -4 & 94 & -92 \\
\hline
   & 4 & -94 & 92 & -47
\end{array}
$$

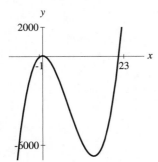

the intervals are $-1 < x < 23, -1 < x < 23$.

79. Multiplying by 100, $t^3 - 8t^2 + 11t + 20 = 0$.

$$
\begin{array}{r|rrrr}
4 & 1 & -8 & 11 & 20 \\
  &   & 4 & -16 & -20 \\
\hline
  & 1 & -4 & -5 & 0
\end{array}
$$

Since $t^2 - 4t - 5 = (t - 5)(t + 1) = 0$, we obtain $t = 4$ hr and $t = 5$ hr.

81. Let x be the radius of the cone.

The volume of the cone is $\dfrac{\pi}{3}x^2 \cdot 2$ and the volume of the cylinder with height $4x$ is $\pi x^2 \cdot (4x)$. So $114\pi = \dfrac{\pi}{3}2x^2 + \pi x^2(4x)$.

Divide by π, multiply by 3, and simplify to obtain $12x^3 + 2x^2 - 342 = 0$.

$$
\begin{array}{r|rrrr}
3 & 12 & 2 & 0 & -342 \\
  &   & 36 & 114 & 342 \\
\hline
  & 12 & 38 & 114 & 0
\end{array}
$$

The radius is $x = 3$ in. since $12x^2 + 38x + 114 = 0$ has no real roots.

83. From the following

$$
\begin{aligned}
\overline{(a + bi) + (c + di)} &= \overline{(a + c) + i(b + d)} \\
&= (a + c) - i(b + d)
\end{aligned}
$$

and

$$
\begin{aligned}
\overline{a + bi} + \overline{c + di} &= (a - bi) + (c - di) \\
&= (a + c) - i(b + d)
\end{aligned}
$$

we obtain that the conjugate of the sum of two complex numbers is equal to the sum of their conjugates.

85. If a is real then $\bar{a} = \overline{a + 0i} = a - 0i = a$.

87. Let $f(x) = a(x - 1)(x - 2)(x - 3)$.

Since $f(0) = -6a = 3$, we find $a = -\dfrac{1}{2}$.

Then substitute $a = -\dfrac{1}{2}$ and multiply out $f(x)$ to obtain $f(x) = -\dfrac{1}{2}x^3 + 3x^2 - \dfrac{11}{2}x + 3$.

For Thought

1. False, since

$$(\sqrt{x-1} + \sqrt{x})^2 = (x-1) + 2\sqrt{x(x-1)} + x.$$

2. False, since -1 is a solution of the first and not of the second equation.

3. False, since -27 is a solution of the first equation but not of the second.

4. False, rather let $u = x^{1/4}$ and $u^2 = x^{1/2}$.

5. True, since $x - 1 = \pm 4^{-3/2}$.

6. False, $\left(-\dfrac{1}{32}\right)^{-2/5} = (-32)^{2/5} = (-2)^2 = 4.$

7. False, $x = -2$ is not a solution.

8. True

9. True

10. False, since $(x^3)^2 = x^6$.

3.4 Exercises

1. Factor: $x^2(x+3) - 4(x+3) = 0$
$(x^2 - 4)(x+3) = (x-2)(x+2)(x+3) = 0$
The solution set is $\{\pm 2, -3\}$.

3. Factor: $2x^2(x+500) - (x+500) = 0$
$(2x^2 - 1)(x + 500) = 0$
The solution set is $\left\{\pm\dfrac{\sqrt{2}}{2}, -500\right\}$.

5. Set the right-hand side to 0 and factor.

$$a(a^2 - 15a + 5) = 0$$
$$a = \frac{15 \pm \sqrt{15^2 - 4(1)(5)}}{2} \quad \text{or} \quad a = 0$$
$$a = \frac{15 \pm \sqrt{205}}{2} \quad \text{or} \quad a = 0$$

The solution set is $\left\{\dfrac{15 \pm \sqrt{205}}{2}, 0\right\}$.

7. Factor: $3y^2(y^2 - 4) = 3y^2(y-2)(y+2) = 0$
The solution set is $\{0, \pm 2\}$.

9. Factor: $(a^2 - 4)(a^2 + 4) =$
$(a-2)(a+2)(a-2i)(a+2i) = 0$.
The solution set is $\{\pm 2, \pm 2i\}$.

11. Squaring each side, we get

$$\begin{aligned} x + 1 &= x^2 - 10x + 25 \\ 0 &= x^2 - 11x + 24 = (x-8)(x-3). \end{aligned}$$

Checking $x = 3$, we get $2 \neq -2$. Then $x = 3$ is an extraneous root. The solution set is $\{8\}$.

13. Isolate the radical and then square each side.

$$\begin{aligned} x &= x^2 - 40x + 400 \\ 0 &= x^2 - 41x + 400 = (x - 25)(x - 16) \end{aligned}$$

Checking $x = 16$, we get $2 \neq -6$. Then $x = 16$ is an extraneous root. The solution set is $\{25\}$.

15. Isolate the radical and then square each side.

$$\begin{aligned} 2w &= \sqrt{1 - 3w} \\ 4w^2 &= 1 - 3w \\ 4w^2 + 3w - 1 &= (4w - 1)(w + 1) = 0 \\ w &= \frac{1}{4}, -1 \end{aligned}$$

Checking $w = -1$, we get $-1 \neq 1$.
Then $w = -1$ is an extraneous root.
The solution set is $\left\{\dfrac{1}{4}\right\}$.

17. Multiply both sides by $z\sqrt{4z+1}$ and square each side.

$$\begin{aligned} \sqrt{4z+1} &= 3z \\ 4z + 1 &= 9z^2 \\ 0 &= 9z^2 - 4z - 1 \end{aligned}$$

By the quadratic formula,

$$z = \frac{4 \pm \sqrt{16 - 4(9)(-1)}}{18}$$
$$z = \frac{4 \pm \sqrt{52}}{18} = \frac{4 \pm 2\sqrt{13}}{18} = \frac{2 \pm \sqrt{13}}{9}$$

Since $z = \dfrac{2 - \sqrt{13}}{9} < 0$ and the right-hand side of the original equation is nonnegative,
$z = \dfrac{2 - \sqrt{13}}{9}$ is an extraneous root.

The solution set is $\left\{\dfrac{2 + \sqrt{13}}{9}\right\}$.

19. Squaring each side, one obtains

$$x^2 - 2x - 15 = 9$$
$$x^2 - 2x - 24 = (x-6)(x+4) = 0.$$

The solution set is $\{-4, 6\}$.

21. Isolate a radical and square each side.

$$\sqrt{x+40} = \sqrt{x} + 4$$
$$x + 40 = x + 8\sqrt{x} + 16$$
$$24 = 8\sqrt{x}$$
$$3 = \sqrt{x}$$
$$9 = x$$

The solution set is $\{9\}$.

23. Isolate a radical and square each side.

$$\sqrt{n+4} = 5 - \sqrt{n-1}$$
$$n + 4 = 25 - 10\sqrt{n-1} + (n-1)$$
$$-20 = -10\sqrt{n-1}$$
$$2 = \sqrt{n-1}$$
$$4 = n - 1$$

The solution set is $\{5\}$.

25. Isolate a radical and square each side.

$$\sqrt{2x+5} = 9 - \sqrt{x+6}$$
$$2x + 5 = 81 - 18\sqrt{x+6} + (x+6)$$
$$x - 82 = -18\sqrt{x+6}$$
$$x^2 - 164x + 6724 = 324(x+6)$$
$$x^2 - 488x + 4780 = 0$$
$$(x-10)(x-478) = 0$$

Checking $x = 478$ we get $53 \neq 9$ and $x = 478$ is an extraneous root. The solution set is $\{10\}$.

27. Raise each side to the power $3/2$.
Then $x = \pm 2^{3/2} = \pm 8^{1/2} = \pm 2\sqrt{2}$.
The solution set is $\left\{\pm 2\sqrt{2}\right\}$.

29. Raise each side to the power $-3/4$.

Thus, $w = \pm (16)^{-3/4} = \pm (2)^{-3} = \pm \dfrac{1}{8}$.

The solution set is $\left\{\pm \dfrac{1}{8}\right\}$.

31. Raise each side to the power -2.
So, $t = (7)^{-2} = \dfrac{1}{49}$. The solution set is $\left\{\dfrac{1}{49}\right\}$.

33. Raise each side to the power -2.

Then $s - 1 = (2)^{-2} = \dfrac{1}{4}$ and $s = 1 + \dfrac{1}{4}$.

The solution set is $\left\{\dfrac{5}{4}\right\}$.

35. Since $(x^2 - 9)(x^2 - 3) = 0$, the solution set is $\left\{\pm 3, \pm\sqrt{3}\right\}$.

37. Let $u = \dfrac{2c-3}{5}$ and $u^2 = \left(\dfrac{2c-3}{5}\right)^2$. Then

$$u^2 + 2u - 8 = 0$$
$$(u+4)(u-2) = 0$$
$$u = -4, 2$$
$$\frac{2c-3}{5} = -4 \quad \text{or} \quad \frac{2c-3}{5} = 2$$
$$2c - 3 = -20 \quad \text{or} \quad 2c - 3 = 10$$
$$c = -\frac{17}{2} \quad \text{or} \quad c = \frac{13}{2}.$$

The solution set is $\left\{-\dfrac{17}{2}, \dfrac{13}{2}\right\}$.

39. Let $u = \dfrac{1}{5x-1}$ and $u^2 = \left(\dfrac{1}{5x-1}\right)^2$. Then

$$u^2 + u - 12 = (u+4)(u-3) = 0$$
$$u = -4, 3$$
$$\frac{1}{5x-1} = -4 \quad \text{or} \quad \frac{1}{5x-1} = 3$$
$$1 = -20x + 4 \quad \text{or} \quad 1 = 15x - 3$$
$$x = \frac{3}{20} \quad \text{or} \quad \frac{4}{15} = x.$$

The solution set is $\left\{\dfrac{3}{20}, \dfrac{4}{15}\right\}$.

41. Let $u = v^2 - 4v$ and $u^2 = (v^2 - 4v)^2$. Then

$$u^2 - 17u + 60 = (u-5)(u-12) = 0$$
$$u = 5, 12$$
$$v^2 - 4v = 5 \quad \text{or} \quad v^2 - 4v = 12$$
$$v^2 - 4v - 5 = 0 \quad \text{or} \quad v^2 - 4v - 12 = 0$$
$$(v-5)(v+1) = 0 \quad \text{or} \quad (v-6)(v+2) = 0.$$

The solution set is $\{-1, -2, 5, 6\}$.

43. Factor the left-hand side.

$$\begin{aligned}(\sqrt{x}-3)(\sqrt{x}-1) &= 0 \\ \sqrt{x} &= 3,1 \\ x &= 9,1\end{aligned}$$

The solution set is $\{1,9\}$.

45. Factor the left-hand side as $(\sqrt{q}-4)(\sqrt{q}-3)=0$. Then $\sqrt{q}=3,4$ and the solution set is $\{9,16\}$.

47. Set the right-hand side to 0 and factor.

$$\begin{aligned}x^{2/3}-7x^{1/3}+10 &= 0 \\ \left(x^{1/3}-5\right)\left(x^{1/3}-2\right) &= 0 \\ x^{1/3} &= 5,2\end{aligned}$$

The solution set is $\{8,125\}$.

49. An equivalent statement is

$$\begin{aligned}w^2-4=3 \quad&\text{or}\quad w^2-4=-3 \\ w^2=7 \quad&\text{or}\quad w^2=1.\end{aligned}$$

The solution set is $\left\{\pm\sqrt{7},\pm 1\right\}$.

51. An equivalent statement assuming $5v\geq 0$ is

$$\begin{aligned}v^2-3v=5v \quad&\text{or}\quad v^2-3v=-5v \\ v^2-8v=0 \quad&\text{or}\quad v^2+2v=0 \\ v(v-8)=0 \quad&\text{or}\quad v(v+2)=0 \\ v &= 0,8,-2.\end{aligned}$$

Since $5v\geq 0$, $v=-2$ is an extraneous root and the solution set is $\{0,8\}$.

53. An equivalent statement is

$$\begin{aligned}x^2-x-6=6 \quad&\text{or}\quad x^2-x-6=-6 \\ x^2-x-12=0 \quad&\text{or}\quad x^2-x=0 \\ (x-4)(x+3)=0 \quad&\text{or}\quad x(x-1)=0.\end{aligned}$$

The solution set is $\{-3,0,1,4\}$.

55. An equivalent statement is

$$\begin{aligned}x+5=2x+1 \quad&\text{or}\quad x+5=-(2x+1) \\ 4=x \quad&\text{or}\quad x=-2.\end{aligned}$$

The solution set is $\{-2,4\}$.

57. Isolate a radical and square both sides.

$$\begin{aligned}\sqrt{16x+1} &= \sqrt{6x+13}-1 \\ 16x+1 &= (6x+13)-2\sqrt{6x+13}+1 \\ 10x-13 &= -2\sqrt{6x+13}\end{aligned}$$

$$\begin{aligned}100x^2-260x+169 &= 4(6x+13) \\ 100x^2-284x+117 &= 0\end{aligned}$$

$$\begin{aligned}x &= \frac{284\pm\sqrt{284^2-4(100)(117)}}{200} \\ x &= \frac{284\pm 184}{200} \\ x &= \frac{1}{2},\frac{117}{50}\end{aligned}$$

Checking $x=\dfrac{117}{50}$ we get $\sqrt{\dfrac{1922}{50}}-\sqrt{\dfrac{1352}{50}}>0$ and so $x=\dfrac{117}{50}$ is an extraneous root.

The solution set is $\left\{\dfrac{1}{2}\right\}$.

59. Factor as a difference of two squares and then as a sum and difference of two cubes.

$$\begin{aligned}(v^3-8)(v^3+8) &= 0 \\ (v-2)(v^2+2v+4)(v+2)(v^2-2v+4) &= 0\end{aligned}$$

Then $v=\pm 2$ or

$$v=\frac{-2\pm\sqrt{2^2-16}}{2} \quad\text{or } v=\frac{2\pm\sqrt{2^2-16}}{2}$$

$$v=\frac{-2\pm 2i\sqrt{3}}{2} \quad\text{or } v=\frac{2\pm 2i\sqrt{3}}{2}$$

The solution set is $\left\{\pm 2,-1\pm i\sqrt{3},1\pm i\sqrt{3}\right\}$.

61. Raise both sides to the power 4. Then

$$\begin{aligned}7x^2-12 &= x^4 \\ 0 &= x^4-7x^2+12=(x^2-4)(x^2-3) \\ x &= \pm 2,\pm\sqrt{3}\end{aligned}$$

Since the left-hand side of the given equation is nonnegative, $x=-2,-\sqrt{3}$ are extraneous roots. The solution set is $\left\{\sqrt{3},2\right\}$.

63. Raise both sides to the power 3.

$$
\begin{aligned}
2 + x - 2x^2 &= x^3 \\
x^3 + 2x^2 - x - 2 &= 0 \\
x^2(x+2) - (x+2) &= (x^2-1)(x+2) = 0 \\
x &= \pm 1, -2
\end{aligned}
$$

The solution set is $\{\pm 1, -2\}$.

65. Let $t = \dfrac{x-2}{3}$ and $t^2 = \left(\dfrac{x-2}{3}\right)^2$. Then

$$
\begin{aligned}
t^2 - 2t + 10 &= 0 \\
t^2 - 2t + 1 &= -10 + 1 \\
(t-1)^2 &= -9 \\
t &= 1 \pm 3i \\
\frac{x-2}{3} &= 1 \pm 3i \\
x - 2 &= 3 \pm 9i \\
x &= 5 \pm 9i.
\end{aligned}
$$

The solution set is $\{5 \pm 9i\}$.

67. Raise both sides to the power 5/2. Then

$$
\begin{aligned}
3u - 1 &= \pm 2^{5/2} \\
3u - 1 &= \pm 32^{1/2} \\
3u &= 1 \pm 4\sqrt{2}
\end{aligned}
$$

The solution set is $\left\{\dfrac{1 \pm 4\sqrt{2}}{3}\right\}$.

69. Factor this quadratic type expression.

$$
\begin{aligned}
(x^2+1) - 11\sqrt{x^2+1} + 30 &= 0 \\
\left(\sqrt{x^2+1} - 5\right)\left(\sqrt{x^2+1} - 6\right) &= 0 \\
\sqrt{x^2+1} = 5 \;\text{ or }\; \sqrt{x^2+1} &= 6 \\
x^2 = 24 \;\text{ or }\; x^2 &= 35 \\
x = \pm 2\sqrt{6}, \pm\sqrt{35}
\end{aligned}
$$

The solution set is $\left\{\pm\sqrt{35}, \pm 2\sqrt{6}\right\}$.

71. An equivalent statement is

$$
\begin{aligned}
x^2 - 2x = 3x - 6 \quad&\text{or}\quad x^2 - 2x = -3x + 6 \\
x^2 - 5x + 6 = 0 \quad&\text{or}\quad x^2 + x - 6 = 0 \\
(x-3)(x-2) = 0 \quad&\text{or}\quad (x+3)(x-2) = 0 \\
x &= 2, \pm 3
\end{aligned}
$$

The solution set is $\{2, \pm 3\}$.

73. Raise both sides to the power $-5/3$. Then

$$
\begin{aligned}
3m + 1 &= \left(-\frac{1}{8}\right)^{-5/3} \\
3m + 1 &= \left(-\frac{1}{2}\right)^{-5} \\
3m + 1 &= -32.
\end{aligned}
$$

The solution set is $\{-11\}$.

75. An equivalent statement assuming $x - 2 \geq 0$ is

$$
\begin{aligned}
x^2 - 4 = x - 2 \quad&\text{or}\quad x^2 - 4 = -x + 2 \\
x^2 - x - 2 = 0 \quad&\text{or}\quad x^2 + x - 6 = 0 \\
(x-2)(x+1) = 0 \quad&\text{or}\quad (x+3)(x-2) = 0 \\
x &= 2, -1, -3.
\end{aligned}
$$

Since $x - 2 \geq 0$, $x = -1, -3$ are extraneous roots and the solution set is $\{2\}$.

77. Solve for S.

$$
\begin{aligned}
21.24 + 1.25S^{1/2} - 9.8(18.34)^{1/3} &= 16.296 \\
1.25S^{1/2} - 25.84396 &\approx -4.944 \\
S^{1/2} &\approx 16.72 \\
S &\approx 279.56
\end{aligned}
$$

The maximum sailing area is 279.56 m^2.

79. Solve for x with $C = 83.50$.

$$
\begin{aligned}
0.5x + \sqrt{8x + 5000} &= 83.50 \\
\sqrt{8x + 5000} &= 83.50 - 0.5x \\
8x + 5000 &= 6972.25 - 83.50x + 0.25x^2 \\
0 &= 0.25x^2 - 91.50x + 1972.25
\end{aligned}
$$

$$
\begin{aligned}
x &= \frac{91.50 \pm \sqrt{(-91.50)^2 - 4(0.25)(1972.25)}}{0.5} \\
x &= \frac{91.50 \pm 80}{0.5} \\
x &= 23, 343
\end{aligned}
$$

Checking $x = 343$, the value of the left-hand side of the first equation exceeds 83.50 and so $x = 343$ is an extraneous root. Thus, 23 loaves cost $83.50.

81. Let x and $x + 6$ be two numbers. Then

$$\sqrt{x+6} - \sqrt{x} = 1$$
$$\sqrt{x+6} = \sqrt{x} + 1$$
$$x + 6 = x + 2\sqrt{x} + 1$$
$$5 = 2\sqrt{x}$$
$$25 = 4x$$
$$x = \frac{25}{4}$$

Since $\dfrac{25}{4} + 6 = \dfrac{49}{4}$, the numbers are $\dfrac{25}{4}$ and $\dfrac{49}{4}$.

83. Let x be the length of the short leg. Since $x+7$ is the other leg, by the Pythagorean Theorem the hypotenuse is $\sqrt{x^2 + (x+7)^2}$. Then

$$x + (x+7) + \sqrt{x^2 + (x+7)^2} = 30$$

$$2x - 23 = -\sqrt{x^2 + (x+7)^2}$$
$$4x^2 - 92x + 529 = 2x^2 + 14x + 49$$
$$2x^2 - 106x + 480 = 0$$
$$x^2 - 53x + 240 = 0$$
$$(x-5)(x-48) = 0$$
$$x = 5, 48$$

Since the perimeter is 30 in., $x = 48$ is an extraneous root. The short leg is $x = 5$ in.

85. Let x be the length of one side of the original square foundation. From the 2100 ft^2 we have

$$(x-10)(x+30) = 2100$$
$$x^2 + 20x - 2400 = 0$$
$$(x+60)(x-40) = 0.$$

Since x is nonnegative, $x = 40$ and the area of the square foundation is $x^2 = 1600$ ft^2.

87. Solving for d, we find

$$598.9\left(\frac{d}{64}\right)^{-2/3} = 14.26$$
$$\left(\frac{d}{64}\right)^{-2/3} = \frac{14.26}{598.9}$$

$$\frac{d}{64} = \left(\frac{14.26}{598.9}\right)^{-3/2}$$
$$d = 64\left(\frac{14.26}{598.9}\right)^{-3/2}$$
$$d \approx 17,419.3 \text{ lbs.}$$

89. Let $x^2(x+2)$ be the volume of shrimp to be shipped. Here, x is the length of one side of the square base and $x+2$ is the height. The volume of the box containing the styrofoam and shrimp is $(x+2)^2(x+4)$. Since the amount of shrimp is one-half of the volume of the box it is to be shipped in, we have

$$2x^2(x+2) = (x+2)^2(x+4)$$
$$2x^3 + 4x^2 = (x^2 + 4x + 4)(x+4)$$
$$x^3 - 4x^2 - 20x - 16 = 0.$$

By using the Rational Zero Theorem and synthetic division, we obtain

$$x^3 - 4x^2 - 20x - 16 = (x+2)(x^2 - 6x - 8).$$

Note, we have to exclude the zero of $x+2$ which is -2 since a dimension is a positive number. Then by using the method of completing the square, we get

$$x^2 - 6x = 8$$
$$x^2 - 6x + 9 = 17$$
$$(x-3)^2 = 17$$
$$x = 3 \pm \sqrt{17}$$
$$x \approx -1.123, 7.123$$

Again, we exclude $x = -1.123$.
So $x = 3 + \sqrt{17}$.
Thus, the volume of shrimp to be shipped is $(3 + \sqrt{17})^2(5 + \sqrt{17}) \approx 462.89$ in.^{3}

91. Let x be the number of hours after 10:00 a.m. so that the distance between Nancy and Edgar is 14 miles greater than the distance between Nancy and William.

By the Pythagorean Theorem, $\sqrt{(5x)^2 + (12x)^2}$ and $\sqrt{(5x)^2 + [4(x+2)]^2}$ are the distances between Nancy and Edgar and

Nancy and William, respectively. Since these distances are equal, we have

$$\sqrt{(5x)^2 + (12x)^2} - 14 = \sqrt{(5x)^2 + [4(x+2)]^2}$$
$$\sqrt{169x^2} - 14 = \sqrt{41x^2 + 64x + 64}$$
$$13x - 14 = \sqrt{41x^2 + 64x + 64}$$
$$169x^2 - 364x + 196 = 41x^2 + 64x + 64$$
$$128x^2 - 428x + 132 = 0$$
$$32x^2 - 107x + 33 = 0$$

$$x = \frac{107 \pm \sqrt{(-107)^2 - 4(32)(33)}}{64} = 3, \frac{11}{32}.$$

Since the left-hand side of the first equation is negative when $x = \frac{11}{32}$, we find that $x = \frac{11}{32}$ is an extraneous root. Thus, $x = 3$ hours and the time is 1:00 p.m.

93. (a) $\sqrt[3]{(2.7)(2.2)(2.6)} \approx \$2.49$ billion

(b) Let p be the net income in the fourth quarter. Then

$$\sqrt[4]{(2.7)(2.2)(2.6)p} = 2.6$$
$$(2.7)(2.2)(2.6)p = 2.6^4$$
$$p = \frac{2.6^4}{(2.7)(2.2)(2.6)}$$
$$p \approx \$2.96 \text{ billion.}$$

95. The weight of a cylindrical tank is its volume times its density. Then

$$\pi \left(\frac{d}{2} \right)^2 \cdot d \cdot 1600 = 25,850,060$$
$$400\pi d^3 = 25,850,060$$
$$d = \sqrt[3]{\frac{25,850,060}{400\pi}} \approx 27.4.$$

The height of the tank is about 27.4 meters.

For Thought

1. False, to be symmetric about the origin one must have $P(-x) = -P(x)$ for *all* x in the domain.

2. True   **3.**   True

4. False, since $f(-x) = -f(x)$.

5. True   **6.**   True   **7.**   False

8. False, y-intercept is $(0, 38)$.   **9.**   True

10. False, only one x-intercept.

3.5 Exercises

1. y-axis, since $f(-x) = f(x)$

3. $x = 3/2$, since $\frac{3}{2}$ is the x-coordinate of the vertex of a parabola

5. None

7. Origin, since $f(-x) = -f(x)$

9. $x = 5$, since 5 is the x-coordinate of the vertex of a parabola

11. Origin, since $f(-x) = -f(x)$

13. It does not cross $(4, 0)$ since $x - 4$ is raised to an even power.

15. It crosses $(1/2, 0)$.

17. It crosses $(1/4, 0)$

19. No x-intercepts since $x^2 - 3x + 10 = 0$ has no real root.

21. It crosses $(3, 0)$ and not $(0, 0)$ since $x^3 - 3x^2 = x^2(x - 3)$.

23. It crosses $(1/2, 0)$ and not $(1, 0)$ since

1	2	-5	4	-1
		2	-3	1
	2	-3	1	0

and $2x^3 - 5x^2 + 4x - 1 = (x - 1)(2x^2 - 3x + 1) = (x - 1)^2(2x - 1)$.

25. It crosses $(2, 0)$ and not $(-3, 0)$ since

-3	-2	-8	6	36
		6	6	-36
	-2	-2	12	0

and $(x + 3)(-2x^2 - 2x + 12) = -2(x + 3)(x^2 + x - 6) = -2(x + 3)(x + 3)(x - 2) = -2(x + 3)^2(x - 2)$.

27. $y \to \infty$ **29.** $y \to -\infty$ **31.** $y \to -\infty$

33. $y \to \infty$

35. $y \to \infty$

37. Neither symmetry, crosses $(-2, 0)$, does not cross $(1, 0)$, $y \to \infty$ as $x \to \infty$, $y \to -\infty$ as $x \to -\infty$

39. Symmetric about y-axis, no x-intercepts, $y \to \infty$ as $x \to \infty$, $y \to \infty$ as $x \to -\infty$

41. The graph of

$$f(x) = (x-1)^2(x+3)$$

is shown below.

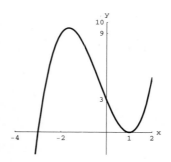

43. The graph of

$$f(x) = -2(2x-1)^2(x+1)^3$$

is given below.

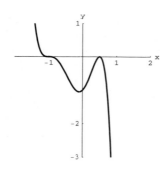

45. e   **47.** g

49. b   **51.** c

53. $f(x) = x - 30$ has x-intercept $(30, 0)$ and y-intercept $(0, -30)$

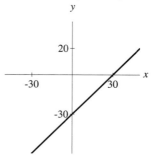

55. $f(x) = (x - 30)^2$ does not cross $(30, 0)$, y-intercept $(0, 900)$, $y \to \infty$ as $x \to \infty$ and as $x \to -\infty$

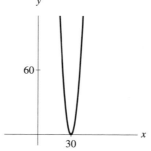

57. $f(x) = x^2(x - 40)$ crosses $(40, 0)$ but does not cross $(0, 0)$, $y \to \infty$ as $x \to \infty$, $y \to -\infty$ as $x \to -\infty$

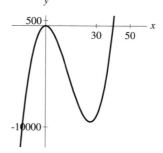

59. $f(x) = (x - 20)^2(x + 20)^2$ does not cross $(20, 0), (-20, 0)$, y-intercept $(0, 160000)$, $y \to \infty$ as $x \to \infty$ and as $x \to -\infty$

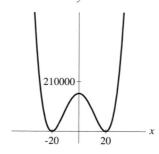

61. Since $f(x) = -x^3 - x^2 + 5x - 3$

$$
\begin{array}{c|cccc}
1 & -1 & -1 & 5 & -3 \\
  &   & -1 & -2 & 3 \\
\hline
  & -1 & -2 & 3 & 0
\end{array}
$$

we find

$$
\begin{aligned}
f(x) &= (x-1)(-x^2 - 2x + 3) \\
     &= -(x-1)(x+3)(x-1)
\end{aligned}
$$

the graph crosses $(-3, 0)$ but not $(1, 0)$, y-intercept $(0, -3)$, $y \to -\infty$ as $x \to \infty$, and $y \to \infty$ as $x \to -\infty$

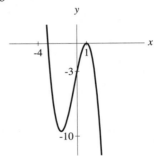

63. Since we have

$$
x^3 - 10x^2 - 600x = x(x-30)(x+20)
$$

the graph crosses $(0,0), (30,0), (-20,0)$, $y \to \infty$ as $x \to \infty$, $y \to -\infty$ as $x \to -\infty$

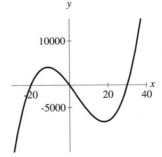

65. $f(x) = x^3 + 18x^2 - 37x + 60$ has only one x-intercept since

$$
\begin{array}{c|cccc}
-20 & 1 & 18 & -37 & 60 \\
    &   & -20 & 40 & -60 \\
\hline
    & 1 & -2 & 3 & 0
\end{array}
$$

and $x^2 - 2x + 3$ has no real root.

The graph of f crosses $(-20, 0)$, y-intercept $(0, 60)$, $y \to \infty$ as $x \to \infty$, and $y \to -\infty$ as $x \to -\infty$

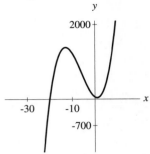

67. Since $-x^2(x^2 - 196) = -x^2(x - 14)(x + 14)$, graph crosses $(\pm 14, 0)$ and does not cross $(0, 0)$, $y \to -\infty$ as $x \to \infty$ and as $x \to -\infty$

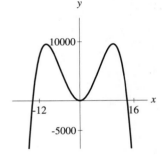

69. Use synthetic division on

$$
f(x) = x^3 + 3x^2 + 3x + 1
$$

with $c = 1$.

$$
\begin{array}{c|cccc}
-1 & 1 & 3 & 3 & 1 \\
   &   & -1 & -2 & -1 \\
\hline
   & 1 & 2 & 1 & 0
\end{array}
$$

Since $x^2 + 2x + 1 = (x+1)^2$, we obtain

$$
f(x) = (x+1)^3.
$$

The graph crosses $(-1, 0)$, y-intercept $(0, 1)$, $y \to \infty$ as $x \to \infty$, and $y \to -\infty$ as $x \to -\infty$.

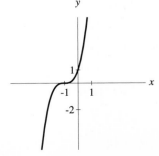

71. The graph crosses $(-7, 0)$, but it does not cross $(3, 0)$ and $(-5, 0)$, y-intercept is $(0, 1575)$, $y \to \infty$ as $x \to \infty$, and $y \to -\infty$ as $x \to -\infty$.

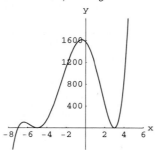

73. The roots of $x(x^2 - 3) = 0$ are $x = 0, \pm\sqrt{3}$.

If $x = 2$, then $(2)^2 - 3(2) > 0$.
If $x = 1$, then $(1)^2 - 3(1) < 0$.
If $x = -1$, then $(-1)^2 - 3(-1) > 0$.
If $x = -2$, then $(-2)^2 - 3(-2) < 0$.

$$
\begin{array}{ccccccc}
- & 0 & + & 0 & - & 0 & + \\
\end{array}
$$

$$
\begin{array}{ccccccc}
-2 & -\sqrt{3} & -1 & 0 & 1 & \sqrt{3} & 2
\end{array}
$$

The solution set is $(-\sqrt{3}, 0) \cup (\sqrt{3}, \infty)$.

75. The roots of $x^2(2 - x^2) = 0$ are $x = 0, \pm\sqrt{2}$.

If $x = 3$, then $2(3)^2 - (3)^4 < 0$.
If $x = 1$, then $2(1)^2 - (1)^4 > 0$.
If $x = -1$, then $2(-1)^2 - (-1)^4 > 0$.
If $x = -3$, then $2(-3)^2 - (-3)^4 < 0$.

$$
\begin{array}{ccccccc}
- & 0 & + & 0 & + & 0 & - \\
\end{array}
$$

$$
\begin{array}{ccccccc}
-3 & -\sqrt{2} & -1 & 0 & -1 & \sqrt{2} & 3
\end{array}
$$

The solution set is $(-\infty, -\sqrt{2}] \cup \{0\} \cup [\sqrt{2}, \infty)$.

77. Let $f(x) = x^3 + 4x^2 - x - 4$. Since

$$x^2(x + 4) - (x + 4) = (x + 4)(x^2 - 1) = 0,$$

the zeros of $f(x)$ are $x = -4, \pm 1$.

If $x = 2$, then $f(2) > 0$.
If $x = 0$, then $f(0) < 0$.
If $x = -2$, then $f(-2) > 0$.
If $x = -5$, then $f(-5) < 0$.

$$
\begin{array}{ccccccc}
- & 0 & + & 0 & - & 0 & + \\
\end{array}
$$

$$
\begin{array}{ccccccc}
-5 & -4 & -2 & -1 & 0 & 1 & 2
\end{array}
$$

The solution set is $(-4, -1) \cup (1, \infty)$.

79. Let $f(x) = x^3 - 4x^2 - 20x + 48$.
Since $f(x) = (x + 4)(x - 2)(x - 6)$, the zeros of $f(x)$ are $x = -4, 2, 6$.

If $x = 7$, then $f(7) > 0$.
If $x = 4$, then $f(4) < 0$.
If $x = 0$, then $f(0) > 0$.
If $x = -5$, then $f(-5) < 0$.

$$
\begin{array}{ccccccc}
- & 0 & + & 0 & - & 0 & + \\
\end{array}
$$

$$
\begin{array}{ccccccc}
-5 & -4 & 0 & 2 & 4 & 6 & 7
\end{array}
$$

The solution set is $[-4, 2] \cup [6, \infty)$.

81. Note, $f(x) = x^3 - x^2 + x - 1 = x^2(x - 1) + (x - 1) = (x - 1)(x^2 + 1) = 0$ has only one solution, namely, $x = 1$.

If $x = 2$, then $f(2) > 0$.
If $x = 0$, then $f(0) < 0$.

$$
\begin{array}{ccc}
- & 0 & + \\
\end{array}
$$

$$
\begin{array}{ccc}
0 & 1 & 2
\end{array}
$$

The solution set is $(-\infty, 1)$.

83. Let $f(x) = x^4 - 19x^2 + 90$. Since

$$f(x) = (x^2 - 10)(x^2 - 9)$$

the zeros of $f(x)$ are

$$x = \pm\sqrt{10}, \pm 3.$$

If $x = 4$, then $f(4) > 0$.
If $x = 3.1$, then $f(3.1) < 0$.
If $x = 0$, then $f(0) > 0$.
If $x = -3.1$, then $f(-3.1) < 0$.
If $x = -4$, then $f(-4) > 0$.

$$
\begin{array}{cccccccc}
+ & 0 & - & 0 & + & 0 & - & 0 & + \\
\end{array}
$$

$$
\begin{array}{cccc}
-\sqrt{10} & -3 & 3 & \sqrt{10}
\end{array}
$$

The solution set is $[-\sqrt{10}, -3] \cup [3, \sqrt{10}]$.

85. d, since the x-intercepts of

$$f(x) = \frac{1}{3}(x + 3)(x - 2)$$

are -3 and 2 and $f(0) = -2$

87. Since $f(-5) = 0 = f(4)$, the quadratic function f may be expressed as

$$f(x) = a(x + 5)(x - 4).$$

Also, we obtain

$$3 = f(0) = a(5)(-4)$$

and so $a = -3/20$. Thus, we have

$$f(x) = -\frac{3}{20}(x + 5)(x - 4).$$

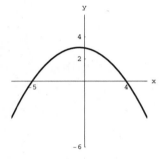

89. Since $f(-1) = 0 = f(3)$, the quadratic function f can be written as

$$f(x) = a(x + 1)(x - 3).$$

Moreover, we find

$$2 = f(1) = a(2)(-2)$$

and $a = -1/2$. Thus, we obtain

$$f(x) = -\frac{1}{2}(x + 1)(x - 3).$$

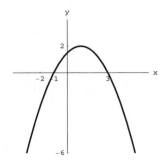

91. Since $f(2) = f(-3) = f(4) = 0$, the cubic function f can be expressed as

$$f(x) = a(x - 2)(x + 3)(x - 4).$$

Also, $6 = f(0) = a(-2)(3)(-4)$ and we get $a = 1/4$. Thus, we obtain

$$f(x) = \frac{1}{4}(x - 2)(x + 3)(x - 4).$$

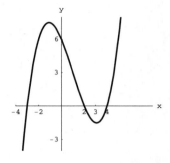

93. Since $f(\pm 2) = 0 = f(\pm 4)$, the quartic function f can be written as

$$f(x) = a(x^2 - 4)(x^2 - 16).$$

Also, $3 = f(1) = a(-3)(-15)$ and we get $a = 1/15$. Thus, we obtain

$$f(x) = \frac{1}{15}(x^2 - 4)(x^2 - 16).$$

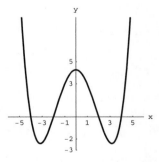

95. Local max. ≈ 3.11, local min. ≈ 0.37

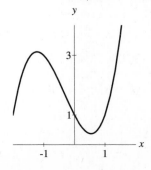

97. Local max. ≈ 23.74, local min. ≈ -163.74

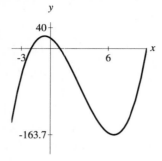

99. Local max. ≈ 21.01, local min. ≈ 13.99

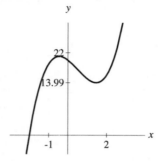

101. From the graph of the profit function, we find a local maximum of \$3400 and a local minimum of \$2600. To get a profit higher than \$3400, the company must spend more than \$2200 on advertising.

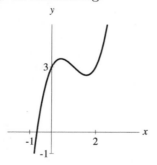

103. Note, $P = \dfrac{x}{10}(x^2 - 60x + 900) = \dfrac{x}{10}(x - 30)^2.$

The graph of P does not cross the x-intercept $(30, 0)$. Thus, the profit decreases to \$0 at $x = 30$ stores, and the profit increases as more stores (greater than 30) are opened.

A closer look at the graph of P shows that the profit is also increasing for $0 < x < 10$, and is decreasing when $10 < x < 30$.

105. Since $3x + 2y = 12$, we get $y = \dfrac{12 - 3x}{2}$.

The volume V of the block is given by

$$
\begin{aligned}
V &= xy\frac{4y}{3} \\
&= \frac{4x}{3}y^2 \\
&= \frac{4x}{3}\left(\frac{12 - 3x}{2}\right)^2 \\
V &= 3x^3 - 24x^2 + 48x.
\end{aligned}
$$

One finds from the graph of

$$
V = \frac{4x}{3}\left(\frac{12 - 3x}{2}\right)^2
$$

that the optimal dimensions of the block are

$$
x = \frac{4}{3} \text{ in. by } y = \frac{12 - 4}{2} = 4 \text{ in. by } \frac{16}{3} \text{ in..}
$$

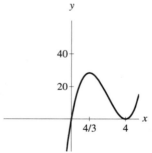

107. If the surface area is 50 square feet, then no paint can be used (since all the paint is used to coat the inside of the can).

Since the amount of paint in a 1-pint can is $\dfrac{1}{8} \cdot \dfrac{1}{7.5}$ or $\dfrac{1}{60}$ feet^3, we get

$$
\frac{1}{60} = \pi r^2 h.
$$

Solving for h, one obtains $h = \dfrac{1}{60\pi r^2}$. Then

$$
\begin{aligned}
2\pi r^2 + 2\pi rh &= 50 \\
2\pi r^2 + 2\pi r\left(\frac{1}{60\pi r^2}\right) &= 50 \\
2\pi r^2 + \frac{1}{30r} - 50 &= 0.
\end{aligned}
$$

With a graphing calculator, one finds

$$
r \approx \pm 2.82 \text{ or } r \approx 6.67 \times 10^{-4}.
$$

If $r \approx 2.82$, then $h \approx 6.67 \times 10^{-4}$ feet since

$h = \dfrac{1}{60\pi r^2}$. Similarly, if $r \approx 6.67 \times 10^{-4}$, then

$h \approx 11,924.69$ feet. Thus, either the radius and height of the can are

$$r \approx 2.82 \text{ ft and } h \approx 6.67 \times 10^{-4} \text{ ft}$$

or

$$r \approx 6.67 \times 10^{-4} \text{ ft and } h \approx 11,924.69 \text{ ft.}$$

For Thought

1. False, $\sqrt{x} - 3$ is not a polynomial.

2. False, domain is $(-\infty, 2) \cup (2, \infty)$.

3. False

4. False, it has three vertical asymptotes.

5. True

6. False, $y = 5$ is the horizontal asymptote.

7. True   **8.**   False

9. True, it is an even function.

10. True, since $x = -3$ is not a vertical asymptote.

3.6 Exercises

1. $(-\infty, -2) \cup (-2, \infty)$

3. $(-\infty, -2) \cup (-2, 2) \cup (2, \infty)$

5. $(-\infty, 3) \cup (3, \infty)$

7. $(-\infty, 0) \cup (0, \infty)$

9. $(-\infty, -1) \cup (-1, 0) \cup (0, 1) \cup (1, \infty)$ since

$$f(x) = \frac{3x^2 - 1}{x(x^2 - 1)}$$

11. $(-\infty, -3) \cup (-3, -2) \cup (-2, \infty)$ since

$$f(x) = \frac{-x^2 + x}{(x + 3)(x + 2)}$$

13. Domain $(-\infty, 2) \cup (2, \infty)$, asymptotes $y = 0$ and $x = 2$

15. Domain $(-\infty, 0) \cup (0, \infty)$, asymptotes $y = x$ and $x = 0$

17. Asymptotes $x = 2$, $y = 0$

19. Asymptotes $x = \pm 3$, $y = 0$

21. Asymptotes $x = 1$, $y = 2$

23. Asymptotes $x = 0$, $y = x - 2$ since

$$f(x) = x - 2 + \frac{1}{x}$$

25. Asymptotes $x = -1$, $y = 3x - 3$ since

$$
\begin{array}{r}
3x - 3 \\
x + 1 \overline{\smash{)}3x^2 + 0x + 4} \\
\underline{3x^2 + 3x} \\
-3x + 4 \\
\underline{-3x - 3} \\
7
\end{array}
$$

and $f(x) = 3x - 3 + \dfrac{7}{x + 1}$

27. Asymptotes $x = -2$, $y = -x + 6$ since

$$
\begin{array}{r}
-x + 6 \\
x + 2 \overline{\smash{)}-x^2 + 4x + 0} \\
\underline{-x^2 - 2x} \\
6x + 0 \\
\underline{6x + 12} \\
-12
\end{array}
$$

and $f(x) = -x + 6 + \dfrac{-12}{x + 2}$

29. Asymptotes $x = 0$, $y = 0$, no x or y-intercept

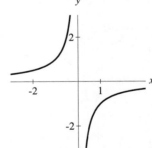

31. Asymptotes $x = 2$, $y = 0$, no x-intercept, y-intercept $(0, -1/2)$

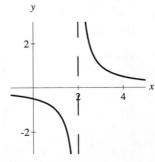

33. Asymptotes $x = \pm 2$, $y = 0$, no x-intercept, y-intercept $(0, -1/4)$

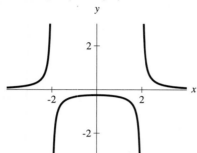

35. Asymptotes $x = -1$, $y = 0$, no x-intercept, y-intercept $(0, -1)$

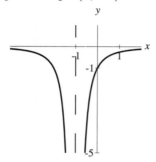

37. Asymptotes $x = 1$, $y = 2$, x-intercept $(-1/2, 0)$, y-intercept $(0, -1)$

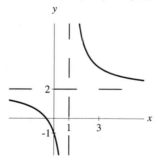

39. Asymptotes $x = -2$, $y = 1$, x-intercept $(3, 0)$, y-intercept $(0, -3/2)$

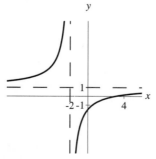

41. Asymptotes $x = \pm 1$, $y = 0$, x- and y-intercept is $(0, 0)$

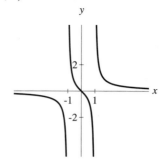

43. Since we have

$$f(x) = \frac{4x}{(x-1)^2}$$

the asymptotes are $x = 1$, $y = 0$, and the x-and y-intercept is $(0, 0)$

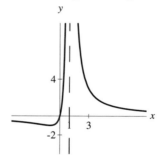

45. Asymptotes are $x = \pm 3$, $y = -1$, x-intercept $(\pm 2\sqrt{2}, 0)$, y-intercept $(0, -8/9)$

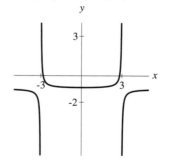

47. Since we have

$$f(x) = \frac{2x^2 + 8x + 2}{(x+1)^2}$$

the asymptotes are $x = -1$ and $y = 2$,
by solving $2x^2 + 8x + 2 = 0$ one gets
x-intercepts $\left(-2 \pm \sqrt{3}, 0\right)$, y-intercept $(0, 2)$

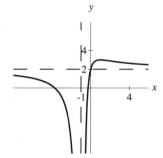

49. Since $f(x) = x + \dfrac{1}{x}$, oblique asymptote is

$y = x$, asymptote $x = 0$, no x-intercept,
no y-intercept, graph goes through
$(1, 2), (-1, -2)$

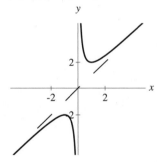

51. Since $f(x) = x - \dfrac{1}{x^2}$, oblique asymptote

is $y = x$, asymptote $x = 0$, x-intercept $(1, 0)$,
no y-intercept, graph goes through $(-1, -2)$

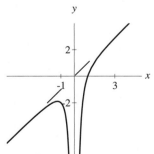

53. $f(x) = x - 1 + \dfrac{1}{x+1}$ since

$$
\begin{array}{r}
x - 1 \\
x+1 \enclose{longdiv}{x^2 + 0x} \\
\underline{x^2 + x} \\
-x + 0 \\
\underline{-x - 1} \\
1
\end{array}
$$

Oblique asymptote $y = x - 1$,
asymptote $x = -1$, x-intercept $(0, 0)$,
graph goes through $(-2, -4)$

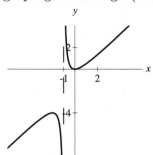

55. $f(x) = 2x + 1 + \dfrac{1}{x-1}$ since

$$
\begin{array}{r}
2x + 1 \\
x-1 \enclose{longdiv}{2x^2 - x + 0} \\
\underline{2x^2 - 2x} \\
x + 0 \\
\underline{x - 1} \\
1
\end{array}
$$

Oblique asymptote $y = 2x + 1$,
asymptote $x = 1$,
x-intercepts $(0, 0), (1/2, 0)$,
graph goes through $(2, 6)$

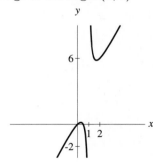

57. e **59.**

61. b **63.** c

65. Domain is $\{x : x \neq \pm 1\}$ since

$$f(x) = \frac{x+1}{(x+1)(x-1)} = \frac{1}{x-1}$$

if $x \neq -1$, there is a 'hole' at $(-1, -1/2)$, asymptotes $x = 1$ and $y = 0$, no x-intercept, y-intercept $(0, -1)$

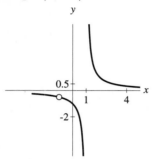

67. The domain is $\{x : x \neq 1\}$ since

$$f(x) = \frac{(x-1)(x+1)}{x-1} = x+1$$

provided $x \neq 1$, a line with a 'hole' at $(1, 2)$

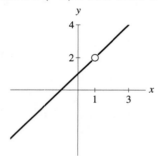

69. Symmetric about y-axis, asymptote $x = 0$, goes through $(0, 2), (1, 1)$

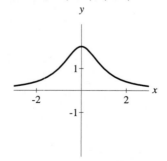

71. Since $f(x) = \dfrac{x-1}{x(x^2-9)}$, asymptotes are

$x = \pm 3, x = 0, y = 0$, crosses $(1, 0)$

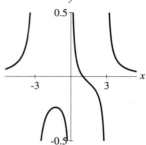

73. Asymptotes $x = 0, y = 0$, x-intercept $(-1, 0)$, no y-intercept, graph goes through $(1, 2)$

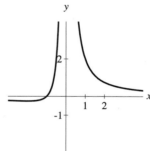

75. The sign graph of $\dfrac{x-4}{x+2} \leq 0$ is

$$
\begin{array}{ccccccccccccc}
- & - & - & - & - & - & - & - & 0 & + & + & + & + \\
- & - & - & - & 0 & + & + & + & + & + & + & + & + \\
\end{array}
$$

$$\xleftarrow{\hspace{2cm}} \quad -2 \qquad\quad 4 \quad \xrightarrow{\hspace{2cm}}$$

The solution set is $(-2, 4]$.

77. The sign graph of $\dfrac{q+8}{q+3} > 0$ is

$$
\begin{array}{ccccccccccccc}
- & - & - & - & - & - & - & - & 0 & + & + & + & + \\
- & - & - & - & 0 & + & + & + & + & + & + & + & + \\
\end{array}
$$

$$\xleftarrow{\hspace{2cm}} \quad -8 \qquad\quad -3 \quad \xrightarrow{\hspace{2cm}}$$

The solution set is $(-\infty, -8) \cup (-3, \infty)$.

79. The sign graph of $\dfrac{(w-3)(w+2)}{w-6} \geq 0$ is

$$
\begin{array}{ccccccccccccccc}
- & - & - & - & - & - & - & - & - & - & - & 0 & + & + & + \\
- & - & - & - & - & - & - & - & 0 & + & + & + & + & + & + \\
- & - & - & - & 0 & + & + & + & + & + & + & + & + & + & + \\
\end{array}
$$

$$\xleftarrow{\hspace{2cm}} \quad -2 \qquad 3 \qquad 6 \quad \xrightarrow{\hspace{1cm}}$$

The solution set is $[-2, 3] \cup (6, \infty)$.

81. The sign graph of $\dfrac{-5}{(x+2)(x-3)} > 0$ is

```
- - - - - - - - - - - - - - - - - - - - - - - - - -
- - - - - - - - - - - - - -  0 + + + + + +
- - - - - - -  0 + + + + + + + + + +
  <--------|----------|-------------->
          -2          3
```

The solution set is $(-2, 3)$.

83. The sign graph of $\dfrac{(x-4)(x+2)}{5-x} > 0$ is

```
+ + + + + + + + + + + + + + + + 0 - - -
- - - - - - - - - - - 0 + + + + + +
- - - - - 0 + + + + + + + + +
  <-----|--------|---------|---------->
       -2        4         5
```

The solution set is

$$(-\infty, -2) \cup (4, 5).$$

85. Let $R(x) = \dfrac{(x-3)(x+1)}{x-5} \geq 0.$

Since $R(-2) < 0, R(0) > 0, R(4) < 0,$
and $R(7) > 0$, we obtain

```
   -    0    +    0    -    U    +
  <--|----|----|----|----|----|----->
    -2   -1    0    3    4    5    7
```

The solution set is

$$[-1, 3] \cup (5, \infty).$$

87. Let $f(x) = \dfrac{(x-\sqrt{7})(x+\sqrt{7})}{(\sqrt{2}-x)(\sqrt{2}+x)}.$

If $x = 3$, then $f(3) < 0.$
If $x = 2$, then $f(2) > 0.$
If $x = 0$, then $f(0) < 0.$
If $x = -2$, then $f(-2) > 0.$
If $x = -3$, then $f(-3) < 0.$

```
    -   0  +   U  -   U  +   0  -
  <----|------|-----|-----|------->
     -√7     -√2    √2    √7
```

The solution set is

$$(-\infty, -\sqrt{7}] \cup (-\sqrt{2}, \sqrt{2}) \cup [\sqrt{7}, \infty).$$

89. Let $R(x) = \dfrac{(x+1)^2}{(x-5)(x+3)} \geq 0.$

Since $R(-4) > 0, R(-2) < 0, R(0) < 0,$
and $R(6) > 0$, we get

```
   +    U    -    0    -    U    +
  <--|----|----|----|----|----|----->
    -4   -3   -2   -1    0    5    6
```

The solution set is $(-\infty, -3) \cup \{-1\} \cup (5, \infty).$

91. Since the sign graph of $\dfrac{w-1}{w^2} > 0$ is

```
- - - - - - - - - 0 + + + + +
+ + + + 0 + + + + + + + + +
  <--------|----------|-------->
          0          1
```

then the solution set is the interval $(1, \infty)$.

93. Since the sign graph of $\dfrac{w^2 - 4w + 5}{w - 3} > 0$ is

```
- - - - - 0 + + + + + + + + +
  <------------|---------------->
              3
```

the solution is the interval $(3, \infty)$.

95. Since $y = 0$ and $x = 1$ are asymptotes, the rational function can be written as

$$y = \frac{a}{x-1}.$$

Since $(3, 1)$ satifies the equation above, we find $1 = a/2$ or $a = 2$. Thus, the function is

$$y = \frac{2}{x-1}.$$

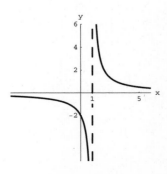

97. Since $y = 1$ and $x = 2$ are asymptotes, the function can be expressed as

$$y = \frac{x - a}{x - 2}.$$

Since $(0, 5)$ satifies the equation above, we obtain $5 = (-a)/(-2)$ or $a = 10$. Thus, the function is

$$y = \frac{x - 10}{x - 2} \text{ or } y = \frac{-8}{x - 2} + 1.$$

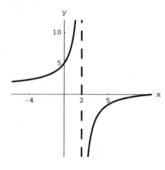

99. Since $y = 0$ and $x = \pm 3$ are asymptotes, the function can be written as

$$y = \frac{ax + b}{x^2 - 9}.$$

Since $(0, 0)$ and $(4, 8/7)$ satifies the equation above, we obtain $0 = (b)/(-9)$ and $8/7 = (4a + b)/7$, respectively. Then $b = 0$ and $8/7 = 4a/7$ or $a = 2$. Thus, the function is

$$y = \frac{2x}{x^2 - 9}.$$

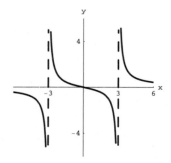

101. Since $y = 2x + 1$ and $x = 1$ are asymptotes, the function can be written as

$$y = (2x + 1) + \frac{a}{x - 1}.$$

Since $(0, 3)$ satifies the equation above, we find

$$3 = 1 + \frac{a}{-1}$$

or $a = -2$. Thus, the function is

$$y = 2x + 1 + \frac{-2}{x - 1}.$$

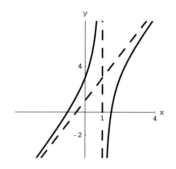

103. The average cost is

$$C = \frac{100 + x}{x}.$$

If she visits the zoo 100 times, then her average cost per visit is

$$C = \frac{200}{100} = \$2.$$

Since $C \to \$1$ as $x \to \infty$, over a long period her average cost per visit is $1.

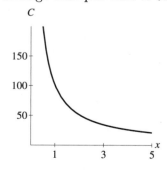

105. Since the second half is 100 miles long and it must be completed in $4 - x$ hours, the average speed for the second half is

$$S(x) = \frac{100}{4 - x}.$$

The asymptote $x = 4$ implies the average speed in the second half goes to ∞ as the completion time in the first half shortens to four hours.

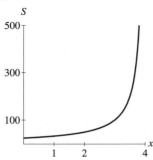

107. a) About 15 minutes

b) About 200 PPM

c) The horizontal asymptote is PPM $= 0$ and the vertical asymptote is $t = 0$. In simple terms, a long exposure to low level carbon monoxide still leads to permanent brain damage, and a very high level carbon monoxide can lead to permanent brain damage in a few minutes.

109.

(a) $h = \dfrac{500}{\pi r^2}$ since $500 = \pi r^2 h$

(b) $S = 2\pi r^2 + 2\pi rh = 2\pi r^2 + 2\pi r\dfrac{500}{\pi r^2}$

or equivalently $S = 2\pi r^2 + \dfrac{1000}{r}$

(c) As can be seen from the graph of $S = 2\pi r^2 + \dfrac{1000}{r}$, S is minimized when $r \approx 4.3$ feet.

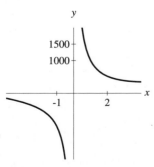

(d) If it costs \$8 per ft^2 to construct the tank in part (c), then the cost of the tank

is given by

$$8\left(2\pi(4.3)^2 + \frac{1000}{4.3}\right) \approx \$2789.87.$$

Chapter 3 Review Exercises

1. By using the method of completing the square, we obtain

$$\begin{aligned} f(x) &= 3\left(x^2 - \frac{2}{3}x + \frac{1}{9}\right) - \frac{1}{3} + 1 \\ &= 3\left(x - \frac{1}{3}\right)^2 + \frac{2}{3} \end{aligned}$$

3. Since $y = 2(x^2 - 2x + 1) - 2 - 1 = 2(x-1)^2 - 3$, the vertex is $(1, -3)$ and axis of symmetry is $x = 1$. Setting $y = 0$, we get $x - 1 = \pm\dfrac{\sqrt{6}}{2}$ and the x-intercepts are $\left(\dfrac{2 \pm \sqrt{6}}{2}, 0\right)$.

The y-intercept is $(0, -1)$.

5. From the x-intercepts, $y = a(x + 1)(x - 3)$. Substitute $(0, 6)$, so $6 = a(-3)$ or $a = -2$. The equation of the parabola is $y = -2(x^2 - 2x - 3)$ or $y = -2x^2 + 4x + 6$.

7. $1/3$ **9.** $\pm 2\sqrt{2}$

11. Factoring, we obtain

$$m(x) = (2x - 1)(4x^2 + 2x + 1).$$

By using the quadratic formula, we obtain the zeros to the second factor. Namely,

$$x = \frac{-2 \pm \sqrt{-12}}{8} = \frac{-2 \pm 2i\sqrt{3}}{8}.$$

The zeros are $\dfrac{1}{2}, \dfrac{-1 \pm i\sqrt{3}}{4}$.

13. Since $P(t) = (t^2 - 10)(t^2 + 10)$, the zeros are $\pm\sqrt{10}, \pm i\sqrt{10}$

15. Factoring, we obtain $R(s) = 4s^2(2s-1) - (2s-1) = (4s^2-1)(2s-1)$.

The zeros are $\dfrac{1}{2}$ (with multiplicity 2) and $-\dfrac{1}{2}$.

17. Find the roots of the second factor of
$f(x) = x(x^2 + 2x - 6)$. Since
$x^2 + 2x - 6 = (x^2 + 2x + 1) - 6 - 1 = (x+1)^2 - 7$, the zeros are $0, -1 \pm \sqrt{7}$.

19. $P(3) = 108 - 27 + 3 - 1 = 83$.
By synthetic division, we get

$$
\begin{array}{r|rrrr}
3 & 4 & -3 & 1 & -1 \\
  &   & 12 & 27 & 84 \\
\hline
  & 4 & 9 & 28 & 83
\end{array}
$$

Remainder is $P(3) = 83$.

21. $P(-1/2) = \dfrac{1}{4} - \dfrac{1}{4} + 3 + 2 = 5$.

By synthetic division, we obtain

$$
\begin{array}{r|rrrrrr}
-1/2 & -8 & 0 & 2 & 0 & -6 & 2 \\
     &   & 4 & -2 & 0 & 0 & 3 \\
\hline
     & -8 & 4 & 0 & 0 & -6 & 5
\end{array}
$$

So the remainder is $P(-1/2) = 5$.

23. $\pm \left\{ 1, \dfrac{1}{3}, 2, \dfrac{2}{3} \right\}$

25. $\pm \left\{ 1, \dfrac{1}{2}, \dfrac{1}{3}, \dfrac{1}{6}, 3, \dfrac{3}{2} \right\}$

27.

$$
\begin{aligned}
2\left(x + \frac{1}{2}\right)(x - 3) &= (2x + 1)(x - 3) \\
&= 2x^2 - 5x - 3
\end{aligned}
$$

An equation is $2x^2 - 5x - 3 = 0$.

29. $(x - (3 - 2i))(x - (3 + 2i)) =$

$$
\begin{aligned}
&= ((x - 3) + 2i)((x - 3) - 2i) \\
&= (x - 3)^2 + 4 \\
&= x^2 - 6x + 13
\end{aligned}
$$

An equation is $x^2 - 6x + 13 = 0$.

31. $(x - 2)(x - (1 - 2i))(x - (1 + 2i)) =$

$$
\begin{aligned}
&= (x - 2)((x - 1) + 2i)((x - 1) - 2i) \\
&= (x - 2)\left((x - 1)^2 + 4\right) \\
&= (x - 2)(x^2 - 2x + 5) \\
&= x^3 - 4x^2 + 9x - 10
\end{aligned}
$$

Thus, an equation is $x^3 - 4x^2 + 9x - 10 = 0$.

33. $\left(x - (2 - \sqrt{3})\right)\left(x - (2 + \sqrt{3})\right) =$

$$
\begin{aligned}
&= \left((x - 2) + \sqrt{3}\right)\left((x - 2) - \sqrt{3}\right) \\
&= \left((x - 2)^2 - 3\right) \\
&= x^2 - 4x + 1
\end{aligned}
$$

Thus, an equation is $x^2 - 4x + 1 = 0$.

35. $P(x) = P(-x) = x^8 + x^6 + 2x^2$ has no sign variation. There are 6 imaginary roots and 0 has multiplicity 2.

37. $P(x) = 4x^3 - 3x^2 + 2x - 9$ has 3 sign variations and $P(-x) = -4x^3 - 3x^2 - 2x - 9$ has no sign variation. There are
(a) 3 positive roots, or
(b) 1 positive and 2 imaginary roots.

39. $P(x) = x^3 + 2x^2 + 2x + 1$ has no sign variation and $P(-x) = -x^3 + 2x^2 - 2x + 1$ has 3 sign variations. There are
(a) 3 negative roots, or
(b) 1 negative root and 2 imaginary roots.

41. Best integral bounds: $-4 < x < 3$. One checks that $1, 2$ are not upper bounds and that 3 is a bound.

$$
\begin{array}{r|rrr}
3 & 6 & 5 & -50 \\
  &   & 18 & 69 \\
\hline
  & 6 & 23 & 19
\end{array}
$$

One checks that $-1, -2, -3$ are not lower bounds and that -4 is a bound since

$$
\begin{array}{r|rrr}
-4 & 6 & 5 & -50 \\
   &   & -24 & 76 \\
\hline
   & 6 & -19 & 26
\end{array}
$$

43. Best integral bounds: $-1 < x < 8$. One checks that $1, 2, 3, 4, 5, 6, 7$ are not upper bounds and that 8 is a bound.

$$
\begin{array}{r|rrrr}
8 & 2 & -15 & 31 & -12 \\
  &   & 16 & 8 & 312 \\
\hline
  & 2 & 1 & 39 & 300
\end{array}
$$

-1 is a lower bound since

$$
\begin{array}{r|rrrr}
-1 & 2 & -15 & 31 & -12 \\
   &   & -2 & 17 & -48 \\
\hline
   & 2 & -17 & 48 & -60
\end{array}
$$

45. Best integral bounds: $-1 < x < 1$

$$
\begin{array}{r|rrrr}
1 & 12 & -4 & -3 & 1 \\
  &   & 12 & 8 & 5 \\
\hline
  & 12 & 8 & 5 & 6
\end{array}
$$

$$
\begin{array}{r|rrrr}
-1 & 12 & -4 & -3 & 1 \\
   &   & -12 & 16 & -13 \\
\hline
   & 12 & -16 & 13 & -12
\end{array}
$$

47. Roots are $1, 2, 3$ since

$$
\begin{array}{r|rrrr}
1 & 1 & -6 & 11 & -6 \\
  &   & 1 & -5 & 6 \\
\hline
  & 1 & -5 & 6 & 0
\end{array}
$$

and $x^2 - 5x + 6 = (x - 3)(x - 2)$.

49. Roots are $\pm i, 1/3, 1/2$ since

$$
\begin{array}{r|rrrrr}
1/2 & 6 & -5 & 7 & -5 & 1 \\
    &   & 3 & -1 & 3 & -1 \\
\hline
    & 6 & -2 & 6 & -2 & 0
\end{array}
$$

$$
\begin{array}{r|rrrr}
1/3 & 6 & -2 & 6 & -2 \\
    &   & 2 & 0 & 2 \\
\hline
    & 6 & 0 & 6 & 0
\end{array}
$$

and the zeros of $6x^2 + 6 = 6(x^2 + 1) = 0$ are $\pm i$.

51. Roots are $3, 3 \pm i$ since

$$
\begin{array}{r|rrrr}
3 & 1 & -9 & 28 & -30 \\
  &   & 3 & -18 & 30 \\
\hline
  & 1 & -6 & 10 & 0
\end{array}
$$

and the zeros of $x^2 - 6x + 10 = (x-3)^2 + 1 = 0$ are $3 \pm i$.

53. Roots are $2, 1 \pm i\sqrt{2}$ since

$$
\begin{array}{r|rrrr}
2 & 1 & -4 & 7 & -6 \\
  &   & 2 & -4 & 6 \\
\hline
  & 1 & -2 & 3 & 0
\end{array}
$$

and the zeros of $x^2 - 2x + 3 = (x-1)^2 + 2 = 0$ are $1 \pm i\sqrt{2}$.

55. Apply synthetic division to the second factor in $x(2x^3 - 5x^2 - 2x + 2) = 0$.

$$
\begin{array}{r|rrrr}
1/2 & 2 & -5 & -2 & 2 \\
    &   & 1 & -2 & -2 \\
\hline
    & 2 & -4 & -4 & 0
\end{array}
$$

By completing the square, the zeros of $2x^2 - 4x - 4 = 2(x^2 - 2x) - 4 = 2(x-1)^2 - 6 = 0$ are $1 \pm \sqrt{3}$. All the roots are $x = 0, 1/2, 1 \pm \sqrt{3}$.

57. Solve an equivalent statement assuming $3v \geq 0$.

$$
\begin{array}{ccc}
2v - 1 = 3v & \text{or} & 2v - 1 = -3v \\
-1 = v & \text{or} & 5v = 1
\end{array}
$$

Since $3v \geq 0$, $v = -1$ is an extraneous root. The solution set is $\{1/5\}$.

59. Let $w = x^2$ and $w^2 = x^4$.

$$
\begin{aligned}
w^2 + 7w &= 18 \\
(w+9)(w-2) &= 0 \\
w &= -9, 2 \\
x^2 = -9 \quad \text{or} \quad x^2 &= 2.
\end{aligned}
$$

Since $x^2 = -9$ has no real solution, the solution set is $\{\pm\sqrt{2}\}$.

61. Isolate a radical and square both sides.

$$
\begin{aligned}
\sqrt{x+6} &= \sqrt{x-5} + 1 \\
x + 6 &= x - 5 + 2\sqrt{x-5} + 1 \\
5 &= \sqrt{x-5} \\
25 &= x - 5
\end{aligned}
$$

The solution set is $\{30\}$.

63. Let $w = \sqrt[4]{y}$ and $w^2 = \sqrt{y}$.

$$
\begin{aligned}
w^2 + w - 6 &= 0 \\
(w + 3)(w - 2) &= 0 \\
w = -3 \quad &\text{or} \quad w = 2 \\
y^{1/4} = -3 \quad &\text{or} \quad y^{1/4} = 2
\end{aligned}
$$

Since $y^{1/4} = -3$ has no real solution, the solution set is $\{16\}$.

65. Let $w = x^2$ and $w^2 = x^4$.

$$
\begin{aligned}
w^2 - 3w - 4 &= 0 \\
(w + 1)(w - 4) &= 0 \\
x^2 = -1 \quad &\text{or} \quad x^2 = 4
\end{aligned}
$$

Since $x^2 = -1$ has no real solution, the solution set is $\{\pm 2\}$.

67. Raise to the power 3/2 and get
$x - 1 = \pm(4^{1/2})^3$. So, $x = 1 \pm 8$.
The solution set is $\{-7, 9\}$.

69. No solution since $(x + 3)^{-3/4}$ is nonnegative.

71. Since $3x - 7 = 4 - x$, we obtain $4x = 11$.
The solution set is $\{11/4\}$.

73. Symmetric about $x = 3/4$ since $\dfrac{-b}{2a} = \dfrac{3}{4}$.

75. Symmetric about y-axis since $f(-x) = f(x)$.

77. Symmetric about the origin for
$f(-x) = -f(x)$.

79. $(-\infty, -2.5) \cup (-2.5, \infty)$

81. $(-\infty, \infty)$

83. Since $f(x) = (x - 2)(x + 1)$, the x-intercepts
are $(2, 0), (-1, 0)$, y-intercept is $(0, -2)$.
Since $\dfrac{-b}{2a} = \dfrac{1}{2}$, the vertex is $(1/2, -9/4)$.

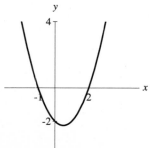

85. Use synthetic division on $f(x) = x^3 - 3x - 2$.

$$
\begin{array}{r|rrrr}
-1 & 1 & 0 & -3 & -2 \\
   &   & -1 & 1 & 2 \\
\hline
   & 1 & -1 & -2 & 0
\end{array}
$$

Since $x^2 - x - 2 = (x - 2)(x + 1)$, we find

$$f(x) = (x + 1)^2(x - 2).$$

The graph crosses $(2, 0)$ but does not cross $(-1, 0)$, the y-intercept is $(0, -2)$.

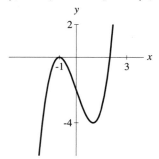

87. Use synthetic division on

$$f(x) = \frac{1}{2}x^3 - \frac{1}{2}x^2 - 2x + 2.$$

$$
\begin{array}{r|rrrr}
1 & 1/2 & -1/2 & -2 & 2 \\
  &   & 1/2 & 0 & -2 \\
\hline
  & 1/2 & 0 & -2 & 0
\end{array}
$$

Since the roots of

$$\frac{1}{2}x^2 - 2 = 0$$

are $x = \pm 2$, the graph crosses the
x-intercepts $(\pm 2, 0)$ and $(1, 0)$, and
the y-intercept is $(0, 2)$.

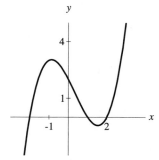

89. Factoring, we get

$$f(x) = \frac{1}{4}(x^4 - 8x^2 + 16) = \frac{1}{4}(x^2 - 4)^2.$$

The graph does not cross the x-intercepts $(\pm 2, 0)$, and the y-intercept is $(0, 4)$.

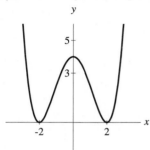

91. $f(x) = \dfrac{2}{x+3}$ has no x-intercept, y-intercept is $(0, 2/3)$, asymptotes are $x = -3$, $y = 0$

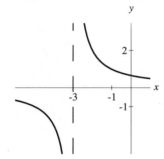

93. $f(x) = \dfrac{2x}{x^2 - 4}$ has x-intercept $(0, 0)$, asymptotes are $x = \pm 2$, $y = 0$. Symmetric about the origin. Graph goes through $(1, -2/3)$, $(3, 6/5)$, and $(-3, -6/5)$.

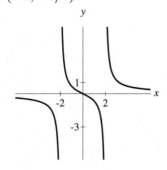

95. $f(x) = \dfrac{(x-1)^2}{x-2}$ has x-intercept $(1, 0)$, y-intercept is $(0, -1/2)$, asymptote $x = 2$, and oblique asymptote $y = x$ since

$$
\begin{array}{c|ccc}
2 & 1 & -2 & 1 \\
  &   & 2 & 0 \\
\hline
  & 1 & 0 & 1
\end{array}
$$

and $f(x) = x + \dfrac{1}{x-2}$.

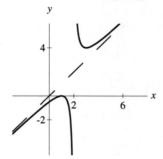

97. $f(x) = \dfrac{2x-1}{2-x}$ has x-intercept $(1/2, 0)$, y-intercept is $(0, -1/2)$, asymptotes are $x = 2$ and $y = -2$, graph goes through $(1, 1), (3, -5)$.

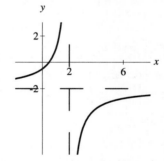

99. Since $f(x) = \dfrac{x^2 - 4}{x - 2} = x + 2$ if $x \neq 2$, x-intercept is $(-2, 0)$, y-intercept is $(0, 2)$, no asymptotes

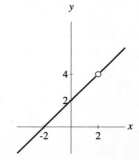

101. Set the right side to zero to obtain

$$8x^2 - 6x + 1 < 0.$$

Factoring, we get $(4x - 1)(2x - 1) < 0$. The sign graph is shown on the next page.

```
- - - - - - - - - 0 + + + +
- - - - - 0 + + + + + + + +
<---------------------------->
       1/4         1/2
```

The solution set is the interval $(1/4, 1/2)$.

103. The sign graph of $(3 - x)(x + 5) \geq 0$ is shown below.

```
+ + + + + + + + 0 - - - - -
- - - - 0 + + + + + + + + +
<---------------------------->
     -5            3
```

The solution set is the interval $[-5, 3]$.

105. Using the Rational Zero Theorem and synthetic division, we obtain

```
1/2 | 4    -400    -1    100
    |        2    -199  -100
    ------------------------------
      4    -398   -200    0
```

Note, $4x^2 - 398x - 200 = (4x + 2)(x - 100)$. Using the zeros, we obtain the sign graph of $4x^3 - 400x^2 - x + 100 \geq 0$. Namely,

```
- - - - - - - - - - - - - - - - 0 + + +
- - - - - - - - - - - 0 + + + + + +
- - - - - 0 + + + + + + + + +
<-------------------------------------->
     -1/2       1/2       100
```

The solution set is $\left[-\dfrac{1}{2}, \dfrac{1}{2}\right] \cup [100, \infty)$.

107. Let $R(x) = \dfrac{x + 10}{x + 2} - 5 = \dfrac{-4x}{x + 2} < 0$. Using the test-point method, we find $R(-3) < 0$, $R(-1) > 0$, and $R(1) < 0$.

```
    -    U    +    0    -
<---------------------------->
  -3   -2   -1    0    1
```

The solution is $(-\infty, -2) \cup (0, \infty)$.

109. $R(x) = \dfrac{12 - 7x}{x^2} + 1 = \dfrac{(x - 3)(x - 4)}{x^2} > 0.$ Using the test-point method, we get $R(-1) > 0$, $R(1) > 0$, $R(3.5) < 0$, and $R(5) > 0$.

```
    +    U    +    0    -    0    +
<------------------------------------>
  -1    0    1    3   3.5   4    5
```

the solution is $(-\infty, 0) \cup (0, 3) \cup (4, \infty)$.

111. Let $R(x) = \dfrac{(x - 1)(x - 2)}{(x - 3)(x - 4)}$. We will use the test-point method. Note, $R(0) > 0$, $R(1.5) < 0$, $R(2.5) > 0$, $R(3.5) < 0$, and $R(5) > 0$.

```
  +    0    -    0    +    U    -    U    +
<------------------------------------------>
  0    1   1.5   2   2.5   3   3.5   4    5
```

The solution is $(-\infty, 1] \cup [2, 3) \cup (4, \infty)$.

113. Quotient $x^2 - 3x$, remainder -15

```
3 | 1    -6     9    -15
  |        3    -9     0
  ------------------------
    1    -3     0    -15
```

115. Since $\dfrac{-b}{2a} = \dfrac{-156}{-32} = 4.875$, the maximum height is $-16(4.875)^2 + 156(4.875) = 380.25$ ft.

117. Let w be the height and let l be the length of the given cross-section of the room. Since the ratios of corresponding sides of similar triangles are equal, we obtain

$$\frac{7}{12} = \frac{w}{(48 - l)/2}$$
$$w = \frac{7}{24}(48 - l)$$

and the area, A, of the cross-section of the room is $A = wl = \dfrac{7}{24}(48 - l)l$. Since $l = 24$ maximizes A, we get $w = \dfrac{7}{24}(48 - 24) = 7$ ft. The dimensions that will maximize the area are 24 ft by 7 ft.

119.

(a) $V(10) = \dfrac{10{,}000}{58} \approx 172.4$ feet per second

(b) $V = 200$

(c) 200 feet per second

Chapter 3 Test

1. Use the method of completing the square.

$$y = 3(x^2 - 4x + 4) + 1 - 12 = 3(x - 2)^2 - 11$$

2. $y = 3(x - 2)^2 - 11$ has vertex $(2, -11)$,
axis of symmetry $x = 2$, y-intercept $(0, 1)$,

the x-intercepts are

$$\left(\frac{6 \pm \sqrt{33}}{3}, 0 \right),$$

and the range is $[-11, \infty)$

3. Minimum value is -11 since vertex is $(2, -11)$

4. Quotient $2x^2 - 6x + 14$, remainder -37

$$
\begin{array}{r|rrrr}
-3 & 2 & 0 & -4 & 5 \\
   &   & -6 & 18 & -42 \\
\hline
   & 2 & -6 & 14 & -37
\end{array}
$$

5. By the Remainder Theorem, remainder is -14.

6. $\pm \left\{ 1, \dfrac{1}{3}, 2, \dfrac{2}{3}, 3, 6 \right\}$

7.

$$
\begin{aligned}
(x + 3)(x - 4i)(x + 4i) &= (x + 3)(x^2 + 16) \\
&= x^3 + 3x^2 + 16x + 48
\end{aligned}
$$

An equation is $x^3 + 3x^2 + 16x + 48 = 0$.

8. $P(x) = x^3 - 3x^2 + 5x + 7$ has 2 sign variations
and $P(-x) = -x^3 - 3x^2 - 5x + 7$ has 1 sign
variation. There are
(a) 2 positive roots and 1 negative root, or
(b) 1 negative root and 2 imaginary roots.

9. Since $\dfrac{-b}{2a} = \dfrac{-128}{-32} = 4$,

the maximum height is $S(4) = 256$ feet.

10. ± 3

11. $\pm 2, \pm 2i$

12.

$$
\begin{array}{r|rrrr}
2 & 1 & -4 & -1 & 10 \\
  &   & 2 & -4 & -10 \\
\hline
  & 1 & -2 & -5 & 0
\end{array}
$$

Since $x^2 - 2x - 5 = (x - 1)^2 - 6$, the
zeros are $2, 1 \pm \sqrt{6}$.

13. Zeros are $x = \pm i$, each with multiplicity 2
since $f(x) = (x^2 + 1)^2 = (x - i)^2(x + i)^2$

14. The zeros are 2, 0 with multiplicity 2, and
$-3/2$ with mulitiplicity 3 since the equation
can be written as $x^2(x - 2)(2x + 3)^3 = 0$.

15.

$$
\begin{array}{r|rrrr}
1/2 & 2 & -9 & 14 & -5 \\
    &   & 1 & -4 & 5 \\
\hline
    & 2 & -8 & 10 & 0
\end{array}
$$

By completing the square, we find that the
roots of $2(x^2 - 4x) + 10 = 2(x - 2)^2 + 2 = 0$
are $2 \pm i$. The zeros are $x = 2 \pm i, 1/2$.

16. Parabola $y = 2(x - 3)^2 + 1$ with vertex $(3, 1)$

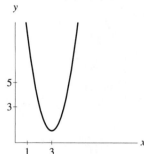

17. $y = (x - 2)^2(x + 1)$ crosses $(-1, 0)$ but does
not cross $(2, 0)$, y-intercept is $(0, 4)$

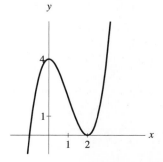

18. $y = x(x-2)(x+2)$ crosses $(0,0), (\pm 2, 0)$, and goes through $(1, -3)$

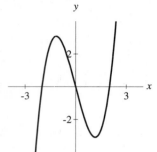

19. $y = \dfrac{1}{x-2}$ has asymptotes $x = 2$, $y = 0$, and goes through $(1, -1), (3, 1), (4, 1/2)$

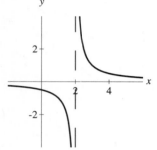

20. $y = \dfrac{2x-3}{x-2}$ has asymptotes $x = 2, y = 2$, x-intercept $(3/2, 0)$, y-intercept $(0, 3/2)$

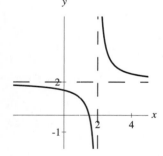

21. $f(x) = x + \dfrac{1}{x}$ has oblique asymptote $y = x$, asymptote $x = 0$, goes through $(1, 2), (-1, -2)$

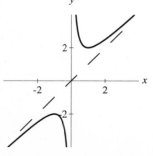

22. $y = \dfrac{4}{x^2 - 4}$ has asymptotes $x = \pm 2, y = 0$, y-intercept $(0, -1)$

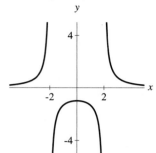

23. Note, $\dfrac{x^2 - 2x + 1}{x - 1} = x - 1$ provided $x \neq 1$; no x-intercept, y-intercept $(0, -1)$

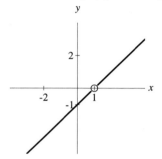

24. The sign graph of $(x - 4)(x + 2) < 0$ is

```
-  -  -  -  -  -  -  -  0  +  +  +  +
-  -  -  -  0  +  +  +  +  +  +  +  +
←─────────────────────────────────→
        -2              4
```

The solution set is the interval $(-2, 4)$.

25. Set the right-hand side to zero to obtain $\dfrac{2x - 1}{x - 3} > 0$. The sign chart is

```
-  -  -  -  -  -  -  -  -  U  +  +  +  +
-  -  -  -  -  0  +  +  +  +  +  +  +  +
←─────────────────────────────────→
        1/2             3
```

The solution set is $(-\infty, 1/2) \cup (3, \infty)$.

26. The sign chart of $\dfrac{x+3}{(4-x)(x+1)} \geq 0$ is

```
+++++++++++++++U- - -
- - - - - - - - - - -U++++++
- - - - -0 ++++++++++
```
$$\xleftarrow{\qquad\qquad\qquad\qquad\qquad}\xrightarrow{}$$
$$\qquad -3 \qquad -1 \qquad 4$$

The solution set is $(-\infty, -3] \cup (-1, 4)$.

27. The sign chart of $x(x^2 - 7) > 0$ is

```
- - - - - - - - - - - - - - 0+++
- - - - - - - - - - 0 ++++++
- - - - - 0 +++++++++++
```
$$\xleftarrow{\qquad\qquad\qquad\qquad\qquad}\xrightarrow{}$$
$$\qquad -\sqrt{7} \qquad 0 \qquad \sqrt{7}$$

The solution set is $(-\sqrt{7}, 0) \cup (\sqrt{7}, \infty)$.

28. If we raise each side of $(x - 3)^{-2/3} = \dfrac{1}{3}$ to

the power -3, then we obtain $(x - 3)^2 = 27$.
Thus, $x = 3 \pm \sqrt{27}$ and the solution set is
$\left\{ 3 \pm 3\sqrt{3} \right\}$.

29. Isolate a radical and square each side.

$$\begin{aligned}
\sqrt{x} - 1 &= \sqrt{x - 7} \\
x - 2\sqrt{x} + 1 &= x - 7 \\
-2\sqrt{x} &= -8 \\
4x &= 64
\end{aligned}$$

The solution set is $\{16\}$.

Tying It All Together

1. Since $2x - 3 = 0$ is equivalent to $2x = 3$, the
solution is $x = \dfrac{3}{2}$.

2. Factoring, we obtain $2x^2 - 3x = x(2x - 3) = 0$.
Thus, the solutions are $x = 0, \dfrac{3}{2}$.

3. Using synthetic division, we obtain

-1	1	3	3	1
		-1	-2	-1
	1	2	1	0

Since $x^2 + 2x + 1 = (x + 1)^2$, we find
$x^3 + 3x^2 + 3x + 1 = (x + 1)^3$.
Thus, the only solution is $x = -1$.

4. Multiply both sides of the equation of $\dfrac{x - 1}{x - 2} = 0$
by $x - 2$. Then we obtain $x - 1 = 0$.
The solution is $x = 1$.

5. If we raise each side of $x^{2/3} = 9$ to

the power 3, then we obtain $x^2 = 729$.
Thus, $x = \pm\sqrt{729}$ and the solution set is
$\{\pm 27\}$.

6. The graph is a straight line with domain

$(-\infty, \infty)$, range $(-\infty, \infty)$, x-intercept $\left(\dfrac{3}{2}, 0 \right)$

(see answer to number 1), and from the graph
we obtain that the function is increasing
in $(-\infty, \infty)$

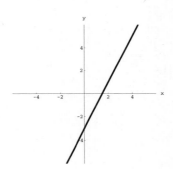

7. The graph is a parabola with domain $(-\infty, \infty)$,

range $\left[-\dfrac{9}{8}, \infty \right)$, x-intercepts are $(0, 0)$ and

$\left(\dfrac{3}{2}, 0 \right)$ (see answer to number 2), and the

function is increasing in $\left(\dfrac{3}{4}, \infty \right)$,

and decreasing in $\left(-\infty, \dfrac{3}{4} \right)$

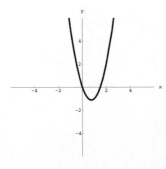

8. domain $(-\infty, \infty)$, range $(-\infty, \infty)$, the x-intercept is $(-1, 0)$ (see answer to number 3), and the function is increasing in $(-\infty, \infty)$

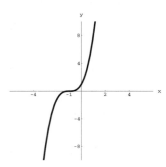

9. domain $(-\infty, 2) \cup (2, \infty)$, range $(-\infty, 1) \cup (1, \infty)$, the x-intercept is $(1, 0)$ (see answer to number 4), and the function is decreasing in $(-\infty, 2)$ and $(2, \infty)$

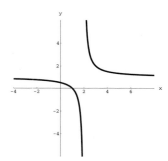

10. domain $(-\infty, \infty)$, range $[-9, \infty)$, the x-intercepts are $(27, 0)$ and $(-27, 0)$ (see answer to number 5), decreasing in $(-\infty, 0)$, increasing in $(0, \infty)$

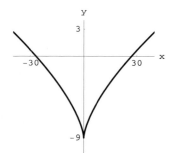

11. Based on the portion of the graph of $y = 2x - 3$ below the x-axis, the solution set of $2x - 3 < 0$ is the interval $\left(-\infty, \dfrac{3}{2}\right)$.

12. By taking the part of the graph of $y = 2x^2 - 3x$ below the x-axis, the solution set of $2x^2 - 3x < 0$ is the interval $\left(0, \dfrac{3}{2}\right)$.

13. By taking the part of the graph of $y = x^3 + 3x^2 + 3x + 1$ above the x-axis, the solution set of $x^3 + 3x^2 + 3x + 1 > 0$ is the interval $(-1, \infty)$.

14. By taking the part of the graph of $y = \dfrac{x-1}{x-2}$ above the x-axis and by considering the x-intercepts, the solution set of $\dfrac{x-1}{x-2} > 0$ is the $(-\infty, 1) \cup (2, \infty)$.

15. By taking the part of the graph of $y = x^{2/3} - 9$ below the x-axis and by conidering the x-intercepts, the solution set of $x^{2/3} - 9 < 0$ is the interval $(-27, 27)$.

16. Let $y = f(x)$, and interchange x and y. Then $x = 2y - 3$. Solving for y, we obtain $y = \dfrac{x+3}{2}$. The inverse is

$$f^{-1}(x) = \frac{x+3}{2}.$$

17. Since the graph of $f(x) = 2x^2 - 3x$ does not satisfy the Horizontal Line Test, the function is not one-to-one. Thus, the inverse does not exist.

18. From the solution of number 3, we see that $f(x) = (x+1)^3$. Let $y = f(x)$, and interchange x and y. Then $x = (y+1)^3$. Taking the cube root of both sides, we get $\sqrt[3]{x} = y + 1$. Thus, the inverse is

$$f^{-1}(x) = \sqrt[3]{x} - 1.$$

19. Let $y = f(x)$, and interchange x and y.

Then $x = \dfrac{y-1}{y-2}$. Solving for y, we obtain

$$\begin{aligned}
x(y-2) &= y-1 \\
xy - 2x &= y - 1 \\
xy - y &= 2x - 1 \\
y(x-1) &= 2x - 1 \\
y &= \frac{2x-1}{x-1}.
\end{aligned}$$

The inverse is

$$f^{-1}(x) = \frac{2x-1}{x-1}.$$

20. Since the graph of $f(x) = x^{2/3} - 9$ does not satisfy the Horizontal Line Test, the function is not one-to-one. Thus, the inverse does not exist.

For Thought

1. False, the base of an exponential function is positive.   **2.**   True

3. True, since $2^{-3} = \dfrac{1}{8}$.

4. True   **5.**   True   **6.**   True   **7.**   True

8. False, since it is decreasing.

9. True, since $0.25 = 4^{-1}$.

10. True, since $\sqrt[100]{2^{173}} = \left(2^{173}\right)^{1/100}$.

4.1 Exercises

1. 27

3. $-\left(2^0\right) = -1$

5. $\dfrac{1}{2^3} = \dfrac{1}{8}$

7. $\left(\dfrac{2}{1}\right)^4 = 2^4 = 16$

9. $\left(8^{1/3}\right)^2 = 2^2 = 4$

11. $-\left(9^{1/2}\right)^{-3} = -(3)^{-3} = -\dfrac{1}{3^3} = -\dfrac{1}{27}$

13. $3^2 = 9$ **15.** $3^{-2} = 1/9$ **17.** $2^{-1} = 1/2$

19. $2^3 = 8$

21. $(1/4)^{-1} = 4$

23. $4^{1/2} = 2$

25. $f(x) = 5^x$ goes through $(-1, 1/5), (0, 1), (1, 5)$, domain is $(-\infty, \infty)$, range is $(0, \infty)$, increasing

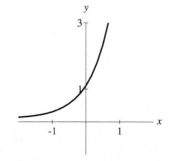

27. $f(x) = 10^{-x}$ goes through $(-1, 10), (0, 1)$, $(1, 1/10)$, domain is $(-\infty, \infty)$, range is $(0, \infty)$, decreasing

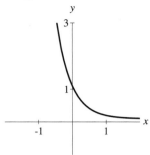

29. $f(x) = (1/4)^x$ goes through $(-1, 4), (0, 1)$, $(1, 1/4)$, domain is $(-\infty, \infty)$, range is $(0, \infty)$, decreasing

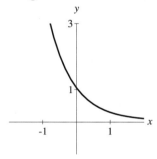

31. Shift $y = 2^x$ down by 3 units; $f(x) = 2^x - 3$ goes through $(-1, -2.5), (0, -2), (2, 1)$, domain $(-\infty, \infty)$, range $(-3, \infty)$, asymptote $y = -3$, increasing

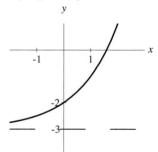

33. Shift $y = 2^x$ to left by 3 units and down by 5 units; $f(x) = 2^{x+3} - 5$ goes through $(-4, -4.5), (-3, -4), (0, 3)$, domain $(-\infty, \infty)$, range $(-5, \infty)$, asymptote $y = -5$, increasing

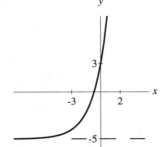

35. Reflect $y = 2^{-x}$ about x-axis, $f(x) = -2^{-x}$ goes through $(-1, -2), (0, -1), (1, -1/2)$, domain $(-\infty, \infty)$, range $(-\infty, 0)$, asymptote $y = 0$, increasing

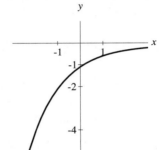

37. Reflect $y = 2^x$ about x-axis and shift up by 1 unit, $f(x) = 1 - 2^x$ goes through $(-1, 0.5), (0, 0), (1, -1)$, domain $(-\infty, \infty)$, range $(-\infty, 1)$, asymptote $y = 1$, decreasing

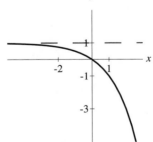

39. Shift $y = 3^x$ to right by 2 and shrink by a factor of 0.5, $f(x) = 0.5 \cdot 3^{x-2}$ goes through $(0, 1/18), (2, 0.5), (3, 1.5)$, domain $(-\infty, \infty)$, range $(0, \infty)$, asymptote $y = 0$, increasing

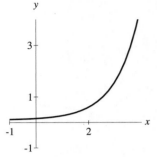

41. Stretch $y = (0.5)^x$ by a factor of 500, $f(x) = 500 \cdot (0.5)^x$ goes through $(0, 500)$, $(1, 250)$, domain $(-\infty, \infty)$, range $(0, \infty)$, asymptote $y = 0$, decreasing

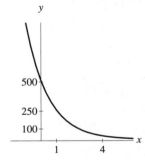

43. $y = 2^{x-5} - 2$

45. $y = -\left(\dfrac{1}{4}\right)^{x-1} - 2$

47. Since $2^x = 2^6$, solution set is $\{6\}$.

49. $\{-1\}$

51. Multiplying the equation by -1, $3^x = 3^3$ and the solution set is $\{3\}$.

53. $\{-2\}$

55. Since $(2^3)^x = 2^{3x} = 2$, $3x = 1$.

The solution set is $\left\{\dfrac{1}{3}\right\}$.

57. $\{-2\}$

59. Since $(2^{-1})^x = 2^{-x} = 2^3$, $-x = 3$. The solution set is $\{-3\}$.

61. Since $10^{x-1} = 10^{-2}$, $x - 1 = -2$. The solution set is $\{-1\}$.

63. Since $2^x = 4$, we find $x = 2$.

65. Since $2^x = \dfrac{1}{2}$, we obtain $x = -1$.

67. Since $\left(\dfrac{1}{3}\right)^x = 1$, we get $x = 0$.

69. Since $\left(\dfrac{1}{3}\right)^x = 3^3$, we find $x = -3$.

71. Since $10^x = 1000$, we obtain $x = 3$.

73. Since $10^x = 0.1 = 10^{-1}$, we find $x = -1$.

75. 1   **77.** -1

79. $(2, 9), (1, 3), (-1, 1/3), (-2, 1/9)$

81. $(0, 1), (-2, 25), (-1, 5), (1, 1/5)$

83. $(4, -16), (-2, -1/4), (-1, -1/2), (5, -32)$

85. When interest is compounded n times a year, the amount at the end of 6 years is

$$A(n) = 5000 \left(1 + \frac{0.08}{n}\right)^{6n}$$

and the interest earned after 6 years is

$$I(n) = A(n) - 5000.$$

a) If $n = 1$, then $A(1) = \$7934.37$ and $I(1) = \$2934.37$.

b) If $n = 4$, then $A(4) = \$8042.19$ and $I(4) = \$3042.19$.

c) If $n = 12$, then $A(12) = \$8067.51$ and $I(12) = \$3067.51$.

d) If $n = 365$, then $A(365) = \$8079.95$ and $I(365) = \$3079.95$.

87. After t years, a deposit of \$5000 will amount to

$$A(t) = 5000e^{0.08t}.$$

We use 30 days per month and 365 days per year.

a) After 6 years, the amount is $A(6) = \$8080.37$.

b) After 8 years and 3 months, the amount is
$$A\left(8 + \frac{90}{365}\right) = \$9671.31.$$

c) After 5 years, 4 months, and 22 days, the amount is

$$A\left(8 + \frac{4(30) + 22}{365}\right) = \$7694.93.$$

d) After 20 years and 321 days, the amount is

$$A\left(20 + \frac{321}{365}\right) = \$26,570.30.$$

89. Assume there are 365 days in a year and 30 days in a month. The present value is

$$3000\left(1 + \frac{0.065}{365}\right)^{-(365)(5 + 120/365)} = \$2121.82.$$

91. $20,000e^{-0.0542(30)} = \$3934.30$

93. a) The interest for first hour is

$$10^6 \cdot e^{0.06\left(\frac{1}{(24)(365)}\right)} - 10^6 = \$6.85.$$

b) The interest for 500th hour is the difference between the amounts in the account at the end of the 500th and 499th hours i.e.

$$10^6 e^{0.06\left(\frac{500}{(24)(365)}\right)} - 10^6 e^{0.06\left(\frac{499}{(24)(365)}\right)} = \$6.87$$

95. If $t = 0$, then $A = 200e^0 = 200$ g.
If $t = 500$, then $A = 200e^{-0.001(500)} \approx 121.3$ g.

97. a) The exponential regression curve is
$y = 5.6837(1.3115)^x$ or about

$$y = 5.68(1.31)^x.$$

b) Yes

c) In 2006 when $t = 16$, the number of subscribers is

$$5.6837(1.3115)^{16} \approx 435.4 \text{ million.}$$

99. $P = 10\left(\dfrac{1}{2}\right)^n$

101. When $t = 31$, the number of damaged O-rings is $n = 644e^{-0.15(31)} \approx 6$.

For Thought

1. True   **2.**   False, since $\log_{100}(10) = 1/2$.

3. True   **4.**   True

5. False, the domain is $(0, \infty)$.   **6.**   True

7. True

8. False, since $\log_a(0)$ is undefined.

9. True   **10.**   True

4.2 Exercises

1. 6   **3.** -4

5. $\dfrac{1}{4}$ **7.** -3

9. Since $2^6 = 64$, $\log_2(64) = 6$.

11. Since $3^{-4} = \dfrac{1}{81}$, $\log_3\left(\dfrac{1}{81}\right) = -4$.

13. Since $16^{1/4} = 2$, $\log_{16}(2) = \dfrac{1}{4}$.

15. Since $\left(\dfrac{1}{5}\right)^{-3} = 125$, $\log_{1/5}(125) = -3$.

17. Since $10^{-1} = 0.1$, $\log(0.1) = -1$

19. Since $10^0 = 1$, $\log(1) = 0$.

21. Since $e^1 = e$, $\ln(e) = 1$.

23. -5

25. $y = \log_3(x)$ goes through $(1/3, -1)$, $(1, 0)$, $(3, 1)$, domain $(0, \infty)$, range $(-\infty, \infty)$

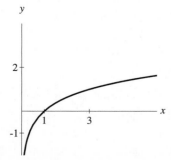

27. $f(x) = \log_5(x)$ goes through $(1/5, -1), (1, 0)$, and $(5, 1)$, domain $(0, \infty)$, range $(-\infty, \infty)$

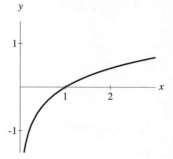

29. $y = \log_{1/2}(x)$ goes through $(2, -1), (1, 0)$, $(1/2, 1)$, domain $(0, \infty)$, range $(-\infty, \infty)$

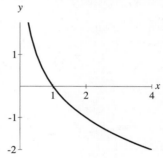

31. $h(x) = \log_{1/5}(x)$ goes through $(5, -1)$, $(1, 0), (1/5, 1)$, domain $(0, \infty)$, range $(-\infty, \infty)$

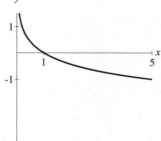

33. $f(x) = \ln(x - 1)$ goes through $\left(1 + \dfrac{1}{e}, -1\right)$, $(2, 0)$, $(1 + e, 1)$, domain $(1, \infty)$, range $(-\infty, \infty)$

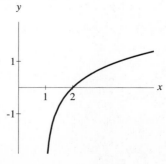

35. $f(x) = -3 + \log(x+2)$ goes through
$(-1.9, -4)$, $(-1, -3)$, $(8, -2)$,
domain $(-2, \infty)$, range $(-\infty, \infty)$

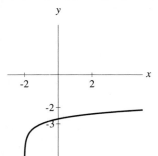

37. $f(x) = -\dfrac{1}{2}\log(x-1)$ goes through $(1.1, 0.5)$,

$(2, 0)$, $(11, -0.5)$, domain $(1, \infty)$,

range $(-\infty, \infty)$

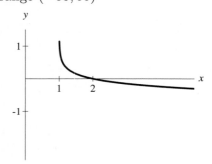

39. $y = \ln(x-3) - 4$

41. $y = -\log_2(x-5) - 1$

43. $2^5 = 32$ **45.** $5^y = x$ **47.** $10^z = 1000$

49. $e^x = 5$

51. $a^m = x$

53. $\log_5(125) = 3$

55. $\ln(y) = 3$

57. $\log(y) = m$

59. $\log_a(y) = z$

61. $\log_a(n) = x - 1$

63. $f^{-1}(x) = \log_2(x)$

65. $f^{-1}(x) = 7^x$

67. Replace $f(x)$ by y, interchange x and y,
solve for y, and replace y by $f^{-1}(x)$.

$$
\begin{aligned}
y &= \ln(x-1) \\
x &= \ln(y-1) \\
e^x &= y - 1 \\
y = f^{-1}(x) &= e^x + 1
\end{aligned}
$$

69. Replace $f(x)$ by y, interchange x and y,
solve for y, and replace y by $f^{-1}(x)$.

$$
\begin{aligned}
y &= 3^{x+2} \\
x &= 3^{y+2} \\
y + 2 &= \log_3(x) \\
y = f^{-1}(x) &= \log_3(x) - 2
\end{aligned}
$$

71. Replace $f(x)$ by y, interchange x and y,
solve for y, and replace y by $f^{-1}(x)$.

$$
\begin{aligned}
x &= \frac{1}{2}10^{y-1} + 5 \\
x - 5 &= \frac{1}{2}10^{y-1} \\
2x - 10 &= 10^{y-1} \\
\log(2x - 10) &= y - 1 \\
\log(2x - 10) + 1 &= y \\
f^{-1}(x) &= \log(2x - 10) + 1
\end{aligned}
$$

73. Since $2^8 = x$, the solution is $x = 256$.

75. Since $3^{1/2} = x$, the solution is $x = \sqrt{3}$.

77. Since $x^2 = 16$ and the base of a logarithm is
positive, $x = 4$.

79. By using the definition of the logarithm, we
get $x = \log_3(77)$.

81. Since $y = \ln(x)$ is one-to-one, $x - 3 = 2x - 9$.
The solution is $x = 6$.

83. Since $x^2 = 18$, we obtain $x = \sqrt{18} = 3\sqrt{2}$.
The base of a logarithm is positive.

85. Since $x + 1 = \log_3(7)$, $x = \log_3(7) - 1$.

87. Since $y = \log(x)$ is one-to-one, we get

$$
\begin{aligned}
x &= 6 - x^2 \\
x^2 + x - 6 &= 0 \\
(x + 3)(x - 2) &= 0 \\
x &= -3, 2.
\end{aligned}
$$

But $\log(-3)$ is undefined, so the solution is $x = 2$.

89. Since $x^{-2/3} = \dfrac{1}{9}$, we find

$$
x = \pm \left(\frac{1}{9} \right)^{-3/2} = \pm \left(\frac{1}{3} \right)^{-3} = \pm 27.
$$

The solution is $x = 27$ since a negative number cannot be a base.

91. Since $(2^2)^{2x-1} = 2^{4x-2} = 2^{-1}$, we get

$4x - 2 = -1$. The solution is $x = \dfrac{1}{4}$.

93. Since $32^x = 64$, we get $2^{5x} = 2^6$.

Thus, the solution is $x = \dfrac{6}{5}$.

95. $x = \log(25) \approx 1.3979$

97. Solving for x, we obtain

$$
\begin{aligned}
e^{2x} &= 3 \\
2x &= \ln(3) \\
x &= \frac{1}{2} \ln(3) \\
x &\approx 0.5493.
\end{aligned}
$$

99. Solving for x, we get

$$
\begin{aligned}
e^x &= \frac{4}{5} \\
x &= \ln\left(\frac{4}{5} \right) \\
x &\approx -0.2231.
\end{aligned}
$$

101. Solving for x, we find

$$
\begin{aligned}
\frac{1}{10^x} &= 2 \\
\frac{1}{2} &= 10^x \\
x &= \log\left(\frac{1}{2} \right) \\
x &\approx -0.3010.
\end{aligned}
$$

103. Solving for the year t when \$10 grows to \$20, we find

$$
\begin{aligned}
20 &= 10e^{rt} \\
2 &= e^{rt} \\
\ln 2 &= rt \\
\frac{\ln 2}{r} &= t.
\end{aligned}
$$

a) If $r = 2\%$, then $t = \dfrac{\ln 2}{0.02} \approx 34.7$ years.

b) If $r = 4\%$, then $t = \dfrac{\ln 2}{0.04} \approx 17.3$ years.

c) If $r = 8\%$, then $t = \dfrac{\ln 2}{0.08} \approx 8.7$ years.

d) If $r = 16\%$, then $t = \dfrac{\ln 2}{0.16} \approx 4.3$ years.

105. Solving for the annual percentage rate r when \$10 becomes \$30, we find

$$
\begin{aligned}
30 &= 10e^{rt} \\
3 &= e^{rt} \\
\ln 3 &= rt \\
\frac{\ln 3}{t} &= r.
\end{aligned}
$$

a) If $t = 5$ years, then $r = \dfrac{\ln 3}{5} \approx 0.2197 \approx 22\%$.

b) If $t = 10$ years, then $r = \dfrac{\ln 3}{10} \approx 0.10986 \approx 11\%$.

c) If $t = 20$ years, then $r = \dfrac{\ln 3}{20} \approx 0.0549 \approx 5.5\%$.

d) If $t = 40$ years, then $r = \dfrac{\ln 3}{40} \approx 0.027465 \approx 2.7\%$.

107. Let t be the number of years.

$$
\begin{aligned}
1000 \cdot e^{0.14t} &= 10^6 \\
e^{0.14t} &= 1000 \\
0.14t &= \ln(1000) \\
t &\approx 49.341 \text{ years}
\end{aligned}
$$

Note, $0.341(365) \approx 125$.
It will take 49 years and 125 days.

109. Since $e^{rt} = A/P$, $rt = \ln(A/P)$ and

$$r = \frac{\ln(A/P)}{t}.$$

Thus, $1000 will double in 3 years if the rate

is $r = \dfrac{\ln(2000/1000)}{3} \approx 0.231$ or 23.1%.

111.

(a) Let t be the number of years.

$$
\begin{aligned}
P \cdot e^{0.1t} &= 2P \\
e^{0.1t} &= 2 \\
0.1t &= \ln(2) \\
t = \frac{\ln(2)}{0.1} &\approx 6.9
\end{aligned}
$$

An investment at 10% doubles every 6.9 years.

(b) If t is the number of years it takes before an investment doubles, then

$$
\begin{aligned}
P \cdot e^{rt} &= 2P \\
e^{rt} &= 2 \\
rt &= \ln(2) \\
t &= \frac{\ln(2)}{r} \\
t &\approx \frac{0.70}{r}.
\end{aligned}
$$

In particular, if $r = 0.07$ then

$$t \approx \frac{0.70}{0.07} = 10 \text{ years.}$$

That is, at 10%, an investment will double in about 10 years.

113. Let r be the interest rate.

$$
\begin{aligned}
4,000 \cdot e^{200r} &= 4,500,000 \\
e^{200r} &= 1,125 \\
200r &= \ln(1,125) \\
r = \frac{\ln(1,125)}{200} &\approx 0.035
\end{aligned}
$$

The rate is 3.5% .

115. Let t be the number of years.

$$
\begin{aligned}
F_o \cdot e^{-0.052t} &= 0.6F_o \\
e^{-0.052t} &= 0.6 \\
-0.052t &= \ln(0.6) \\
t = \frac{\ln(0.6)}{-0.052} &\approx 9.8
\end{aligned}
$$

Only 60% of the present forest will remain after 9.8 years.

117. Let r be the annual rate from 1950 to 1987.

$$
\begin{aligned}
2.5 \cdot e^{37r} &= 5 \\
e^{37r} &= 2 \\
37r &= \ln(2) \\
r = \frac{\ln(2)}{37} &\approx 0.0187
\end{aligned}
$$

The annual rate is 1.87%.

If the annual rate is 1.63% and the initial population is 5 billion in 1987, the world population in year 2010 will be

$$5 \cdot e^{0.0163(23)} \approx 7.3 \text{ billion.}$$

119.

(a) $0.1e^{0.46(15)} \approx 99.2$ billion gigabits per year.

(b) Solving for t, we find

$$
\begin{aligned}
.1e^{.46t} &= 14 \\
e^{.46t} &= 140 \\
.46t &= \ln(140) \\
t &= \frac{\ln(140)}{.46} \\
t &\approx 11.
\end{aligned}
$$

In 2005 ($\approx 1994 + 11$), the data transmission is expected to be 14 billion gigabits/yr.

121.

a) The function is given by

$$p - 100 = \frac{100 - 10}{2 - 4}(x - 2)$$
$$p - 100 = -45(x - 2)$$
$$p = -45x + 90 + 100$$
$$p = -45x + 190$$
$$p = -45\log(I) + 190.$$

b) $p = -45\log(100,000) + 190 = -35\%$ or 0% of the population is expected to be without safe drinking water, i.e., everyone is expected to have safe water.

123. $pH = -\log\left(10^{-4.1}\right) = 4.1$

125. $pH = -\log\left(10^{-3.7}\right) = 3.7$

127. By substituting $x = 1$, we find a formula for c.

$$a \cdot b^x = a \cdot e^{cx}$$
$$b^x = e^{cx}$$
$$b = e^c$$
$$c = \ln b.$$

By using this formula, we find

$$y = 500(1.036)^x = 500e^{\ln(1.036)x}$$

and the continuous growth rate is

$$\ln(1.036) \cdot 100 \approx 3.54\%.$$

For Thought

1. False, since $\log(8) - \log(3) = \log(8/3) \neq \dfrac{\log(8)}{\log(3)}$.

2. True, since $\ln(3^{1/2}) = \dfrac{1}{2} \cdot \ln(3) = \dfrac{\ln(3)}{2}$.

3. True, since $\dfrac{\log_{19}(8)}{\log_{19}(2)} = \log_2(8) = 3 = \log_3(27)$.

4. True, because of the base-change formula.

5. False

6. False, since $\log(x) - \log(2) = \log(x/2)$.

7. False, since the solution of the first equation is $x = -2$ and the second equation is not defined when $x = -2$.

8. True   **9.**   False, since x can be negative.

10. False, since a can be negative and so $\ln(a)$ will not be a real number.

4.3 Exercises

1. $\log(15)$

3. $\log_2((x-1)x) = \log_2(x^2 - x)$

5. $\log_4(6)$ **7.** $\ln\left(\dfrac{x^8}{x^3}\right) = \ln(x^5)$

9. $\log_2(3x) = \log_2(3) + \log_2(x)$

11. $\log\left(\dfrac{x}{2}\right) = \log(x) - \log(2)$

13. $\log((x-1)(x+1)) = \log(x-1) + \log(x+1)$

15. $\ln\left(\dfrac{x-1}{x}\right) = \ln(x-1) - \ln(x)$

17. $\log_a(5^3) = 3\log_a(5)$

19. $\log_a(5^{1/2}) = \dfrac{1}{2} \cdot \log_a(5)$

21. $\log_a(5^{-1}) = -\log_a(5)$

23. $\sqrt{y}$ **25.** $y + 1$ **27.** 999

29. $\log_a(2) + \log_a(5)$

31. $\log_a(5/2) = \log_a(5) - \log_a(2)$

33. $\log_a(\sqrt{2^2 \cdot 5}) = \dfrac{1}{2}(2\log_a(2) + \log_a(5)) = \log_a(2) + \dfrac{1}{2}\log_a(5)$

35. $\log_a(4) - \log_a(25) = \log_a(2^2) - \log_a(5^2) = 2\log_a(2) - 2\log_a(5)$

37. $\log_3(5) + \log_3(x)$

39. $\log_2(5) - \log_2(2y) = \log_2(5) - \log_2(2) - \log_2(y)$

41. $\log(3) + \dfrac{1}{2}\log(x)$

43. $\log(3) + (x-1)\log(2)$

45. $\frac{1}{3} \cdot \ln(xy) - \frac{4}{3}\ln(t) = \frac{1}{3}\cdot\ln(x) + \frac{1}{3}\cdot\ln(y) - \frac{4}{3}\ln(t)$

47. $\ln(6\sqrt{x-1}) - \ln(5x^3) =$

$\ln(6) + \frac{1}{2}\cdot\ln(x-1) - \ln(5) - 3\cdot\ln(x)$

49. $\log_2(5x^3)$

51. $\log_7(x^5) - \log_7(x^8) = \log(x^5/x^8) = \log_7(x^{-3})$

53. $\log(2xy/z)$

55. $\log\left(\frac{\sqrt{x}}{y}\right) + \log\left(\frac{z}{\sqrt[3]{w}}\right) = \log\left(\frac{z\sqrt{x}}{y\sqrt[3]{w}}\right)$

57. $\log_4(x^6) + \log_4(x^{12}) + \log_4(x^2) = \log_4(x^{20})$

59. Since $2^x = 9$, we get $x = \frac{\log 9}{\log 2} \approx 3.1699$.

61. Since $0.56^x = 8$, we get

$x = \frac{\log 8}{\log 0.56} \approx -3.5864$.

63. Since $1.06^x = 2$, we get

$x = \frac{\log 2}{\log 1.06} \approx 11.8957$.

65. Since $0.73^x = 0.5$, we get

$x = \frac{\log 0.5}{\log 0.73} \approx 2.2025$.

67. $\frac{\ln(9)}{\ln(4)} \approx \frac{2.1972246}{1.3862944} \approx 1.5850$

69. $\frac{\ln(2.3)}{\ln(9.1)} \approx \frac{0.8329091}{2.2082744} \approx 0.3772$

71. $\frac{\ln(12)}{\ln(1/2)} \approx -3.5850$

73. Since $4t = \log_{1.02}(3) = \frac{\ln(3)}{\ln(1.02)}$,

we find $t = \frac{\ln(3)}{4\cdot\ln(1.02)} \approx 13.8695$.

75. Since $365t = \log_{1.0001}(3.5) = \frac{\ln(3.5)}{\ln(1.0001)}$,

we get $t = \frac{\ln(3.5)}{365\cdot\ln(1.0001)} \approx 34.3240$.

77. $1 + r = \sqrt[3]{2.3}$, so $r = \sqrt[3]{2.3} - 1 \approx 0.3200$

79.

$$\left(1 + \frac{r}{12}\right)^{360} = 4.2$$
$$1 + \frac{r}{12} = \pm\sqrt[360]{4.2}$$
$$r = 12\left(\pm\sqrt[360]{4.2} - 1\right)$$
$$r \approx 0.0479, -24.0479$$

81. Since $x^5 = 33.4$, we get $x = \sqrt[5]{33.4} \approx 2.0172$.

83. Since $x^{-1.3} = 0.546$, we have $x = 0.546^{1/(-1.3)}$ or $x \approx 1.5928$.

85. Let t be the number of years.

$$800\left(1 + \frac{0.08}{365}\right)^{365t} = 2000$$
$$\left(1 + \frac{0.08}{365}\right)^{365t} = 2.5$$
$$(1.0002192)^{365t} \approx 2.5$$
$$365t \approx \log_{1.0002192}(2.5)$$
$$t \approx \frac{1}{365}\cdot\frac{\ln(2.5)}{\ln(1.0002192)}$$
$$t \approx 11.454889$$
$$t \approx 11 \text{ years}, 166 \text{ days}$$

87. Let t be the number of years.

$$W\left(1 + \frac{0.1}{4}\right)^{4t} = 3W$$
$$(1.025)^{4t} = 3$$
$$4t \approx \log_{1.025}(3)$$
$$t \approx \frac{1}{4}\cdot\frac{\ln(3)}{\ln(1.025)}$$
$$t \approx 11.123 \text{ years}$$
$$t \approx 11.123(4) \approx 44 \text{ quarters}$$

89. Let t be the number of years.

$$500(1+r)^{25} = 2000$$
$$(1+r)^{25} = 4$$
$$1 + r \approx \sqrt[25]{4}$$
$$r = \sqrt[25]{4} - 1$$
$$r \approx 0.057 \text{ or } 5.7\%$$

91. Let t be the number of years.

$$
\begin{aligned}
4000\,(1+r)^{200} &= 4.5 \times 10^6 \\
(1+r)^{200} &= 1125 \\
r &= \sqrt[200]{1125} - 1 \\
r &\approx 0.035752 \text{ or } 3.58\%
\end{aligned}
$$

93. The Richter scale rating is

$$
\log(I) - \log(I_o) = \log\left(\frac{I}{I_o}\right).
$$

When $I = 1000 \cdot I_o$, the Richter scale rating is

$$
\log\left(\frac{1000 \cdot I_o}{I_o}\right) = \log(1000) = 3.
$$

95. $t = \dfrac{1}{r}\ln(P/P_o) = \dfrac{1}{r}\ln(P) - \dfrac{1}{r}\ln(P_o)$

97.

(a) p decreases as n increases

(b) Solving for n, one can take the logarithm of both sides and to note that $y = \log(x)$ is an increasing function.

$$
\left(\frac{7,059,051}{7,059,052}\right)^n > \frac{1}{2}
$$

$$
\log\left(\left(\frac{7,059,051}{7,059,052}\right)^n\right) > \log\left(\frac{1}{2}\right)
$$

$$
n\log\left(\frac{7,059,051}{7,059,052}\right) > \log\left(\frac{1}{2}\right)
$$

$$
n < \frac{\log(1/2)}{\log(7,059,051/7,059,052)}
$$

$$
n < 4,892,962
$$

If at most $4,892,961$ tickets are purchsed, then the probability of a rollover is greater than 50%.

99. We note that $MR(x) = R(x+1) - R(x) = 500 \cdot \log(x+2) - 500 \cdot \log(x+1) =$

$$
500 \cdot \log\left(\frac{x+2}{x+1}\right) = \log\left(\left(\frac{x+2}{x+1}\right)^{500}\right).
$$

Hence, as $x \to \infty$, then $\dfrac{x+2}{x+1} \to 1$ and $MR(x) \to 0$.

101.

a) Let x be the number of years since 1985 and let y be the number of computers per 1000 people. With the aid of a calculator, we find that an exponential regression curve is

$$
y = a \cdot b^x
$$

where

$$
a \approx 369.18888803
$$

and

$$
b \approx 1.10777161.
$$

Approximately, the exponential curve is

$$
y = 369.2(1.108^x).
$$

b) Using the constant b in part a), we note

$$
\ln(b) \approx 0.1023504387.
$$

Thus, approximately we find

$$
y = 369.2(e^{\ln b})^x \approx 369.2e^{0.102x}.
$$

c) Assuming continuous growth, the annual percentage rate is $\ln b \approx 0.102 = 10.2\%$.

d) Let $y = 1500$. Using the constants a and b from part a), we find

$$
\begin{aligned}
1500 &= a \cdot b^x \\
\frac{1500}{a} &= b^x \\
\log_b\left(\frac{1500}{a}\right) &= x \\
13.7 &\approx x
\end{aligned}
$$

In the year 2010 ($\approx 1996 + 14$), there will be 1500 computers per 1000 people.

e) No, the data does not look exponential, rather it looks linear.

103. Let $a > 0$, $a \neq 1$. Using the definition of a logarithm and properties of exponents, we have

$$
\frac{M}{N} = \frac{a^{\log_a(M)}}{a^{\log_a(N)}} = a^{\log_a(M) - \log_a(N)}
$$

and

$$\frac{M}{N} = a^{\log_a(M/N)}$$

Thus, $a^{\log_a(M/N)} = a^{\log_a(M)-\log_a(N)}$
and $\log_a(M) - \log_a(N) = \log_a(M/N)$.

105.

a) The graphs of $y_1 = \log_3(\sqrt{3}x)$ and $y_2 = 0.5 + \log_3(x)$ are identical.

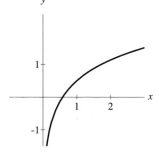

b) The graphs of $y_1 = \log_2(1/x)$ and $y_2 = -\log_2(x)$ are identical.

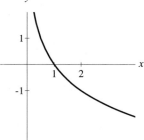

c) $y_1 = 3^{x-1}$ and $y_2 = \log_3(x) + 1$ are inverse functions of each other

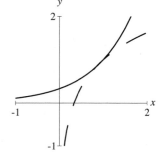

d) $y_1 = 3 + 2^{x-4}$ and $y_2 = \log_2(x-3) + 4$ are inverse functions of each other

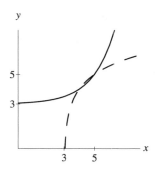

For Thought

1. True, since $(1.02)^x = 7$ is equivalent to $x = \log_{1.02}(7)$.

2. True, for $x(1 - \ln(3)) = 8$ implies $x = \dfrac{8}{1 - \ln(3)}$.

3. False, $\ln(1 - \sqrt{6})$ is undefined.

4. True, by the definition of a logarithm.

5. False, the exact solution is $x = \log_3(17)$ which is not the same as 2.5789.

6. False, since $x = -2$ is not a solution of the first equation but is a solution of the second one.

7. True, since $4^x = 2^{2x} = 2^{x-1}$ is equivalent to $2x = x - 1$.

8. True, since we may take the ln of both sides of $1.09^x = 2.3$.

9. True, since $\dfrac{\ln(2)}{\ln(7)} = \dfrac{\log(2)}{\log(7)}$.

10. True, since $\log(e) \cdot \ln(10) = \ln\left(10^{\log(e)}\right) = \ln(e) = 1$.

4.4 Exercises

1. Since $x = 2^3$, we get $x = 8$.

3. Since $10^2 = x + 20$, $x = 80$.

5. Since $10^1 = x^2 - 15$, we get $x^2 = 25$. The solutions are $x = \pm 5$.

7. Note, $x^2 = 9$ and $x = \pm 3$. Since the base of a logarithm is positive, the solution is $x = 3$.

9. Note, $x^{-2} = 4$ or $\dfrac{1}{4} = x^2$. Since the base of a logarithm is positive, the solution is $x = \dfrac{1}{2}$.

11. Since $x^3 = 10$, the solution is $x = \sqrt[3]{10}$.

13. Since $x = \left(8^{-1/3}\right)^2$, the solution is $x = \dfrac{1}{4}$.

15.

$$
\begin{aligned}
\log_2(x^2 - 4) &= 5 \\
x^2 - 4 &= 2^5 \\
x^2 &= 36 \\
x &= \pm 6
\end{aligned}
$$

Checking $x = -6$, one gets $\log_2(-6 + 2)$ which is undefined. The solution is $x = 6$.

17.

$$
\begin{aligned}
\log(5) + \log(x) &= 2 \\
\log(5x) &= 2 \\
5x &= 10^2 \\
x &= 20
\end{aligned}
$$

19. Since $\ln(x(x + 2)) = \ln(8)$ and $y = \ln(x)$ is one-to-one, we get

$$
\begin{aligned}
x^2 + 2x &= 8 \\
x^2 + 2x - 8 &= 0 \\
(x + 4)(x - 2) &= 0 \\
x &= -4, 2
\end{aligned}
$$

Since $\ln(-4)$ is undefined, the solution is $x = 2$.

21. Since $\log(4x) = \log\left(\dfrac{5}{x}\right)$ and $y = \log(x)$ is one-to-one, we obtain

$$
\begin{aligned}
4x &= \frac{5}{x} \\
4x^2 &= 5 \\
x^2 &= \frac{5}{4} \\
x &= \pm\frac{\sqrt{5}}{2}.
\end{aligned}
$$

But $\log\left(-\dfrac{\sqrt{5}}{2}\right)$ is undefined, so $x = \dfrac{\sqrt{5}}{2}$.

23. Since $\log_2\left(\dfrac{x}{3x - 1}\right) = 0$, we get

$$
\begin{aligned}
\frac{x}{3x - 1} &= 1 \\
x &= 3x - 1 \\
1 &= 2x \\
x &= \frac{1}{2}.
\end{aligned}
$$

25.

$$
\begin{aligned}
x \cdot \ln(3) + x \cdot \ln(2) &= 2 \\
x(\ln(3) + \ln(2)) &= 2 \\
&= \frac{2}{\ln(3) + \ln(2)} \\
x &= \frac{2}{\ln(6)}
\end{aligned}
$$

27. Since $x - 1 = \log_2(7)$, $x = \dfrac{\ln(7)}{\ln(2)} + 1 \approx 3.8074$.

29. Since $4x = \log_{1.09}(3.4)$, we find

$$
x = \frac{1}{4} \cdot \frac{\ln(3.4)}{\ln(1.09)} \approx 3.5502.
$$

31. Since $-x = \log_3(30)$, we obtain

$$
x = -\frac{\ln(30)}{\ln(3)} \approx -3.0959.
$$

33. Note, $-3x^2 = \ln(9)$. There is no solution since the left-hand side is non-negative and the right-hand side is positive.

35.

$$
\begin{aligned}
\ln(6^x) &= \ln(3^{x+1}) \\
x \cdot \ln(6) &= (x + 1) \cdot \ln(3) \\
x \cdot \ln(6) &= x \cdot \ln(3) + \ln(3) \\
x(\ln(6) - \ln(3)) &= \ln(3) \\
x &= \frac{\ln(3)}{\ln(6) - \ln(3)} \\
x &\approx 1.5850
\end{aligned}
$$

37.

$$\begin{aligned}
\ln(e^{x+1}) &= \ln(10^x) \\
(x+1)\cdot\ln(e) &= x\cdot\ln(10) \\
x+1 &= x\cdot\ln(10) \\
1 &= x(\ln(10)-1) \\
x &= \frac{1}{\ln(10)-1} \\
x &\approx 0.7677
\end{aligned}$$

39.

$$\begin{aligned}
2^{x-1} &= (2^2)^{3x} \\
2^{x-1} &= 2^{6x} \\
x-1 &= 6x \\
-1 &= 5x \\
x &= -0.2
\end{aligned}$$

41.

$$\begin{aligned}
\ln(6^{x+1}) &= \ln(12^x) \\
(x+1)\cdot\ln(6) &= x\cdot\ln(12) \\
x\cdot\ln(6)+\ln(6) &= x\cdot\ln(12) \\
\ln(6) &= x(\ln(12)-\ln(6)) \\
x &= \frac{\ln(6)}{\ln(12)-\ln(6)} \\
x &\approx 2.5850
\end{aligned}$$

43. Since $3 = e^{-\ln(w)} = e^{\ln(1/w)} = 1/w$, we

have $\dfrac{1}{w} = 3$ and $w = \dfrac{1}{3}$.

45.

$$\begin{aligned}
(\log(z))^2 &= 2\cdot\log(z) \\
(\log(z))^2 - 2\cdot\log(z) &= 0 \\
\log(z)\cdot(\log(z)-2) &= 0 \\
\log(z) = 0 \quad &\text{or} \quad \log(z) = 2 \\
z = 10^0 \quad &\text{or} \quad z = 10^2 \\
z = 1 \quad &\text{or} \quad z = 100
\end{aligned}$$

The solutions are $z = 1, 100$.

47. Divide the equation by $4(1.03)^x$.

$$\begin{aligned}
\left(\frac{1.02}{1.03}\right)^x &= \frac{3}{4} \\
\ln\left(\left(\frac{1.02}{1.03}\right)^x\right) &= \ln\left(\frac{3}{4}\right) \\
x\cdot\ln\left(\frac{1.02}{1.03}\right) &= \ln\left(\frac{3}{4}\right) \\
x &= \frac{\ln\left(\frac{3}{4}\right)}{\ln\left(\frac{1.02}{1.03}\right)} \\
x &\approx 29.4872
\end{aligned}$$

49. Note that

$$\begin{aligned}
e^{\ln((x^2)^3)-\ln(x^2)} &= e^{\ln(x^6)-\ln(x^2)} = e^{\ln(x^6/x^2)} \\
&= e^{\ln(x^4)} = x^4.
\end{aligned}$$

Thus, $x^4 = 16$ and $x = \pm 2$.
But $\ln(-2)$ is undefined, so $x = 2$.

51. Since $\left(\dfrac{1}{2}\right)^2 = \dfrac{1}{4}$, we find

$$\begin{aligned}
\left(\frac{1}{2}\right)^{2x-1} &= \left(\frac{1}{2}\right)^{6x+4} \\
2x-1 &= 6x+4 \\
-5 &= 4x \\
x &= -\frac{5}{4}.
\end{aligned}$$

53. By approximating the x-intercepts of the graph $y = 2^x - 3^{x-1} - 5^{-x}$, we find that the solutions are $x \approx 0.194, 2.70$.

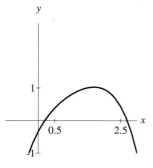

55. By approximating the x-intercept of the graph $y = \ln(x + 51) - \log(-48 - x)$, we obtain that the solution is $x \approx -49.73$.

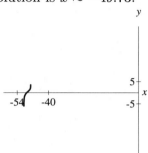

57. By approximating the x-intercepts of the graph $y = x^2 - 2^x$, we find that the solutions are $x \approx -0.767, 2, 4$.

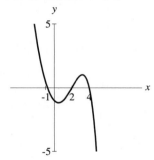

59. Solving for the rate of decay r, we find

$$
\begin{aligned}
A_0/2 &= A_0 e^{10,000r} \\
1/2 &= e^{10,000r} \\
\ln(1/2) &= 10,000r \\
\frac{\ln(1/2)}{10,000} &= r
\end{aligned}
$$

Approximately, $r \approx -6.93 \times 10^{-5}$.

61. Using $A = A_o e^{rt}$ with $A_o = 1$ and the half-life, we obtain $\dfrac{1}{2} = e^{5730r}$. Thus, $5730r = \ln\left(\dfrac{1}{2}\right)$ and $r \approx -0.000120968$. When $A = 0.1$, we get

$$
\begin{aligned}
0.1 &= e^{-0.000120968t} \\
\ln(0.1) &= -0.000120968t \\
t = \frac{\ln(0.1)}{-0.000120968} &\approx 19,035 \text{ years}
\end{aligned}
$$

63. From Number 61, $r \approx -0.000120968$. When $A = 10$ and $A_o = 12$, we obtain

$$
\begin{aligned}
10 &= 12 \cdot e^{-0.000120968t} \\
\ln(10/12) &= -0.000120968t \\
t = \frac{\ln(5/6)}{-0.000120968} &\approx 1507 \text{ years.}
\end{aligned}
$$

65. Let $A = A_o e^{rt}$ where $A_o = 25$, $A = 20$, and $t = 8000$. Then

$$
\begin{aligned}
20 &= 25 \cdot e^{8000r} \\
\ln(0.8) &= 8000r \\
r &\approx -0.000027893.
\end{aligned}
$$

To find the half-life, let $A = 12.5$. Thus, we have

$$
\begin{aligned}
12.5 &= 25 \cdot e^{-0.000027893t} \\
\ln(0.5) &= -0.000027893t \\
t &\approx 24,850 \text{ years.}
\end{aligned}
$$

67. $\dfrac{2.5(0.5)^{24/14}}{2.5} \times 100 \approx 30.5\%$, the percentage of the last dosage that remains before the next dosage is taken

69. Since the half-life is $t = 5730$ years and $A_o = 1$ and $A = 0.5$, we get

$$
\begin{aligned}
0.5 &= e^{(5730) \cdot r} \\
\ln(0.5) &= 5730 \cdot r \\
r &\approx -0.000121.
\end{aligned}
$$

If 79.3% of the carbon is still present, then

$$
\begin{aligned}
0.793 &= e^{(-0.000121) \cdot t} \\
\ln(0.793) &= -0.000121 \cdot t \\
t &\approx 1917.
\end{aligned}
$$

The scrolls were made in the year $1951 - 1917 = 34$ AD.

71. The initial difference in temperature is $325 - 35 = 290$, and after $t = 3$ hours the difference is $325 - 140 = 185$. Then

$$185 = 290 \cdot e^{3k}$$
$$\ln\left(\frac{185}{290}\right) = 3k$$
$$k \approx -0.1498417.$$

The difference in temperature when the roast is well-done is $325 - 170 = 155$. Thus,

$$155 = 290 \cdot e^{(-0.1498417) \cdot t}$$
$$\ln\left(\frac{155}{290}\right) = (-0.1498417) \cdot t$$
$$t \approx 4.18 \text{ hr}$$
$$t \approx 4 \text{ hr and } 11 \text{ min}.$$

James must wait 1 hour and 11 minutes longer.

If the oven temperature is set at 170°, then the initial and final differences are 135 and 0, respectively. Since $0 = 135 \cdot e^{(-0.1498417) \cdot t}$ has no solution, James has to wait forever.

73. At 7:00 a.m., the difference in temperature is $80 - 40 = 40$, and $t = 1$ hour later the difference in temperature is $72 - 40 = 32$. Then

$$32 = 40 \cdot e^{1 \cdot k}$$
$$\ln\left(\frac{32}{40}\right) = k$$
$$k \approx -0.2231436.$$

Let n be the number of hours before 7:00 a.m. when death occured. At the time of death, the difference in temperature is $98 - 40 = 58$. Then

$$40 = 58 \cdot e^{-0.2231436 \cdot n}$$
$$\ln\left(\frac{40}{58}\right) = -0.2231436 \cdot n$$
$$n \approx 1.665 \text{ hr}$$
$$n \approx 1 \text{ hr and } 40 \text{ min}.$$

Death occured at 5:20 a.m.

75. Since $R = P\dfrac{i}{1 - (1+i)^{-nt}}$, we obtain

$$1 - (1+i)^{-nt} = Pi/R$$
$$1 - Pi/R = (1+i)^{-nt}$$
$$\ln(1 - Pi/R) = -nt\ln(1+i)$$
$$\frac{-\ln(1 - Pi/R)}{n\ln(1+i)} = t.$$

Let $i = 0.09/12 = 0.0075$, $P = 100,000$, and $R = 1250$. Then

$$t = \frac{-\ln(1 - Pi/R)}{n\ln(1+i)} \approx 10.219.$$

It will take 10 yr, 3 mo to pay off the loan.

77. The future values of the $1000 and $1100 investments are equal. Then

$$1,000 \cdot e^{0.06t} = 1100\left(1 + \frac{0.06}{365}\right)^{365t}$$
$$e^{0.06t} = 1.1\left(1 + \frac{0.06}{365}\right)^{365t}$$
$$.06t = \ln(1.1) + 365t\ln\left(1 + \frac{0.06}{365}\right)$$

$$.06t - 365t\ln\left(1 + \frac{0.06}{365}\right) = \ln(1.1)$$

$$t = \frac{\ln(1.1)}{.06 - 365\ln\left(1 + \frac{0.06}{365}\right)}$$
$$t \approx 19,328.84173 \text{ years}$$

They will be equal after $19,328$ yr, 307 days.

79. a) Let $t = 0$. The present number of rabbits is

$$P = 12,300 + 1000 \cdot \ln(1) = 12,300 + 0 = 12,300.$$

b) In about 15 years.

c) The number of years before there will be $15,000$ rabbits is given by

$$12,300 + 1000 \cdot \ln(t+1) = 15,000$$
$$1000 \cdot \ln(t+1) = 2,700$$
$$\ln(t+1) = 2.7$$
$$t + 1 = e^{2.7}$$
$$t \approx 13.9 \text{ yr}.$$

81. a) Let $n = 2500$ and $A = 400$.
Since $n = k\log(A)$, $2500 = k\log(400)$.
Then $k = \dfrac{2500}{\log(400)}$. When $A = 200$, the number of species left is

$$n = \frac{2500}{\log(400)}\log(200) \approx 2211 \text{ species}.$$

b) Let $n = 3500$ and $A = 1200$.

Since $n = k \log(A)$, $3500 = k \log(1200)$.

Then $k = \dfrac{3500}{\log(1200)}$. When $n = 1000$,

the remaining forest area is given by

$$1000 = \frac{3500}{\log(1200)} \log(A)$$

$$\frac{1000}{3500} \log(1200) = \log(A)$$

$$\frac{2}{7} \log(1200) = \log(A)$$

$$\log\left(1200^{2/7}\right) = \log(A)$$

$$1200^{2/7} = A$$

Thus, the percentage of forest that has

been destroyed is $100 - \dfrac{A}{1200}(100) \approx 99\%$.

83. Since $m = 0$ and $M_v = 4.39$, the distance to α Centauri is given by

$$4.39 - 5 + 5 \cdot \log(d) = 0$$

$$5 \cdot \log(d) = 0.61$$

$$\log(d) = 0.122$$

$$d = 10^{0.122} \approx 1.32 \text{ parsecs.}$$

85.

a) Let $y = \log(P)$. A formula for P is

$$y - 4.5 = \frac{4.5 - 0.3}{0 - 23}(x - 0)$$

$$y = -\frac{21}{115}x + 4.5$$

$$\log(P) = -\frac{21}{115}x + 4.5$$

$$P = 10^{(-21x/115 + 4.5)}$$

$$P \approx 10^{-0.1826x + 4.5}$$

b) In 1991, $x = 9$ and

$$P \approx 10^{-0.1826(9) + 4.5} = \$718.79.$$

c) Solving for x, we derive

$$0.25 = 10^{(-21x/115 + 4.5)}$$

$$\log(0.25) = -\frac{21}{115}x + 4.5$$

$$\frac{\log(0.25) - 4.5}{-21/115} = x$$

$$x \approx 28$$

In the year $2010 = (1982 + 28)$, according to this model a 1-gigabit hard drive will cost $0.25.

87. If the sound level is 90 db, then the intensity of the sound is given by

$$10 \cdot \log(I \times 10^{12}) = 90$$

$$\log(I) + \log(10^{12}) = 9$$

$$\log(I) + 12 = 9$$

$$\log(I) = -3$$

$$I = 10^{-3} \text{ watts per/m}^2.$$

89. Since $P \cdot e^{(0.06)18} = 20{,}000$, the investment will grow to $P = \dfrac{20{,}000}{e^{(0.06)18}} \approx \$6791.91.$

91.

a) The logarithmic regression line is

$$y = 114.7865206 - 12.75457939 \ln(x)$$

or approximately is

$$y = 114.8 - 12.8 \ln(x)$$

where the year is $1960 + x$.

b) Let $x = 50$. Since

$$114.7865206 - 12.75457939 \ln(50) \approx 65,$$

the percentage of households in 2010 with two parents is 65%.

c) Let $y = 50$. Then

$$50 = 114.7865206 - 12.75457939 \ln(x)$$

$$\ln(x) = \frac{114.7865206 - 50}{12.75457939}$$

$$x \approx 161.$$

Since $1960 + 161 = 2121$, the percentage of two-parent families will reach 50% in the year 2121.

d) The data looks like a quadratic or exponential model.

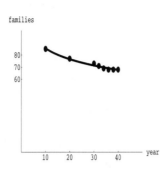

93. By using the first five terms of the formula,

$$e^{0.1} \approx 1 + 0.1 + \frac{(0.1)^2}{2} + \frac{(0.1)^3}{6} + \frac{(0.1)^4}{24}$$
$$\approx 1.105170833$$

From a calculator, $e^{0.1} \approx 1.105170918$.

Chapter 4 Review Exercises

1. 64   **3.**   6   **5.**   0

7. 17   **9.** $2^5 = 32$ **11.** $\log(10^3) = 3$

13. $\log_2(2^9) = 9$ **15.** $\log(1000) = 3$

17. $\log_2(1) - \log_2(8) = 0 - 3 = -3$ **19.** $\log_2(8) = 3$

21. $\log((x-3)x) = \log(x^2 - 3x)$

23. $\ln(x^2) + \ln(3y) = \ln(3x^2y)$

25. $\log(3) + \log(x^4) = \log(3) + 4 \cdot \log(x)$

27. $\log_3(5) + \log_3(x^{1/2}) - \log_3(y^4) =$
$\log_3(5) + \dfrac{1}{2} \cdot \log_3(x) - 4 \cdot \log_3(y)$

29. $\ln(2 \cdot 5) = \ln(2) + \ln(5)$

31. $\ln(5^2 \cdot 2) = \ln(5^2) + \ln(2) = 2 \cdot \ln(5) + \ln(2)$

33. Since $\log_{10}(x) = 10$, we get $x = 10^{10}$.

35. Since $x^4 = 81$ and $x > 0$, $x = 3$.

37. Since $\log_{1/3}(27) = -3 = x + 2$, we get $x = -5$.

39. Since $3^{x+2} = 3^{-2}$, $x + 2 = -2$. So $x = -4$.

41. Since $x - 2 = \ln(9)$, we obtain $x = 2 + \ln(9)$.

43. Since $(2^2)^{x+3} = 2^{2x+6} = 2^{-x}$, we get $2x + 6 = -x$. Then $6 = -3x$ and so $x = -2$.

45.

$$
\begin{aligned}
\log(2x^2) &= 5 \\
2x^2 &= 10^5 \\
x^2 &= 50{,}000 \\
x &= \pm 100\sqrt{5}
\end{aligned}
$$

Since $\log(-100\sqrt{5})$ is undefined, $x = 100\sqrt{5}$.

47.

$$
\begin{aligned}
\log_2\left(x^2 - 4x\right) &= \log_2(x + 24) \\
x^2 - 4x &= x + 24 \\
x^2 - 5x - 24 &= 0 \\
(x - 8)(x + 3) &= 0 \\
x &= 8, -3
\end{aligned}
$$

Since $\log(-3)$ is undefined, $x = 8$.

49. Since $\ln((x + 2)^2) = \ln(4^3)$ and $y = \ln(x)$ is a one-to-one function, we obtain

$$
\begin{aligned}
(x + 2)^2 &= 64 \\
x + 2 &= \pm 8 \\
x &= -2 \pm 8 \\
x &= 6, -10.
\end{aligned}
$$

Checking $x = -10$ one gets $2 \ln(-8)$ which is undefined. So $x = 6$.

51.

$$
\begin{aligned}
x \cdot \log(4) + x \cdot \log(25) &= 6 \\
x(\log(4) + \log(25)) &= 6 \\
x \cdot \log(100) &= 6 \\
x \cdot 2 &= 6 \\
x &= 3
\end{aligned}
$$

53. The missing coordinates are

(i) 3 since $\left(\dfrac{1}{3}\right)^{-1} = 3$,

(ii) -3 since $\left(\dfrac{1}{3}\right)^{-3} = 27$,

(iii) $\sqrt{3}$ since $\left(\dfrac{1}{3}\right)^{-1/2} = \sqrt{3}$, and

(iv) 0 since $\left(\dfrac{1}{3}\right)^{0} = 1$.

55. c

57. b

59. d

61. e

63. Domain $(-\infty, \infty)$, range $(0, \infty)$, increasing, asymptote $y = 0$

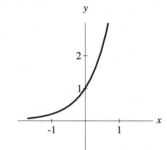

65. Domain $(-\infty, \infty)$, range $(0, \infty)$, decreasing, asymptote $y = 0$

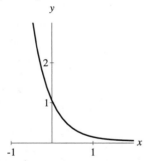

67. Domain $(0, \infty)$, range $(-\infty, \infty)$, increasing, asymptote $x = 0$

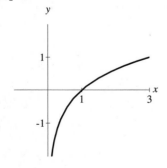

69. Domain $(-3, \infty)$, range $(-\infty, \infty)$, increasing, asymptote $x = -3$

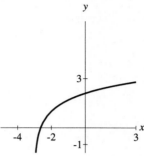

71. Domain $(-\infty, \infty)$, range $(1, \infty)$, increasing, asymptote $y = 1$

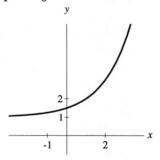

73. Domain $(-\infty, 2)$, range $(-\infty, \infty)$, decreasing, asymptote $x = 2$

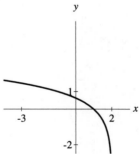

75. $f^{-1}(x) = \log_7(x)$

77. $f^{-1}(x) = 5^x$

79. Replace $f(x)$ by y, interchange x and y, solve for y, and replace y by $f^{-1}(x)$.

$$
\begin{aligned}
y &= 3\log(x - 1) \\
x &= 3\log(y - 1) \\
\frac{x}{3} &= \log(y - 1) \\
10^{x/3} &= y - 1 \\
10^{x/3} + 1 &= y \\
f^{-1}(x) &= 10^{x/3} + 1
\end{aligned}
$$

81. Replace $f(x)$ by y, interchange x and y, solve for y, and replace y by $f^{-1}(x)$.

$$
\begin{aligned}
y &= e^{x+2} - 3 \\
x &= e^{y+2} - 3 \\
x + 3 &= e^{y+2} \\
\ln(x+3) &= y + 2 \\
\ln(x+3) - 2 &= y \\
f^{-1}(x) &= \ln(x+3) - 2
\end{aligned}
$$

83. Since $3^x = 10$, $x = \log_3(10) = \dfrac{\ln(10)}{\ln(3)} \approx 2.0959$

.

85. Since $\log_3(x) = 1.876$, $x = 3^{1.876} \approx 7.8538$.

87. After taking the natural logarithm of both sides,
we have

$$
\begin{aligned}
\ln(5^x) &= \ln(8^{x+1}) \\
x \cdot \ln(5) &= (x+1) \cdot \ln(8) \\
x \cdot \ln(5) &= x \cdot \ln(8) + \ln(8) \\
x \cdot (\ln(5) - \ln(8)) &= \ln(8) \\
x &= \dfrac{\ln(8)}{\ln(5) - \ln(8)} \\
x &\approx -4.4243.
\end{aligned}
$$

89. True, since $\log_3(81) = 4$ and $2 = \log_3(9)$.

91. False, since $\ln(3^2) = 2 \cdot \ln(3) \neq (\ln(3))^2$.

93. True, since $4 \cdot \log_2(8) = 4 \cdot 3 = 12$.

95. False, since $3 + \log(6) = \log(10^3) + \log(6) = \log(6000)$.

97. False, since $\log_2(16) = 4$, $\log_2(8) = 3$, and $\dfrac{3}{4} \neq 3 - 4$.

99. False, since $\log_2(25) = 2\log_2(5) = 2 \cdot \dfrac{\log(5)}{\log(2)} \neq 2 \cdot \log(5)$.

101. True, because of the base-changing formula.

103. If H_B^+ is the hydrogen ion concentration of liquid B then $10 \cdot H_B^+$ is the hydrogen ion concentration of liquid A. The pH of A is

$$
-\log(10 \cdot H_B^+) = -1 - \log(H_B^+)
$$

i.e. the pH of A is one less than the pH of B.

105. The value at the end of 18 years is

$$
50{,}000 \left(1 + \frac{0.05}{4}\right)^{18 \cdot 4} \approx \$122{,}296.01.
$$

107. Let t be the number of years.

$$
\begin{aligned}
50{,}000 \left(1 + \frac{0.05}{4}\right)^{4t} &= 100{,}000 \\
(1.0125)^{4t} &= 2 \\
4t &= \log_{1.0125}(2) \\
t &= \frac{1}{4} \cdot \frac{\ln(2)}{\ln(1.0125)} \\
t &\approx 13.9 \ \text{years}
\end{aligned}
$$

It doubles in $4 \cdot 13.9 \approx 56$ quarters.

109. The present amount is $A = 25 \cdot e^0 = 25$ g. After $t = 1000$ years, the amount left is $A = 25 \cdot e^{-0.32} \approx 18.15$ g.

To find the half-life, let $A = 12.5$.

$$
\begin{aligned}
25 \cdot e^{-0.00032t} &= 12.5 \\
e^{-0.00032t} &= 0.5 \\
-0.00032t &= \ln(0.5) \\
t &\approx 2166
\end{aligned}
$$

The half-life is 2166 years.

111. Let $f(t) = 10{,}000$.

$$
\begin{aligned}
40{,}000 \cdot (1 - e^{-0.0001t}) &= 10{,}000 \\
1 - e^{-0.0001t} &= 0.25 \\
0.75 &= e^{-0.0001t} \\
\ln(0.75) &= -0.0001t \\
t &\approx 2877 \ \text{hr}
\end{aligned}
$$

It takes 2877 hours to learn $10{,}000$ words.

113. In the following equations, we solve for x.

$$
\begin{aligned}
1026 \left(\frac{25{,}005}{64}\right)^x &= 19.2 \\
\left(\frac{25{,}005}{64}\right)^x &= \frac{19.2}{1026} \\
x \ln\left(\frac{25{,}005}{64}\right) &= \ln\left(\frac{19.2}{1026}\right) \\
x &= \frac{\ln\left(\frac{19.2}{1026}\right)}{\ln\left(\frac{25{,}005}{64}\right)} \\
x \approx -0.667 \quad &\text{or} \quad x \approx -\frac{2}{3}
\end{aligned}
$$

In the following equations, we obtain y.

$$\frac{25005}{2240}\left(\frac{35+\frac{1}{12}}{100}\right)^y = 258.51$$

$$\left(\frac{35+\frac{1}{12}}{100}\right)^y = \frac{258.51(2240)}{25005}$$

$$y = \frac{\ln\left(\frac{258.51(2240)}{25005}\right)}{\ln\left(\frac{35+\frac{1}{12}}{100}\right)}$$

$$y \approx -3$$

Next, we solve for z.

$$13.5\left(\frac{25005}{64}\right)^z = 1.85$$

$$\left(\frac{25005}{64}\right)^z = \frac{1.85}{13.5}$$

$$z\ln\left(\frac{25005}{64}\right) = \ln\left(\frac{1.85}{13.5}\right)$$

$$z = \frac{\ln\left(\frac{1.85}{13.5}\right)}{\ln\left(\frac{25005}{64}\right)}$$

$$z \approx -0.333 \quad \text{or } z \approx -\frac{1}{3}$$

Chapter 4 Test

1. 3   **2.** -2 **3.**   6.47   **4.** $\sqrt{2}$

5. $f^{-1}(x) = e^x$

6. Replace $f(x)$ by y, interchange x and y, solve for y, and replace y by $f^{-1}(x)$.

$$\begin{aligned}
y &= 8^{x+1} - 3 \\
x &= 8^{y+1} - 3 \\
x + 3 &= 8^{y+1} \\
\log_8(x+3) &= y + 1 \\
\log_8(x+3) - 1 &= y \\
f^{-1}(x) &= \log_8(x+3) - 1
\end{aligned}$$

7. $\log(x) + \log(y^3) = \log(xy^3)$

8. $\ln(\sqrt{x-1}) - \ln(33) = \ln\left(\frac{\sqrt{x-1}}{33}\right)$

9. $\log_a(2^2 \cdot 7) = \log_a(2^2) + \log_a(7) = 2\cdot\log_a(2) + \log_a(7)$

10. $\log_a\left(\frac{7}{2}\right) = \log_a(7) - \log_a(2)$

11.

$$\begin{aligned}
\log_2(x^2 - 2x) &= 3 \\
x^2 - 2x &= 2^3 \\
x^2 - 2x - 8 &= 0 \\
(x-4)(x+2) &= 0 \\
x &= 4, -2
\end{aligned}$$

But $\log_2(-2)$ is undefined, so $x = 4$.

12.

$$\begin{aligned}
\log\left(\frac{10x}{x+2}\right) &= \log(3^2) \\
\frac{10x}{x+2} &= 9 \\
10x &= 9x + 18 \\
x &= 18
\end{aligned}$$

13.

$$\begin{aligned}
\ln(3^x) &= \ln(5^{x-1}) \\
x\cdot\ln(3) &= (x-1)\cdot\ln(5) \\
x\cdot\ln(3) &= x\cdot\ln(5) - \ln(5) \\
x(\ln(3) - \ln(5)) &= -\ln(5) \\
x(\ln(5) - \ln(3)) &= \ln(5) \\
x &= \frac{\ln(5)}{\ln(5) - \ln(3)} \\
x &\approx 3.1507
\end{aligned}$$

14. By the definition of a logarithm, we obtain $x - 1 = 3^{5.46}$. Then $x = 1 + 3^{5.46} \approx 403.7931$.

15. Domain $(-\infty, \infty)$, range $(1, \infty)$, increasing, asymptote $y = 1$

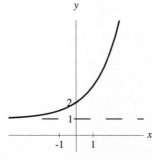

16. Domain $(1, \infty)$, range $(-\infty, \infty)$, decreasing, asymptote $x = 1$

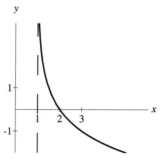

17. $(1, 0)$

18. Compounded quarterly, the investment is worth

$$2000 \left(1 + \frac{0.08}{4}\right)^{80} \approx \$9750.88.$$

Compounded continuously, the investment is worth

$$2000 \cdot e^{0.08(20)} \approx \$9906.06.$$

19. The amount of power at the end of $t = 200$ days is $P = 50 \cdot e^{-200/250} \approx 22.5$ watts.

To find the half-life, let $P = 25$.

$$\begin{aligned}
50 \cdot e^{-t/250} &= 25 \\
e^{-t/250} &= 0.5 \\
-\frac{t}{250} &= \ln(0.5) \\
t &\approx 173.3
\end{aligned}$$

The half-life is 173.3 days.

The operational life for a power of $P = 9$ watts is given by

$$\begin{aligned}
50 \cdot e^{-t/250} &= 9 \\
e^{-t/250} &= \frac{9}{50} \\
-\frac{t}{250} &= \ln\left(\frac{9}{50}\right) \\
t &\approx 428.7 \text{ days.}
\end{aligned}$$

20.

$$\begin{aligned}
4,000 \cdot \left(1 + \frac{0.06}{4}\right)^{4t} &= 10,000 \\
(1.015)^{4t} &= 2.5 \\
4t &= \log_{1.015}(2.5) \\
t &\approx 15.38576 \text{ years} \\
t &\approx 61.5 \text{ quarters}
\end{aligned}$$

21. Substituting $t = 100$, we obtain

$$\begin{aligned}
-50 \cdot \ln(1 - p) &= 100 \\
\ln(1 - p) &= -2 \\
1 - p &= e^{-2} \\
p &= 1 - e^{-2} \\
p &\approx 0.86.
\end{aligned}$$

The level reached after 100 hr is $p = 0.86$. When $p = 1$, then $t = -50 \ln(0)$ which is undefined, so it is impossible to master MGM.

Tying It All Together

1. Since $x - 3 = \pm 2$, we get $x = 3 \pm 2 = 1, 5$.

2. Since $\log((x - 3)^2) = \log(4)$, we obtain

$$\begin{aligned}
(x - 3)^2 &= 4 \\
x - 3 &= \pm 2 \\
x &= 3 \pm 2 \\
x &= 5, 1.
\end{aligned}$$

Checking $x = 1$, one gets $2\log(-2)$ which is undefined, so $x = 5$.

3. Using the definition of a logarithm, we obtain $x - 3 = 2^4$. The solution is $x = 16 + 3 = 19$.

4. Since $2^{x-3} = 2^2$, we have $x - 3 = 2$. So $x = 5$.

5. Square both sides of $\sqrt{x - 3} = 4$. Then $x - 3 = 16$. The solution is $x = 19$.

6. An equivalent equation is $x - 3 = \pm 4$. Thus, $x = 3 \pm 4 = -1, 7$.

7. Completing the square, we obtain

$$\begin{aligned}
x^2 - 4x + 4 &= -2 + 4 \\
(x - 2)^2 &= 2 \\
x - 2 &= \pm\sqrt{2} \\
x &= 2 \pm \sqrt{2}.
\end{aligned}$$

8. Since $2^{x-3} = 2^{2x}$, $x - 3 = 2x$. So $x = -3$.

9. Raise both sides to the third power.

$$\begin{aligned}
(\sqrt[3]{x-5})^3 &= 5^3 \\
x - 5 &= 125 \\
x &= 130
\end{aligned}$$

10. By using the definition of a logarithm, we have $x = \log_2(3)$.

11.

$$\begin{aligned}
\log(4x - 12) &= \log(x) \\
4x - 12 &= x \\
3x &= 12 \\
x &= 4
\end{aligned}$$

12. Use synthetic division with $c = -1$.

$$
\begin{array}{r|rrrr}
-1 & 1 & -4 & 1 & 6 \\
   &   & -1 & 5 & -6 \\
\hline
   & 1 & -5 & 6 & 0
\end{array}
$$

The quotient factors as $x^2 - 5x + 6 = (x - 3)(x - 2)$. The solutions are $x = -1, 2, 3$.

13. Parabola $y = x^2$ goes through $(0,0), (\pm 1, 1)$

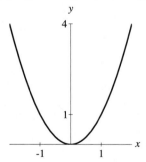

14. Parabola $y = (x - 2)^2$ goes through $(0, 4)$, $(1, 1)$, $(3, 1)$

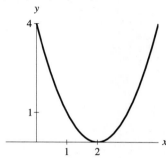

15. $y = 2^x$ goes through $\left(-1, \dfrac{1}{2}\right)$, $(0, 1)$, $(1, 2)$

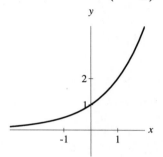

16. $y = x^{-2}$ goes through $(\pm 1, 1)$, $\left(\pm 2, \dfrac{1}{4}\right)$, and $\left(\pm \dfrac{1}{2}, 4\right)$

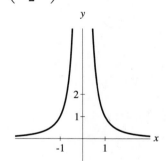

17. $y = \log_2(x - 2)$ goes through $(3, 0)$, $(4, 1)$, and $(2.5, -1)$

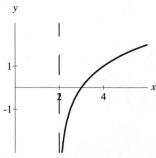

18. Line $y = x - 2$ goes through $(0, -2)$, $(2, 0)$

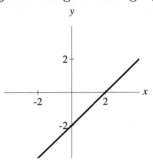

19. Line $y = 2x$ goes through $(0, 0)$, $(1, 2)$

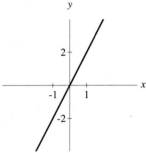

20. Line $y = x \cdot \log(2)$ goes through $(0, 0)$, $(1, \log(2))$

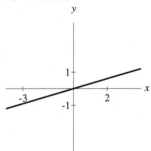

21. Horizontal line $y = e^2$ goes through $(0, e^2)$

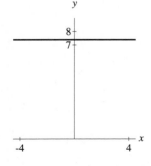

22. Parabola $y = 2 - x^2$ goes through $(0, 2)$, $(\pm 1, 1)$

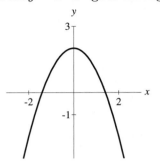

23. $y = \dfrac{2}{x}$ is symmetric about the origin and

goes through $(1, 2)$, $(2, 1)$, $\left(\dfrac{1}{2}, 4\right)$, $\left(4, \dfrac{1}{2}\right)$

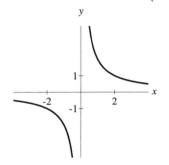

24. The graph of

$$y = \frac{1}{x - 2}$$

is obtained by shifting the graph

$$y = \frac{1}{x}$$

to the right by 2 units.

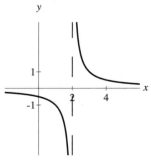

25. $f^{-1}(x) = 3x$

26. $f^{-1}(x) = -\log_3(x)$ since $f(x) = 3^{-x}$

27. $f^{-1}(x) = x^2 + 2$ for $x \geq 0$

28. $f^{-1}(x) = \sqrt[3]{x - 2} + 5$

29. $f^{-1}(x) = (10^x + 3)^2$ **30.** $\{(1,3),(4,5)\}$

31. $f^{-1}(x) = 5 + \dfrac{1}{x-3}$

32. $f^{-1}(x) = (\ln(3-x))^2$ for $x \leq 2$

33. $(p \circ m)(x) = e^{x+5}$, domain $(-\infty, \infty)$, range $(0, \infty)$

34. $(p \circ q)(x) = e^{\sqrt{x}}$, domain $[0, \infty)$, range $[1, \infty)$

35. $(q \circ p \circ m)(x) = q\left(e^{x+5}\right) = \sqrt{e^{x+5}}$, domain $(-\infty, \infty)$, range $(0, \infty)$

36. $(m \circ r \circ q)(x) = m\left(\ln(\sqrt{x})\right) = \ln(\sqrt{x}) + 5$, domain $(0, \infty)$, range $(-\infty, \infty)$

37. $(p \circ r \circ m)(x) = p\left(\ln(x+5)\right) = e^{\ln(x+5)}$, domain $(-5, \infty)$, range $(0, \infty)$

38. $(r \circ q \circ p)(x) = r\left(\sqrt{e^x}\right) = \ln\left(\sqrt{e^x}\right) = \dfrac{1}{2}x$, domain $(-\infty, \infty)$, range $(-\infty, \infty)$

39. $F(x) = (f \circ g \circ h)(x)$

40. $H(x) = (g \circ h \circ f)(x)$

41. $G(x) = (h \circ g \circ f)(x)$

42. $M(x) = (h \circ f \circ g)(x)$

For Thought

1. True, since if $x = 2$ and $y = 3$ we get $2 + 3 = 5$.

2. False, since $(2, 3)$ does not satisfy $x - y = 1$.

3. False, it is independent since the two lines are perpendicular and intersect at only one point.

4. True. Multiply $x - 2y = 4$ by -3 and add to the second equation.

$$\begin{aligned} -3x + 6y &= -12 \\ 3x - 6y &= 8 \\ \hline 0 &= -4 \end{aligned}$$

Since $0 = -4$ is false, there is no solution.

5. True, adding gives $2x = 6$.   **6.**   True

7. False, it is dependent because substituting $x = 5 + 3y$ results in an identity.

$$\begin{aligned} 9y - 3(5 + 3y) &= -15 \\ 9y - 15 - 9y &= -15 \\ -15 &= -15 \end{aligned}$$

8. False, it is dependent and the solution set is $\{(5 + 3y, y) \mid y$ is any real number$\}$.

9. False, it is dependent, there are an infinite number of solutions.

10. True, since both lines have slope $1/2$.

5.1 Exercises

1. Note, $x = 1$ and $y = 3$ satisfies both $x + y = 4$ and $x - y = -2$. Yes, $(1, 2)$ is a solution.

3. Note, $x = -1$ and $y = 5$ does not satisfy $x - 2y = -9$. Thus, $(-1, 5)$ is not a solution.

5. $\{(1, 2)\}$

7. No solution since the lines are parallel. The solution set is $\emptyset$.

9. $\{(3, 2)\}$

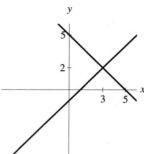

11. $\{(3, 1)\}$

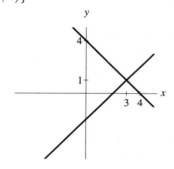

13. No solution since the lines are parallel. The solution set is $\emptyset$.

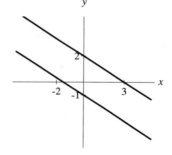

15. Since the lines are identical, the solution set is $\{(x, y) \mid x - 2y = 6\}$.

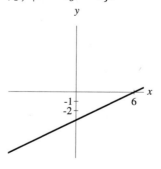

17. Substitute $y = 2x + 1$ into $3x - 4y = 1$.

$$
\begin{aligned}
3x - 4(2x + 1) &= 1 \\
3x - 8x - 4 &= 1 \\
-5x &= 5 \\
x &= -1
\end{aligned}
$$

From $y = 2x + 1$, $y = 2(-1) + 1 = -1$.
Independent and solution set is $\{(-1, -1)\}$.

19. Substitute $y = 1 - x$ into $2x - 3y = 8$.

$$
\begin{aligned}
2x - 3(1 - x) &= 8 \\
2x - 3 + 3x &= 8 \\
5x &= 11 \\
x &= 11/5
\end{aligned}
$$

From $y = 1 - x$, $y = 1 - 11/5 = -6/5$.
Independent and solution set is
$\{(11/5, -6/5)\}$.

21. Substitute $y = 3x + 5$ into $3(x + 1) = y - 2$.

$$
\begin{aligned}
3x + 3 &= (3x + 5) - 2 \\
3x + 3 &= 3x + 3 \\
3 &= 3
\end{aligned}
$$

Dependent and solution set is
$\{(x, y) \mid y = 3x + 5\}$.

23. Multiplying $\frac{1}{2}x + \frac{1}{3}y = 3$ by 6, we get $3x + 2y = 18$. Then substitute $2y = 6 - 3x$ into $3x + 2y = 18$. So

$$
\begin{aligned}
3x + (6 - 3x) &= 18 \\
6 &= 18
\end{aligned}
$$

Inconsistent and the solution set is $\emptyset$.

25. Multiplying $0.05x + 0.06y = 10.50$ by 100, we obtain $5x + 6y = 1050$. Then substitute $y = 200 - x$ into $5x + 6y = 1050$.

$$
\begin{aligned}
5x + 6(200 - x) &= 1050 \\
5x + 1200 - 6x &= 1050 \\
-x &= -150 \\
x &= 150
\end{aligned}
$$

From $y = 200 - x$, $y = 200 - 150 = 50$.
Independent and solution set is $\{(150, 50)\}$.

27. Since $3x + 1 = 3x - 7$ leads to $1 = -7$, the system is inconsistent and the solution set is $\emptyset$.

29. Multiplying the first and second equations by 6 and 4, respectively, we have $3x - 2y = 72$ and $x - 2y = 4$.
Substitute $2y = x - 4$ into $3x - 2y = 72$.

$$
\begin{aligned}
3x - (x - 4) &= 72 \\
2x + 4 &= 72 \\
2x &= 68 \\
x &= 34
\end{aligned}
$$

From $2y = x - 4$, $y = \dfrac{34 - 4}{2} = 15$.
Independent and solution set is $\{(34, 15)\}$.

31. Adding the two equations, we get $2x = 26$.
So $x = 13$ and from $x + y = 20$,
we obtain $13 + y = 20$ or $y = 7$.
Independent and solution set is $\{(13, 7)\}$.

33. Multiplying $x - y = 5$ by 2 and by adding to the second equation, we obtain

$$
\begin{aligned}
2x - 2y &= 10 \\
3x + 2y &= 10 \\
\hline
5x &= 20 \\
x &= 4
\end{aligned}
$$

From $x - y = 5$, $4 - y = 5$ or $y = -1$.
Independent and solution set is $\{(4, -1)\}$.

35. Adding the two equations leads to $0 = 12$.
Inconsistent and the solution set is $\emptyset$.

37. Multiply $2x + 3y = 1$ by -3 and $3x - 5y = -8$ by 2. Then add the equations.

$$
\begin{aligned}
-6x - 9y &= -3 \\
6x - 10y &= -16 \\
\hline
-19y &= -19 \\
y &= 1
\end{aligned}
$$

Since $2x + 3y = 1$, $2x + 3 = 1$ or $x = -1$.
Independent and solution set is $\{(-1, 1)\}$.

39. Multiply $0.05x + 0.1y = 0.6$ by -100 and $x + 2y = 12$ by 5. Then add the equations.

$$
\begin{aligned}
-5x - 10y &= -60 \\
5x + 10y &= 60 \\
\hline
0 &= 0
\end{aligned}
$$

Dependent and the solution set is
$\{(x, y) \mid x + 2y = 12\}$.

41. Multiplying $\dfrac{x}{2} + \dfrac{y}{2} = 5$ by 2 and

$\dfrac{3x}{2} - \dfrac{2y}{3} = 2$ by 6, we have $x + y = 10$
and $9x - 4y = 12$, respectively. Then multiply
$x + y = 10$ by 4 and add to the second equation.

$$\begin{array}{rcl}
4x + 4y & = & 40 \\
9x - 4y & = & 12 \\
\hline
13x & = & 52 \\
x & = & 4
\end{array}$$

Since $x + y = 10$, $4 + y = 10$ and $y = 6$.
Independent and the solution set is $\{(4, 6)\}$.

43. Multiply $3x - 2.5y = -4.2$ by -4 and
$0.12x + 0.09y = 0.4932$ by 100.
Then add the equations.

$$\begin{array}{rcl}
-12x + 10y & = & 16.8 \\
12x + 9y & = & 49.32 \\
\hline
19y & = & 66.12 \\
y & = & 3.48
\end{array}$$

From $3x - 2.5y = -4.2$, we find

$$\begin{array}{rcl}
3x - 2.5(3.48) & = & -4.2 \\
3x - 8.7 & = & -4.2 \\
3x & = & 4.5 \\
x & = & 1.5.
\end{array}$$

Independent and the solution set is
$\{(1.5, 3.48)\}$.

45. Independent

47. Dependent

49. The point of intersection is $(-1000, -497)$.

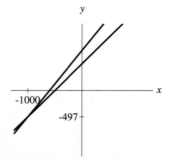

51. The point of intersection is approximately
$(6.18, -0.54)$

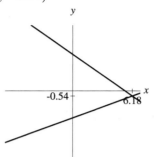

53. Let x and y be Althea's and Vaughn's incomes,
respectively. Then $x + y = 82,000$ and $x - y = 16,000$. By adding these two equations, we get
$2x = 98,000$ or $x = 49,000$. Thus, Althea's
income is $49,000 and Vaughn's income
is $33,000.

55. Let x and y be the amounts invested at 10%
and 8%, respectively.

$$\begin{array}{rcl}
x + y & = & 25,000 \\
0.1x + 0.08y & = & 2200
\end{array}$$

Multiply the second equation by -10 and
add to the first.

$$\begin{array}{rcl}
x + y & = & 25,000 \\
-x - 0.8y & = & -22,000 \\
\hline
0.2y & = & 3000 \\
y & = & 15,000
\end{array}$$

Carmen invested $15,000 at 8% and $10,000
at 10%.

57. Let x and y be the prices of an adult ticket
and child ticket, respectively.

$$\begin{array}{rcl}
2x + 5y & = & 33 \\
x + 3y & = & 18.50
\end{array}$$

Multiply the second equation by -2 and add
to the first equation.

$$\begin{array}{rcl}
2x + 5y & = & 33 \\
-2x - 6y & = & -37 \\
\hline
-y & = & -4 \\
y & = & 4
\end{array}$$

A child's ticket costs $4. Since $x + 3y = 18.50$,
$x + 12 = 18.50$ and so an adult's ticket
costs $x = \$6.50$.

59. If m and f are the number of male and female memberships, respectively, then a system of equations is

$$\begin{aligned} m + f &= 12 \\ 500m + 500f &= 6,000. \end{aligned}$$

Dividing the second equation by 500, one finds $m + f = 12$; the system of equations is dependent. Therefore, one cannot conclude the number of female memberships and male memberships.

All we know is that m and $12 - m$ are the number of male and female memberships where $0 \leq m \leq 12$.

61. Let x and y be the number of cows and ostriches, respectively.

$$\begin{aligned} 2x + 2y &= 84 \\ 4x + 2y &= 122 \end{aligned}$$

Subtract the first equation from the second equation.

$$\begin{aligned} 2x &= 38 \\ x &= 19 \end{aligned}$$

Consequently, we have

$$\begin{aligned} 2x + 2y &= 84 \\ 38 + 2y &= 84 \\ 2y &= 46 \\ y &= 23 \end{aligned}$$

Hence, there are $x = 19$ cows and $y = 23$ ostriches.

63. Let x and y be the number of cows and horses, respectively.

$$\begin{aligned} 4x + 4y &= 96 \\ x + y &= 24 \end{aligned}$$

Divide the first equation by 4.

$$\begin{aligned} x + y &= 24 \\ x + y &= 24 \end{aligned}$$

Since the two equations are identical, the system of equations is dependent and there are infinitely many possible solutions.

65. If c and m are the prices of a coffee and a muffin, respectively, then a system of equations is

$$\begin{aligned} 3c + 7m &= 7.77 \\ 6c + 14m &= 14.80. \end{aligned}$$

But if one multiplies the second equation by two, then one obtains $6c + 14m = 15.54$; this last equation contradicts the second equation in the system. Therefore, the system of equations is inconsistent and has no solution.

67. Let x and y be the number of male and female students, respectively.

$$\begin{aligned} 0.5x + 0.3y &= 230 \\ 0.2x + 0.6y &= 260 \end{aligned}$$

Multiply the first equation by -2 and the second by 5. Then add the resulting equations.

$$\begin{aligned} -x - 0.6y &= -460 \\ \underline{x + 3y} &= \underline{1300} \\ 2.4y &= 840 \\ y &= 350 \end{aligned}$$

From $0.2x + 0.6y = 260$, $0.2x + 210 = 260$ and so $x = 250$. There are $250 + 350 = 600$ students at CHS.

69. Let x and y be the number of nickels and pennies, respectively. We obtain

$$\begin{aligned} x + y &= 87 \\ 0.05x + 0.01y &= 1.75. \end{aligned}$$

Multiply the second equation by -100 and then add it to the first equation.

$$\begin{aligned} x + y &= 87 \\ \underline{-5x - y} &= \underline{-175} \\ -4x &= -88 \\ x &= 22 \end{aligned}$$

From $x + y = 87$, $22 + y = 87$ and $y = 65$. Isabelle has 22 nickels and 65 pennies.

71. The weights x and y must satisfy

$$\begin{aligned} 5x &= 3y \\ 3(4 + x + y) &= 6y. \end{aligned}$$

The second equation can be written as $12 + 3x = 3y$. Substituting into the first equation one finds

$$
\begin{aligned}
12 + 3x &= 5x \\
12 &= 2x \\
6 &= x.
\end{aligned}
$$

Then $x = 6$ oz and $y = 10$ oz since

$$
y = \frac{5x}{3} = \frac{5(6)}{3}.
$$

73. Let x be the number of months. Plan A costs $150x + 800$ and Plan B costs $200x + 200$. Thus, Plan A is cheaper in the long run. The number of months for which the costs are the same is given by

$$
\begin{aligned}
150x + 800 &= 200x + 200 \\
600 &= 50x \\
x &= 12 \ \text{ months}.
\end{aligned}
$$

75. If we set the formulas equal to each other, then we find

$$
\begin{aligned}
0.08aD &= \frac{D(a+1)}{24} \\
0.08a &= \frac{(a+1)}{24} \\
1.92a &= a + 1 \\
a &= \frac{1}{0.92} \\
a &\approx 1.09.
\end{aligned}
$$

The dosage is the same if the age is 1.1 yr.

77. Suppose the fronts of the trucks are at the same points. Let t be the number of hours before the trucks pass each other. They will be passing each other when the ends of the trucks are at the same position, i.e., when the total distance driven by the trucks is 100 feet.

$$
\begin{aligned}
40(5280)t + 50(5280)t &= 100 \\
40t + 50t &= \frac{100}{5280} \\
90t &= \frac{100}{5280}
\end{aligned}
$$

$$
\begin{aligned}
t &= \frac{100}{(90)5280} \ \text{hour} \\
t &= \frac{100}{(90)5280(3600)} \ \text{sec} \\
t &= \frac{25}{33} \ \text{sec} \\
t &\approx 0.76 \ \text{sec}
\end{aligned}
$$

Hence, the trucks pass each other in 0.76 sec, approximately.

79. Solve for a and b.

$$
\begin{aligned}
-3a + b &= 9 \\
2a + b &= -1.
\end{aligned}
$$

Multiply the second equation by -1 and add to the first equation.

$$
\begin{aligned}
-3a + b &= 9 \\
-2a - b &= 1 \\
\hline
-5a &= 10 \\
a &= -2
\end{aligned}
$$

Substituting $a = -2$ into $2a + b = -1$, one finds $b = 3$. An equation of the line is $y = -2x + 3$.

81. Solve for a and b.

$$
\begin{aligned}
-2a + b &= 3 \\
4a + b &= -7
\end{aligned}
$$

Multiply the first equation by -1 and add to the second equation.

$$
\begin{aligned}
2a - b &= -3 \\
4a + b &= -7 \\
\hline
6a &= -10 \\
a &= -\frac{5}{3}
\end{aligned}
$$

Substituting $a = -\frac{5}{3}$ into $-2a + b = 3$, one gets $b = -\frac{1}{3}$. An equation of the line is

$$
y = -\frac{5}{3}x - \frac{1}{3}.
$$

85.

$$x + y = -1$$
$$x - y = 5$$

For Thought

1. True

2. False, since $(1, 1, 0)$ does not satisfy
$-x - y + z = 4$.

3. True, adding the first two equations gives
$0 = 6$ which is false.

4. False, since $(2, 3, -1)$ does not satisfy
$x - y - z = 8$.

5. True

6. True, adding the first two equations gives
an identity. Also, multiplying the first
by -2 and then adding to the third
equation gives an identity.

$$
\begin{array}{rcr}
x - y + z &=& 1 \\
-x + y - z &=& -1 \\
\hline
0 &=& 0
\end{array}
$$

$$
\begin{array}{rcr}
-2x + 2y - 2z &=& -2 \\
2x - 2y + 2z &=& 2 \\
\hline
0 &=& 0
\end{array}
$$

7. True. The calculations in Exercise 6 above
show system (c) is dependent.

8. True

9. True, if $x = 1$ then $(x + 2, x, x - 1) = (3, 1, 0)$.

10. False, the value is $0.05x + 0.10y + 0.25z$ dollars.

5.2 Exercises

1. Points on the plane are $(5, 0, 0)$, $(0, 5, 0)$, and
$(0, 0, 5)$.

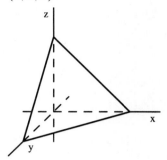

3. Points on the plane are $(3, 0, 0)$, $(0, 3, 0)$,
and $(0, 0, -3)$.

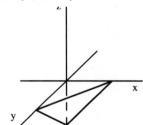

5. Note, $x = 1$, $y = 3$, and $z = 2$ satisfy all the
three equations in the system . Yes, $(1, 3, 2)$ is
a solution.

7. Note, $x = -1$, $y = 5$, and $z = 2$ does not satisfy
$x - y - 2z = -8$. No, $(-1, 5, 2)$ is not a solution.

9. Add the first and second equations and add
the first and third equations.

$$
\begin{array}{rcr}
x + y + z &=& 6 \\
2x - 2y - z &=& -5 \\
\hline
3x - y &=& 1
\end{array}
$$

$$
\begin{array}{rcr}
x + y + z &=& 6 \\
3x + y - z &=& 2 \\
\hline
4x + 2y &=& 8
\end{array}
$$

Divide $4x + 2y = 8$ by 2 and add to $3x - y = 1$.

$$
\begin{array}{rcr}
2x + y &=& 4 \\
3x - y &=& 1 \\
\hline
5x &=& 5
\end{array}
$$

So $x = 1$. From $3x - y = 1$, $3 - y = 1$ and
$y = 2$. From $x + y + z = 6$, $1 + 2 + z = 6$ and
$z = 3$. The solution set is $\{(1, 2, 3)\}$.

11. Multiply the first equation by 2 and add to the second. Then multiply the first equation by -3 and add to the third equation.

$$\begin{array}{rcr} 6x + 4y + 2z &=& 2 \\ x + y - 2z &=& -4 \\ \hline 7x + 5y &=& -2 \end{array}$$

$$\begin{array}{rcr} -9x - 6y - 3z &=& -3 \\ 2x - 3y + 3z &=& 1 \\ \hline -7x - 9y &=& -2 \end{array}$$

Add $7x + 5y = -2$ to $-7x - 9y = -2$.

$$\begin{array}{rcr} 7x + 5y &=& -2 \\ -7x - 9y &=& -2 \\ \hline -4y &=& -4 \end{array}$$

So $y = 1$. From $7x + 5y = -2$, $7x + 5 = -2$ and $x = -1$. From $3x + 2y + z = 1$, $-3 + 2 + z = 1$ and $z = 2$.
The solution set is $\{(-1, 1, 2)\}$.

13. Add first equation to third equation. Multiply second equation by 2 and add to the first.

$$\begin{array}{rcr} 2x + y - 2z &=& -15 \\ x + 3y + 2z &=& -5 \\ \hline 3x + 4y &=& -20 \end{array}$$

$$\begin{array}{rcr} 8x - 4y + 2z &=& 30 \\ 2x + y - 2z &=& -15 \\ \hline 10x - 3y &=& 15 \end{array}$$

Multiply $3x + 4y = -20$ by 3 and $10x - 3y = 15$ by 4 and then add the equations.

$$\begin{array}{rcr} 9x + 12y &=& -60 \\ 40x - 12y &=& 60 \\ \hline 49x &=& 0 \end{array}$$

So $x = 0$. From $3x + 4y = -20$, $4y = -20$ and $y = -5$. From $2x + y - 2z = -15$, we get $-5 - 2z = -15$ and $z = 5$.
The solution set is $\{(0, -5, 5)\}$.

15. If we substitute $x = 1, 2, 3$ in $(x, x + 3, x - 5)$, we obtain

$$(1, 4, -4), \ (2, 5, -3), \ \text{and} \ (3, 6, -2),$$

respectively.

17. If we substitute $y = 1, 2, 3$ in $(2y, y, y - 7)$, we obtain

$$(2, 1, -6), \ (4, 2, -5), \ \text{and} \ (6, 3, -4),$$

respectively.

19. Let $y = x + 3$. Then $x = y - 3$, $x - 5 = y - 8$, and

$$(x, x + 3, x - 5) = (y - 3, y, y - 8).$$

21. Let $z = x - 1$. Then $x = z + 1$, $x + 1 = z + 2$, and

$$(x, x + 1, x - 1) = (z + 1, z + 2, z).$$

23. Let $y = 2x + 1$. Then $x = (y - 1)/2$, $3x - 1 = (3y - 5)/2$, and

$$(x, 2x + 1, 3x - 1) = \left(\frac{y - 1}{2}, y, \frac{3y - 5}{2}\right).$$

25. Adding the two equations, we have $4x - 4z = -20$ or $z = x + 5$.
From $x + 2y - 3z = -17$, we obtain

$$\begin{array}{rcr} x + 2y - 3(x + 5) &=& -17 \\ 2y - 2x - 15 &=& -17 \\ 2y &=& 2x - 2 \\ y &=& x - 1. \end{array}$$

The solution set is
$\{(x, x - 1, x + 5) | x \text{ is any real number}\}$.

27. Adding the first two equations, we have $0 = 0$, which is an identity. Multiply the second equation by 2 and add to the third equation.

$$\begin{array}{rcr} -2x - 4y + 6z &=& -10 \\ 2x + 4y - 6z &=& 10 \\ \hline 0 &=& 0 \end{array}$$

The system is dependent and the solution set is $\{(x, y, z) | x + 2y - 3z = 5\}$.

29. Multiply first equation by -2 and add to the second equation.

$$\begin{array}{rcr} -2x + 4y - 6z &=& -10 \\ 2x - 4y + 6z &=& 3 \\ \hline 0 &=& -7 \end{array}$$

This is false, so the solution set is $\emptyset$.

31. Adding the two equations, we get $3x = 6$ or $x = 2$. Substitute into the two equations. Then

$$\begin{aligned} 2 + y - z &= 2 \\ y - z &= 0 \end{aligned}$$

and

$$\begin{aligned} 4 - y + z &= 4 \\ -y + z &= 0. \end{aligned}$$

In any case $y = z$. The solution set is $\{(2, y, y) | y \text{ is any real number}\}$.

33. If the second equation is multiplied by -1 and added to the first equation, the result is the third equation.

$$\begin{aligned} x + y &= 5 \\ -y + z &= -2 \\ \hline x + z &= 3 \end{aligned}$$

There ae infinitely number many solutions. Since $y = 5 - x$ and $z = 3 - x$, the solution set $\{(x, 5 - x, 3 - x) | x \text{ is any real number}\}$.

35. Multiply the first equation by 2 and add to the second equation.

$$\begin{aligned} 2x - 2y + 2z &= 14 \\ 2y - 3z &= -13 \\ \hline 2x - z &= 1 \end{aligned}$$

Multiply $2x - z = 1$ by -2 and add to the third equation.

$$\begin{aligned} -4x + 2z &= -2 \\ 3x - 2z &= -3 \\ \hline -x &= -5 \end{aligned}$$

So $x = 5$. From $2x - z = 1$, we get $10 - z = 1$ and $z = 9$. Since $2y - 3z = -13$, $2y - 27 = -13$ and $y = 7$. The solution set is $\{(5, 7, 9)\}$.

37. Multiply first equation by -2 and add to third equation. Then multiply first equation by -4 and add to the second.

$$\begin{aligned} -2x - 2y - 4z &= -15 \\ 5x + 2y + 5z &= 21 \\ \hline 3x + z &= 6 \end{aligned}$$

$$\begin{aligned} -4x - 4y - 8z &= -30 \\ 3x + 4y + z &= 12 \\ \hline -x - 7z &= -18 \end{aligned}$$

Multiply $-x - 7z = -18$ by 3 and add to $3x + z = 6$.

$$\begin{aligned} -3x - 21z &= -54 \\ 3x + z &= 6 \\ \hline -20z &= -48 \end{aligned}$$

So $z = 48/20 = 2.4$. From $3x + z = 6$, we get $3x + 2.4 = 6$ and $x = 1.2$. From $x + y + 2z = 7.5$, $1.2 + y + 4.8 = 7.5$ and $y = 1.5$. The solution set is $\{(1.2, 1.5, 2.4)\}$.

39. Multiply the first equation by -5 and add to 100 times the second one.

$$\begin{aligned} -5x - 5y - 5z &= -45,000 \\ 5x + 6y + 9z &= 71,000 \\ \hline y + 4z &= 26,000 \end{aligned}$$

Substitute $z = 3y$ into $y + 4z = 26,000$.

$$\begin{aligned} y + 12y &= 26,000 \\ 13y &= 26,000 \\ y &= 2,000 \end{aligned}$$

From $z = 3y$, we obtain $z = 6000$. From $x + y + z = 9,000$, we have $x + 2000 + 6000 = 9000$ and $x = 1000$. The solution set is $\{(1000, 2000, 6000)\}$.

41. Substitute $x = 2y - 1$ into $z = 2x - 3$ to get $z = 4y - 5$. Then substitute $z = 4y - 5$ into $y = 3z + 2$.

$$\begin{aligned} y &= 3(4y - 5) + 2 \\ y &= 12y - 13 \\ -11y &= -13 \\ y &= 13/11 \end{aligned}$$

From $z = 4y - 5$, we obtain $z = 52/11 - 5 = -3/11$. Since $x = 2y - 1$, $x = 26/11 - 1 = 15/11$. The solution set is $\{(15/11, 13/11, -3/11)\}$.

43. Substitute $(-1, -2)$, $(2, 1)$, $(-2, 1)$ into $y = ax^2 + bx + c$. So

$$\begin{aligned} a - b + c &= -2 \\ 4a + 2b + c &= 1 \\ 4a - 2b + c &= 1. \end{aligned}$$

Multiply first equation by -1 and add to the second and third equations.

$$
\begin{array}{rcl}
-a + b - c &=& 2 \\
4a + 2b + c &=& 1 \\
\hline
3a + 3b &=& 3
\end{array}
$$

$$
\begin{array}{rcl}
-a + b - c &=& 2 \\
4a - 2b + c &=& 1 \\
\hline
3a - b &=& 3
\end{array}
$$

Multiply $3a + 3b = 3$ by -1 and add to $3a - b = 3$.

$$
\begin{array}{rcl}
-3a - 3b &=& -3 \\
3a - b &=& 3 \\
\hline
-4b &=& 0
\end{array}
$$

So $b = 0$. From $3a - b = 3$, we get $3a = 3$ and $a = 1$. From $a - b + c = -2$, $1 + c = -2$ and $c = -3$. Since the solution is $(a, b, c) = (1, 0, -3)$, the parabola is $y = x^2 - 3$.

45. Substitute $(0,0)$, $(1,3)$, $(2,2)$ into $y = ax^2 + bx + c$. Then

$$
\begin{array}{rcl}
c &=& 0 \\
a + b + c &=& 3 \\
4a + 2b + c &=& 2.
\end{array}
$$

Multiply second one by -2 and add to third equation.

$$
\begin{array}{rcl}
-2a - 2b - 2c &=& -6 \\
4a + 2b + c &=& 2 \\
\hline
2a - c &=& -4
\end{array}
$$

Substituting $c = 0$ into $2a - c = -4$, we get $2a = -4$ and $a = -2$. From $a + b + c = 3$, $-2 + b = 3$ and $b = 5$. Since $(a, b, c) = (-2, 5, 0)$, the parabola is $y = -2x^2 + 5x$.

47. Substitute $(0,4)$, $(-2,0)$, $(-3,1)$ into $y = ax^2 + bx + c$. Then

$$
\begin{array}{rcl}
c &=& 4 \\
4a - 2b + c &=& 0 \\
9a - 3b + c &=& 1.
\end{array}
$$

Multiply second and third equations by 3 and -2, respectively, then add the equations.

$$
\begin{array}{rcl}
12a - 6b + 3c &=& 0 \\
-18a + 6b - 2c &=& -2 \\
\hline
-6a + c &=& -2
\end{array}
$$

Substituting $c = 4$ into $-6a + c = -2$, $-6a + 4 = -2$ and $a = 1$. From $4a - 2b + c = 0$, $4 - 2b + 4 = 0$ and $b = 4$. Since $(a, b, c) = (1, 4, 4)$, the parabola is $y = x^2 + 4x + 4$.

49. By substituting the coordinates of the points $(1,0,0)$, $(0,1,0)$, and $(0,0,1)$ into $ax + by + cz = 1$, we get $a = 1$, $b = 1$, and $c = 1$. A linear equation satisfied by the ordered triples is $x + y + z = 1$.

51. By substituting the coordinates of the points $(1,1,1)$, $(0,2,0)$, and $(1,0,0)$ into $ax + by + cz = 1$, we get $a + b + c = 1$, $2b = 1$, and $a = 1$. Then $b- = \dfrac{1}{2}$ and $c = -\dfrac{1}{2}$. A linear equation satisfied by the ordered triples is

$$
x + \frac{1}{2}y - \frac{1}{2}z = 1 \text{ or } 2x + y - z = 2
$$

53. Let x, y, z be the three numbers listed in increasing order.

$$
\begin{array}{rcl}
x + y + z &=& 40 \\
-x \phantom{{}+ y} + z &=& 12 \\
x + y - z &=& 0.
\end{array}
$$

If we add the first two equations and add the last two equations, we obtain (by subtracting the 2nd sum from the 1st sum)

$$
\begin{array}{rcl}
y + 2z &=& 52 \\
y &=& 12 \\
\hline
2z &=& 40
\end{array}
$$

Then $z = 20$. Since $-x + z = 12$, we find $-x + 20 = 12$ or $x = 8$. Thus, the numbers are $8, 12,$ and 20.

55. Let x, y, z be the scores in the 1st, 2nd, and 3rd quizzes, respectively.

$$
\begin{array}{rcl}
x + y + z &=& 21 \\
x - y &=& -1 \\
y - z &=& -4.
\end{array}
$$

If we add the first two equations and add the last two equations, we obtain

$$
\begin{array}{rcl}
2x + z &=& 20 \\
x - z &=& -5 \\
\hline
3x &=& 15
\end{array}
$$

Since $x = 5$, the scores on the quizzes are 5, 6, and 10.

57. Let x, y, and z be the amounts invested in stocks, bonds, and a mutual fund. Then

$$\begin{aligned} x + y + z &= 25,000 \\ 0.08x + 0.10y + 0.06z &= 1,860 \\ 2y &= z. \end{aligned}$$

Multiply the first equation by -8 and add to 100 times the second.

$$\begin{array}{r} -8x - 8y - 8z = -200,000 \\ 8x + 10y + 6z = 186,000 \\ \hline 2y - 2z = -14,000 \end{array}$$

Substitute $z = 2y$ into $2y - 2z = -14,000$.

$$\begin{aligned} 2y - 4y &= -14,000 \\ -2y &= -14,000 \\ y &= 7,000 \end{aligned}$$

Since $z = 2y$, we obtain $z = 14,000$.
Since $x + y + z = 25,000$, $x = 4000$.
Marita invested $4,000 in stocks,
$7000 in bonds, and
$14,000 in a mutual fund.

59. Let x, y, and z be the prices last year of a hamburger, fries, and a Coke, respectively. So

$$\begin{aligned} x + y + z &= 1.90 \\ 1.1x + 1.2y + 1.5z &= 2.37 \\ 1.5z &= 1.1x + 0.09. \end{aligned}$$

Multiply first equation by -12 and add to 100 times the second equation.

$$\begin{array}{r} -12x - 12y - 12z = -22.8 \\ 11x + 12y + 15z = 23.7 \\ \hline -x + 3z = 0.9 \end{array}$$

Substitute $x = 3z - 0.9$ into $1.5z = 1.1x + 0.09$.

$$\begin{aligned} 1.5z &= 1.1(3z - 0.9) + 0.09 \\ 1.5z &= 3.3z - 0.99 + 0.09 \\ 0.90 &= 1.8z \\ 0.50 &= z \end{aligned}$$

Since $x = 3z - 0.9$, we find $x = 1.50 - 0.90 = 0.60$. From $x + y + z = 1.90$, we have $y = 0.80$. The prices last year of a hamburger, fries and a Coke are $0.60, $0.80, and $0.50, respectively.

61. Let L_f and L_r be the weights on the left front tire and left rear tire, respectively. Let R_f and R_r be the weights on the right front tire and right rear tire, respectively. Since $1200(0.51) = 612$ and $1200(0.48) = 576$, we obtain

$$\begin{aligned} L_f + L_r &= 612 \\ R_r + L_r &= 576 \\ L_f, L_r, R_f, R_r &\geq 280. \end{aligned}$$

Three possible weight distributions are
$(L_r, L_f, R_r, R_f) = (280, 332, 296, 292)$,
$(L_r, L_f, R_r, R_f) = (285, 327, 291, 297)$,
and $(L_r, L_f, R_r, R_f) = (290, 322, 286, 302)$.

63. Let x, y, and z be the number of pennies, nickels, and dimes, respectively. Then

$$\begin{aligned} x + y + z &= 232 \\ y + z &= x \\ 0.01x + 0.05y + 0.10z &= 10.36. \end{aligned}$$

Multiply first equation by -1 and add to 10 times the third equation. Also combine first two equations.

$$\begin{array}{r} -x - y - z = -232 \\ 0.1x + 0.5y + z = 103.6 \\ \hline -0.9x - 0.5y = -128.4 \end{array}$$

$$\begin{array}{r} -x - y - z = -232 \\ -x + y + z = 0 \\ \hline -2x = -232 \end{array}$$

Then $x = 116$. Substituting into $-0.9x - 0.5y = -128.4$, we obtain

$$\begin{aligned} -0.9(116) - 0.5y &= -128.4 \\ -104.4 - 0.5y &= -128.4 \\ 24 &= 0.5y \\ 48 &= y. \end{aligned}$$

Since $x + y + z = 232$, $116 + 48 + z = 232$ and $z = 68$. Emma used 116 pennies, 48 nickels, and 68 dimes.

65. Let x, y, and z be the prices of a carton of milk, a cup of coffee, and a doughnut, respectively. So

$$3x + 4y + 7z = 5.45$$
$$4x + 2y + 8z = 5.30$$
$$2x + 5y + 6z = 5.15.$$

Multiply third equation by -2 and add to second equation. Also, multiply first equation by -4 and add to 3 times the second equation.

$$\begin{aligned} -4x - 10y - 12z &= -10.30 \\ 4x + 2y + 8z &= 5.30 \\ \hline -8y - 4z &= -5 \end{aligned}$$

$$\begin{aligned} -12x - 16y - 28z &= -21.80 \\ 12x + 6y + 24z &= 15.90 \\ \hline -10y - 4z &= -5.90 \end{aligned}$$

Multiply $-8y - 4z = -5$ by -1 and add to $-10y - 4z = -5.90$.

$$\begin{aligned} 8y + 4z &= 5 \\ -10y - 4z &= -5.90 \\ \hline -2y &= -0.90 \end{aligned}$$

Then $y = 0.45$. Substitute into $-8y - 4z = -5$ to get $-3.60 - 4z = -5$ and $z = 0.35$. Since $3x + 4y + 7z = 5.45$, we have $3x + 1.80 + 2.45 = 5.45$ and $x = 0.40$.

Alphonse's bill was $5(0.40) + 2(0.45) + 9(0.35) = \$6.05$. His change is \$3.95.

67. A system of equations is

$$\begin{aligned} 4x &= 6y \\ 2(x + y) &= 15(8) \\ 6(x + y + 15 + 10) &= 10z. \end{aligned}$$

Rewriting the first two equations, one obtains

$$\begin{aligned} 2x - 3y &= 0 \\ x + y &= 60. \end{aligned}$$

Solving this smaller system, one finds $x = 36$ lb, $y = 24$ lb. Substituting into the third equation, one finds

$$z = \frac{6(x + y + 25)}{10} = \frac{6(60 + 25)}{10} = 51 \text{ lb}.$$

69.

a) Substitute $(0,0)$, $(10,40)$, $(20,70)$ into $y = ax^2 + bx + c$. Consequently, we obtain the
following system of equations:

$$\begin{aligned} c &= 0 \\ 100a + 10b + c &= 40 \\ 400a + 20b + c &= 70. \end{aligned}$$

Multiply the second equation by -2 and add to third equation.

$$\begin{aligned} -200a - 20b - 2c &= -80 \\ 400a + 20b + c &= 70 \\ \hline 200a - c &= -10 \end{aligned}$$

Substitute $c = 0$ into $200a - c = -10$. Then $a = -\dfrac{1}{20}$. From $100a + 10b + c = 40$, we obtain $-5 + 10b = 40$ and $b = \dfrac{9}{2}$. The parabola is $y = -\dfrac{1}{20}x^2 + \dfrac{9}{2}x$.

b) Since $-b/(2a) = 45$, the maximum height is

$$-\frac{1}{20}(45)^2 + \frac{9}{2}(45) = 101.25 \text{ m}.$$

c) Since the zeros of $y = -\dfrac{x}{20}(x - 90)$ are $x = 0, 90$, the missile will strike 90 m from the origin.

For Thought

1. True, since the line $y = x$ passes through the center of the circle.

2. False, when a line is tangent to a circle it intersects the circle at only one point.

3. True, the parabola $y = 2x^2 - 4$ intersects the circle $x^2 + y^2 = 16$ at three points.

4. False, they intersect at $(\pm 1, 0)$.

5. False, they intersect at $(1, 1)$ and $(-1, -1)$.

6. False, since three noncollinear points determine a unique circle through the points.

7. True, since either leg can serve as base and the other leg as altitude.

8. True   **9.**   True

10. False, two such numbers are $\frac{1}{2}\left(7 \pm \sqrt{45}\right)$.

5.3 Exercises

1. Since $x = -1$ and $y = 4$ satisfy both equations in the system , $(-1, 4)$ is a solution.

3. Since $x = 4$ and $y = -5$ does not satisfy $x - y = 1$, $(4, -5)$ is not a solution.

5. Since $y = x$ and $y = x^2$, we obtain

$$
\begin{aligned}
x &= x^2 \\
x - x^2 &= 0 \\
x(x - 1) &= 0.
\end{aligned}
$$

Then $x = 0, 1$ and the solution set is

$$\{(0, 0),\ (1, 1)\}.$$

The points of intersection are seen in the graphs on the next column.

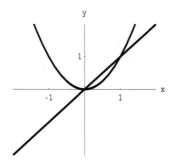

7. Substitute $y = x^2$ into $5x - y = 6$.

$$
\begin{aligned}
5x - x^2 &= 6 \\
x^2 - 5x &= -6 \\
x^2 - 5x + 6 &= 0 \\
(x - 3)(x - 2) &= 0
\end{aligned}
$$

If $x = 3, 2$ in $y = x^2$, then $y = 9, 4$. The solution set is $\{(2, 4),\ (3, 9)\}$.

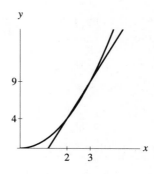

9. Substitute $y = |x|$ into $y = x + 3$.

$$
\begin{aligned}
|x| &= x + 3 \\
x = x + 3 \quad &\text{or} \quad -x = x + 3 \\
0 = 3 \quad &\text{or} \quad -2x = 3
\end{aligned}
$$

Then $x = -3/2$. Since $y = |x|$, $y = 3/2$. The solution set is $\{(-3/2, 3/2)\}$.

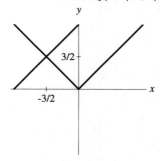

11. Substitute $y = x^2$ into $y = |x|$.

$$
\begin{aligned}
x^2 &= |x| \\
x^2 = x \quad &\text{or} \quad x^2 = -x \\
x^2 - x = 0 \quad &\text{or} \quad x^2 + x = 0 \\
x(x - 1) = 0 \quad &\text{or} \quad x(x + 1) = 0 \\
x = 0, -1 \quad &\text{or} \quad x = 0, -1
\end{aligned}
$$

Using $x = 0, \pm 1$ in $y = x^2$, one finds $y = 0, 1$. The solution set is $\{(-1, 1), (0, 0), (1, 1)\}$.

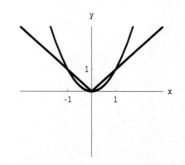

13. Substitute $y = \sqrt{x}$ into $y = 2x$ to obtain

$$
\begin{aligned}
\sqrt{x} &= 2x \\
x &= 4x^2 \\
x - 4x^2 &= 0 \\
x(1 - 4x) &= 0.
\end{aligned}
$$

Using $x = 0, 1/4$ in $y = 2x$, $y = 0, 1/2$.
The solution set is

$$\{(0,0), (1/4, 1/2)\}.$$

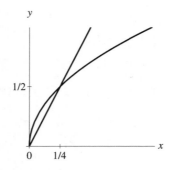

15. Substitute $y = x^3$ into $y = 4x$ to get

$$
\begin{aligned}
x^3 &= 4x \\
x^3 - 4x &= 0 \\
x(x^2 - 4) &= 0 \\
x &= 0, \pm 2.
\end{aligned}
$$

Substituting $x = 0, 2, -2$ into $y = 4x$,
we get $y = 0, 8, -8$.
The solution set is

$$\{(0,0), (2,8), (-2,-8).$$

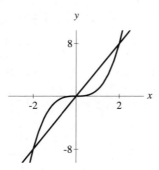

17. Substitute $y = x$ into $y = x^3 - x$.

$$
\begin{aligned}
x &= x^3 - x \\
2x - x^3 &= 0 \\
x(2 - x^2) &= 0 \\
x &= 0, \sqrt{2}, -\sqrt{2}
\end{aligned}
$$

Since $y = x$, the solution set is

$$\{(0,0), (\sqrt{2}, \sqrt{2}), (-\sqrt{2}, -\sqrt{2})\}.$$

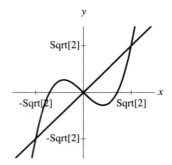

19. Substitute $y = x$ into $x^2 + y^2 = 1$.

$$
\begin{aligned}
x^2 + x^2 &= 1 \\
2x^2 &= 1 \\
x^2 &= \frac{1}{2} \\
x &= \pm\frac{\sqrt{2}}{2}
\end{aligned}
$$

Since $y = x$, the solution set is

$$\left\{\left(\frac{\sqrt{2}}{2}, \frac{\sqrt{2}}{2}\right), \left(-\frac{\sqrt{2}}{2}, -\frac{\sqrt{2}}{2}\right)\right\}.$$

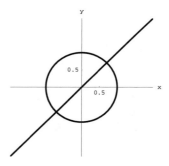

21. Substitute $y = -x - 4$ into $xy = 1$.

$$
\begin{aligned}
x(-x - 4) &= 1 \\
-x^2 - 4x &= 1 \\
x^2 + 4x &= -1 \\
x^2 + 4x + 4 &= 3 \\
(x + 2)^2 &= 3 \\
x &= -2 \pm \sqrt{3}
\end{aligned}
$$

Using $x = -2 + \sqrt{3}, -2 - \sqrt{3}$ in $y = -x - 4$,
we find $y = -2 - \sqrt{3}, -2 + \sqrt{3}$.
The solution set is

$$\left\{(-2 + \sqrt{3}, -2 - \sqrt{3}), (-2 - \sqrt{3}, -2 + \sqrt{3})\right\}.$$

23. Substitute $x^2 = 2y^2 - 1$ into $2x^2 - y^2 = 1$.

$$
\begin{aligned}
2(2y^2 - 1) - y^2 &= 1 \\
3y^2 - 2 &= 1 \\
3y^2 &= 3 \\
y^2 &= 1 \\
y &= \pm 1
\end{aligned}
$$

Using $y = 1$ in $x^2 = 2y^2 - 1$, we get $x^2 = 1$ or $x = \pm 1$. Also, if $y = -1$ then $x = \pm 1$. The solution set is $\{(1, \pm 1), (-1, \pm 1)\}$.

25. Since $2x - y = 1$, we obtain $y = 2x - 1$. Then substitute into $xy - 2x = 2$.

$$
\begin{aligned}
x(2x - 1) - 2x &= 2 \\
2x^2 - 3x - 2 &= 0 \\
(2x + 1)(x - 2) &= 0 \\
x &= -\frac{1}{2}, 2
\end{aligned}
$$

Since $y = 2x - 1$, we get $y = 2 \cdot \left(-\frac{1}{2}\right) - 1 = -2$

and $x = 2(2) - 1 = 3$. The solution set is $\left\{\left(-\frac{1}{2}, -2\right), (2, 3)\right\}$.

27. Multiply the first equation by 2 and add to the second.

$$
\begin{aligned}
\frac{6}{x} - \frac{2}{y} &= \frac{26}{10} \\
\frac{1}{x} + \frac{2}{y} &= \frac{9}{10} \\
\hline
\frac{7}{x} &= \frac{35}{10}
\end{aligned}
$$

So $70 = 35x$ and $x = 2$.

Substituting into $\dfrac{3}{x} - \dfrac{1}{y} = \dfrac{13}{10}$, we obtain

$$
\begin{aligned}
\frac{3}{2} - \frac{1}{y} &= \frac{13}{10} \\
-\frac{1}{y} &= \frac{13}{10} - \frac{15}{10} \\
-\frac{1}{y} &= -\frac{2}{10} \\
y &= 5.
\end{aligned}
$$

The solution set is $\{(2, 5)\}$.

29. Substitute $y = 1 - x$ into $x^2 + xy - y^2 = -5$.

$$
\begin{aligned}
x^2 + x(1 - x) - (1 - x)^2 &= -5 \\
x^2 + x - x^2 - (1 - 2x + x^2) &= -5 \\
-x^2 + 3x - 1 &= -5 \\
x^2 - 3x + 1 &= 5 \\
x^2 - 3x - 4 &= 0 \\
(x - 4)(x + 1) &= 0 \\
x &= 4, -1
\end{aligned}
$$

Using $x = 4, -1$ in $y = 1 - x$, we get $y = -3, 2$. The solution set is $\{(4, -3), (-1, 2)\}$.

31. Add the two given equations to obtain $xy = -2$. Substitute $y = -2/x$ into $x^2 + 2xy - 2y^2 = -11$.

$$
\begin{aligned}
x^2 + 2x\left(-\frac{2}{x}\right) - 2 \cdot \frac{4}{x^2} &= -11 \\
x^2 - 4 - \frac{8}{x^2} &= -11 \\
x^4 + 7x^2 - 8 &= 0 \\
(x^2 + 8)(x^2 - 1) &= 0 \\
x &= \pm 1
\end{aligned}
$$

Using $x = 1, -1$ in $y = -2/x$, $y = -2, 2$. The solution set is $\{(1, -2), (-1, 2)\}$.

33. Substitute $y = 2^{x+1}$ into $y = 4^{-x}$.

$$
\begin{aligned}
2^{x+1} &= \left(2^2\right)^{-x} \\
2^{x+1} &= 2^{-2x} \\
x + 1 &= -2x \\
3x &= -1
\end{aligned}
$$

Using $x = -1/3$ in $y = 2^{x+1}$, we get

$y = 2^{2/3}$. The solution set is $\left\{\left(-\frac{1}{3}, 2^{2/3}\right)\right\}$.

35. Substitute $y = \log_2(x)$ into $y = \log_4(x + 2)$ and use the base-changing formula.

$$
\begin{aligned}
\log_2(x) &= \log_4(x + 2) \\
\log_2(x) &= \frac{\log_2(x + 2)}{\log_2(4)} \\
\log_2(x) &= \frac{\log_2(x + 2)}{2} \\
2 \cdot \log_2(x) &= \log_2(x + 2)
\end{aligned}
$$

$$2 \cdot \log_2(x) - \log_2(x+2) = 0$$
$$\log_2\left(\frac{x^2}{x+2}\right) = 0$$
$$\frac{x^2}{x+2} = 1$$
$$x^2 = x+2$$
$$x^2 - x - 2 = 0$$
$$(x-2)(x+1) = 0$$
$$x = 2, -1$$

But $x = -1$ is an extraneous root since $\log_2(-1)$ is undefined. Using $x = 2$ in $y = \log_2(x)$, we have $y = 1$. The solution set is $\{(2, 1)\}$.

37. Substitute $y = \log_2(x+2)$ into $y = 3 - \log_2(x)$.

$$\log_2(x+2) = 3 - \log_2(x)$$
$$\log_2(x+2) + \log_2(x) = 3$$
$$\log_2(x^2 + 2x) = 3$$
$$x^2 + 2x = 2^3$$
$$x^2 + 2x - 8 = 0$$
$$(x+4)(x-2) = 0$$
$$x = -4, 2$$

But $x = -4$ is an extraneous root since $\log_2(-4)$ is undefined. Using $x = 2$ in $y = \log_2(x+2)$, we get $y = \log_2(4) = 2$. The solution set is $\{(2, 2)\}$.

39. Substitute $y = 3^x$ into $y = 2^x$.

$$3^x = 2^x$$
$$\log(3^x) = \log(2^x)$$
$$x \cdot \log(3) = x \cdot \log(2)$$
$$x\left[\log(3) - \log(2)\right] = 0$$
$$x = 0$$

Using $x = 0$ in $y = 3^x$, we obtain $y = 3^0 = 1$. The solution set is $\{(0, 1)\}$.

41. From the graphs, the solution set is $\{(2, 1), (0.3, -1.8)\}$.

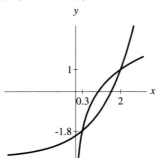

43. From the graphs, the solution set is $\{(1.9, 0.6), (0.1, -2.0)\}$.

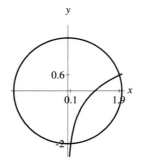

45. From the graphs, the solution set is $\{(-0.8, 0.6), (2, 4), (4, 16)\}$.

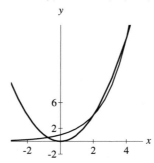

47. Let x and $6 - x$ be the two numbers.

$$x(6-x) = -16$$
$$x^2 - 6x = 16$$
$$x^2 - 6x + 9 = 25$$
$$(x-3)^2 = 25$$
$$x - 3 = \pm 5$$
$$x = 3 \pm 5$$
$$x = -2, 8$$

The numbers are -2 and 8.

49. Let x and y be the lengths of the legs of the triangle.

$$x^2 + y^2 = 15^2$$
$$\frac{1}{2}xy = 54$$

Substitute $y = \dfrac{108}{x}$ into $x^2 + y^2 = 225$ and use the quadratic formula.

$$x^2 + \frac{11664}{x^2} = 225$$
$$x^4 - 225x^2 + 11,664 = 0$$

$$x^2 = \frac{225 \pm \sqrt{225^2 - 4(11,664)}}{2}$$

$$x^2 = \frac{225 \pm 63}{2}$$
$$x^2 = 144, 81$$
$$x = 12, 9$$

Using $x = 12, 9$ in $y = \dfrac{108}{x}$, we get $y = 9, 12$. The sides of the right triangle are 9 m and 12 m.

51. If x is the length of the hypotenuse, then $\dfrac{x}{2}$ and $\dfrac{x\sqrt{3}}{2}$ are the lengths of the sides opposite the 30° and 60° angles.

$$x + \frac{x}{2} + \frac{x\sqrt{3}}{2} = 12$$
$$2x + x + \sqrt{3}x = 24$$
$$(3 + \sqrt{3})x = 24$$
$$x = \frac{24}{3 + \sqrt{3}} \cdot \frac{3 - \sqrt{3}}{3 - \sqrt{3}}$$
$$x = 12 - 4\sqrt{3}$$

Substituting $x = 12 - 4\sqrt{3}$ in $\dfrac{x}{2}$ and $\dfrac{x\sqrt{3}}{2}$, we find $6 - 2\sqrt{3}$ ft and $6\sqrt{3} - 6$ ft are the lengths of the two sides and the hypotenuse is $12 - 4\sqrt{3}$ ft.

53. The values of x and y must satisfy

$$6y = 6x$$
$$x(6 + y) = 7(4 + 12).$$

Since $x = y$ as seen from the first equation, upon substitution into the second equation one obtains

$$6x + x^2 = 112$$
$$x^2 + 6x - 112 = 0$$
$$(x + 14)(x - 8) = 0.$$

Then $x = 8$ in. and $y = 8$ oz. One must exclude the negative value $x = -14$.

55. Let x and y be the number of minutes it takes for pump A and pump B, respectively, to fill the vat.

$$\frac{1}{x} + \frac{1}{y} = \frac{1}{8}$$
$$\frac{1}{x} - \frac{1}{y} = \frac{1}{12}$$

Adding the two equations, we get

$$\frac{2}{x} = \frac{3}{24} + \frac{2}{24}$$
$$\frac{2}{x} = \frac{5}{24}$$
$$5x = 48$$
$$x = 9.6.$$

Substituting $x = \dfrac{48}{5}$ into $\dfrac{1}{x} + \dfrac{1}{y} = \dfrac{1}{8}$, we find

$$\frac{5}{48} + \frac{1}{y} = \frac{1}{8}$$
$$\frac{1}{y} = \frac{6}{48} - \frac{5}{48}$$
$$\frac{1}{y} = \frac{1}{48}$$
$$y = 48.$$

Pump A can fill the vat by itself in $x = 9.6$ min while Pump B will take $y = 48$ min.

57. Let x and y be two numbers satisfying

$$x + y = 6$$
$$xy = 10.$$

Substituting $y = 6 - x$ into $xy = 10$,

$$x(6 - x) = 10$$
$$6x - x^2 = 10$$
$$x^2 - 6x = -10$$
$$x^2 - 6x + 9 = -1$$
$$(x - 3)^2 = -1$$
$$x = 3 \pm i.$$

If $x = 3 + i$, then $y = 6 - (3 + i) = 3 - i$.
If $x = 3 - i$, then $y = 6 - (3 - i) = 3 + i$.
The two numbers are $3 + i$ and $3 - i$.

59. Let x and y be the length and width.

$$20xy = 36,000$$
$$40x + 40y + 2xy = 7200$$

Substitute $x = \dfrac{1800}{y}$ into $40x + 40y + 2xy = 7200$.

$$40 \cdot \frac{1800}{y} + 40y + 2 \cdot \frac{1800}{y} \cdot y = 7200$$
$$\frac{72,000}{y} + 40y + 3600 = 7200$$
$$\frac{72,000}{y} + 40y = 3600$$
$$40y^2 - 3600y + 72,000 = 0$$
$$y^2 - 90y + 1800 = 0$$
$$(y - 60)(y - 30) = 0$$

Using $y = 30, 60$ in $x = \dfrac{1800}{w}$, we get $x = 60, 30$. The length is 60 ft and width is 30 ft.

61. From the graphs, the two models give the same population for $t = 0$ and $t = 9.65$ years. The exponential population model is twice the linear model when $t \approx 29.5$ years.

63. Let x be the time of the sunrise or the number of hours since midnight. Let S and B be Sally's and Bob's speeds. The number of miles Sally and Bob drove are $(16 - x)S$ and $(21 - x)B$, respectively. Since they met at noon, the distance between Sally's house and Bob's house is $(12 - x)S + (12 - x)B$; or equivalently $(12 - x)(S + B)$. Then

$$(16 - x)S = (12 - x)(S + B)$$
$$(16 - x)S = (16 - x)(S + B) - 4(S + B)$$
$$0 = (16 - x)B - 4(S + B)$$
$$(16 - x)B = 4(S + B).$$

Likewise,

$$(21 - x)B = (21 - x)(S + B) - 9(S + B)$$
$$(21 - x)S = 9(S + B).$$

Combining, one obtains

$$\frac{(16 - x)B}{(21 - x)S} = \frac{4(S + B)}{9(S + B)}. \text{ So, } \frac{(16 - x)9B}{(21 - x)4S} = 1.$$

Furthermore, since $(16 - x)S = (21 - x)B$ one finds $\dfrac{B}{S} = \dfrac{16 - x}{21 - x}$. Then

$$\frac{9(16 - x)}{4(21 - x)} \cdot \frac{B}{S} = 1$$
$$\frac{9}{4}\left(\frac{16 - x}{21 - x}\right)^2 = 1$$
$$\frac{16 - x}{21 - x} = \frac{2}{3} \text{ since } 16 - x > 0$$
$$48 - 3x = 42 - 2x$$
$$x = 6.$$

The sunrise was at 6:00 A.M.

65. By the time the mouse reaches the northeast corner (50 feet from the southwest corner) of the train, the train would have traveled 20 feet which is one-half the distance the train travels before the mouse returns to the southeast corner. So, when the mouse reaches the northwest corner (40 feet from the southwest corner) the train (in proportion to 20 feet) would have traveled 16 feet.

As the distance traveled by the train ranges from 16 feet to 20 feet, the path the mouse takes will be the hypotenuse of a right triangle whose sides are 10 feet and 4 feet. So, in this range, the mouse travels a distance of

$\sqrt{10^2 + 4^2}$ feet on the ground, or equivalently $2\sqrt{29}$ feet on the ground.

Thus, the total diagonal ground distance (including the diagonal ground distance as the mouse moves from the southeast corner to the southwest corner) traveled by the mouse is twice of $2\sqrt{29}$ feet, or $4\sqrt{29}$ feet. Since the north-south distance traveled by the mouse is 80 feet on the ground, the total distance on the ground traveled by the mouse is $80 + 4\sqrt{29}$ feet, or about 101.54 ft.

For Thought

1. True, $\dfrac{1}{x} + \dfrac{3}{x+1} = \dfrac{(x+1) + 3x}{x(x+1)} =$

   $\dfrac{4x+1}{x(x+1)}$.

2. True, $x + \dfrac{3x}{x^2 - 1} = \dfrac{x(x^2 - 1) + 3x}{x^2 - 1} =$

   $\dfrac{x^3 + 2x}{x^2 - 1}$.

3. False, by using long division we obtain

   $\dfrac{x^2}{x^2 - 9} = 1 + \dfrac{9}{x^2 - 9} = 1 + \dfrac{A}{x - 3} + \dfrac{B}{x + 3}$

4. True, since $\dfrac{1}{2} + \dfrac{1}{2^3} = \dfrac{2^2 + 1}{2^3} = \dfrac{5}{8}$.

5. False, since $\dfrac{3x - 1}{x^3 + x} = \dfrac{A}{x} + \dfrac{Bx + C}{x^2 + 1}$.

6. False, since $\dfrac{1}{x^2 - 1} = \dfrac{1/2}{x - 1} - \dfrac{1/2}{x + 1}$.

7. True, by using long division we get

   $$\begin{array}{r} x - 1 \\ x^2 + x - 2 \overline{)x^3 + 0x^2 + 0x + 1} \\ \underline{x^3 + x^2 - 2x} \\ -x^2 + 2x + 1 \\ \underline{-x^2 - x + 2} \\ 3x - 1 \end{array}$$

   So $\dfrac{x^3 + 1}{x^2 + x - 2} = x - 1 + \dfrac{3x - 1}{x^2 + x - 2}$.

8. False, since $x^3 - 8 = (x - 2)(x^2 + 2x + 4)$.

9. True, since $\dfrac{1}{x - 1} + \dfrac{1}{x^2 + x + 1} =$

   $\dfrac{(x^2 + x + 1) + (x - 1)}{x^3 - 1} = \dfrac{x^2 + 2x}{x^3 - 1}$.

10. True, it is already in the form $\dfrac{Ax + B}{x^2 + 9}$.

5.4 Exercises

1. $\dfrac{3(x + 1) + 4(x - 2)}{(x - 2)(x + 1)} = \dfrac{7x - 5}{(x - 2)(x + 1)}$

3. $\dfrac{(x^2 + 2) - 3(x - 1)}{(x - 1)(x^2 + 2)} = \dfrac{x^2 - 3x + 5}{(x - 1)(x^2 + 2)}$

5.

   $\dfrac{(2x + 1)(x^2 + 3) + (x^3 + 2x + 2)}{(x^2 + 3)^2} =$

   $\dfrac{(2x^3 + x^2 + 6x + 3) + (x^3 + 2x + 2)}{(x^2 + 3)^2} =$

   $\dfrac{3x^3 + x^2 + 8x + 5}{(x^2 + 3)^2}$

7.

   $\dfrac{(x - 1)^2 + (2x + 3)(x - 1) + (x^2 + 1)}{(x - 1)^3} =$

   $\dfrac{(x^2 - 2x + 1) + (2x^2 + x - 3) + (x^2 + 1)}{(x - 1)^3} =$

   $\dfrac{4x^2 - x - 1}{(x - 1)^3}$

9. Multiply the equation by $(x - 3)(x + 3)$.

   $$\begin{aligned} 12 &= A(x + 3) + B(x - 3) \\ 12 &= (A + B)x + (3A - 3B) \\ A + B = 0 \quad &\text{and} \quad 3A - 3B = 12 \end{aligned}$$

   Divide $3A - 3B = 12$ by 3 and add to $A + B = 0$.

   $$\begin{aligned} A - B &= 4 \\ A + B &= 0 \\ \hline 2A &= 4 \end{aligned}$$

   Using $A = 2$ in $A + B = 0$, $B = -2$. Then $A = 2$ and $B = -2$.

11. $\dfrac{5x - 1}{(x + 1)(x - 2)} = \dfrac{A}{x + 1} + \dfrac{B}{x - 2}$

   $$\begin{aligned} 5x - 1 &= A(x - 2) + B(x + 1) \\ 5x - 1 &= (A + B)x + (-2A + B) \\ A + B = 5 \quad &\text{and} \quad -2A + B = -1 \end{aligned}$$

Multiply $-2A + B = -1$ by -1 and add to $A + B = 5$.

$$\begin{aligned} 2A - B &= 1 \\ A + B &= 5 \\ \hline 3A &= 6 \end{aligned}$$

Using $A = 2$ in $A + B = 5$, $B = 3$.

The answer is $\dfrac{2}{x + 1} + \dfrac{3}{x - 2}$.

13. $\dfrac{2x + 5}{(x + 4)(x + 2)} = \dfrac{A}{x + 4} + \dfrac{B}{x + 2}$

$$\begin{aligned} 2x + 5 &= A(x + 2) + B(x + 4) \\ 2x + 5 &= (A + B)x + (2A + 4B) \\ A + B = 2 \quad &\text{and} \quad 2A + 4B = 5 \end{aligned}$$

Multiply $A + B = 2$ by -2 and add to $2A + 4B = 5$.

$$\begin{aligned} -2A - 2B &= -4 \\ 2A + 4B &= 5 \\ \hline 2B &= 1 \end{aligned}$$

Using $B = 1/2$ in $A + B = 2$, we find $A = 3/2$.

The answer is $\dfrac{3/2}{x + 4} + \dfrac{1/2}{x + 2}$.

15. $\dfrac{2}{(x - 3)(x + 3)} = \dfrac{A}{x - 3} + \dfrac{B}{x + 3}$

$$\begin{aligned} 2 &= A(x + 3) + B(x - 3) \\ 2 &= (A + B)x + (3A - 3B) \\ A + B = 0 \quad &\text{and} \quad 3A - 3B = 2 \end{aligned}$$

Multiply $A + B = 0$ by 3 and add to $3A - 3B = 2$.

$$\begin{aligned} 3A + 3B &= 0 \\ 3A - 3B &= 2 \\ \hline 6A &= 2 \end{aligned}$$

Using $A = 1/3$ in $A + B = 0$, we get $B = -1/3$.

The answer is $\dfrac{1/3}{x - 3} + \dfrac{-1/3}{x + 3}$.

17.

$$\begin{aligned} \frac{1}{x(x - 1)} &= \frac{A}{x} + \frac{B}{x - 1} \\ 1 &= A(x - 1) + Bx \\ 1 &= (A + B)x - A \\ A + B = 0 \quad &\text{and} \quad -A = 1 \end{aligned}$$

Using $A = -1$ in $A + B = 0$, we find $B = 1$.

The answer is $\dfrac{-1}{x} + \dfrac{1}{x - 1}$.

19. Multiplying the equation by $(x + 3)^2(x - 2)$, we obtain $x^2 + x - 31 =$

$$\begin{aligned} &= A(x + 3)(x - 2) + B(x - 2) + C(x + 3)^2 \\ &= A(x^2 + x - 6) + B(x - 2) + C(x^2 + 6x + 9) \\ &= (A+C)x^2 + (A+B+6C)x + (-6A-2B+9C). \end{aligned}$$

Equate the coefficients and solve the system.

$$\begin{aligned} A + C &= 1 \\ A + B + 6C &= 1 \\ -6A - 2B + 9C &= -31 \end{aligned}$$

Multiply $A + B + 6C = 1$ by 2 and add to $-6A - 2B + 9C = -31$.

$$\begin{aligned} 2A + 2B + 12C &= 2 \\ -6A - 2B + 9C &= -31 \\ \hline -4A + 21C &= -29 \end{aligned}$$

Multiply $A + C = 1$ by 4 and add to $-4A + 21C = -29$.

$$\begin{aligned} 4A + 4C &= 4 \\ -4A + 21C &= -29 \\ \hline 25C &= -25 \end{aligned}$$

Using $C = -1$ in $A + C = 1$, we obtain $A = 2$. From $A + B + 6C = 1$, $2 + B - 6 = 1$ and $B = 5$. So $A = 2, B = 5$, and $C = -1$.

21. $\dfrac{4x - 1}{(x - 1)^2(x + 2)} = \dfrac{A}{x - 1} + \dfrac{B}{(x - 1)^2} + \dfrac{C}{x + 2}$

$$\begin{aligned} 4x - 1 &= A(x - 1)(x + 2) + B(x + 2) + C(x - 1)^2 \\ 4x - 1 &= A(x^2 + x - 2) + B(x + 2) + C(x^2 - 2x + 1) \\ 4x - 1 &= (A + C)x^2 + (A + B - 2C)x + \\ &\quad (-2A + 2B + C) \end{aligned}$$

If we equate the coefficients of x, we obtain

$$\begin{aligned} A + C &= 0 \\ A + B - 2C &= 4 \\ -2A + 2B + C &= -1. \end{aligned}$$

Solving the system, we get $A = 1$, $B = 1$, and $C = -1$. The answer is

$$\frac{1}{x - 1} + \frac{1}{(x - 1)^2} + \frac{-1}{x + 2}.$$

23. $\dfrac{20 - 4x}{(x - 2)^2(x + 4)} = \dfrac{A}{x - 2} + \dfrac{B}{(x - 2)^2} + \dfrac{C}{x + 4}$

$$20 - 4x = A(x - 2)(x + 4) + B(x + 4) +$$
$$C(x - 2)^2$$
$$20 - 4x = A(x^2 + 2x - 8) + B(x + 4) +$$
$$C(x^2 - 4x + 4)$$
$$20 - 4x = (A + C)x^2 + (2A + B - 4C)x +$$
$$(-8A + 4B + 4C)$$

If we equate the coefficients of x, we obtain

$$A + C = 0$$
$$2A + B - 4C = -4$$
$$-8A + 4B + 4C = 20$$

Solving the system, we get $A = -1$, $B = 2$, and $C = 1$. The answer is

$$\dfrac{-1}{x - 2} + \dfrac{2}{(x - 2)^2} + \dfrac{1}{x + 4}.$$

25. Note, $\dfrac{3x^2 + 3x - 2}{(x + 1)^2(x - 1)} = \dfrac{A}{x + 1} + \dfrac{B}{(x + 1)^2} + \dfrac{C}{x - 1}$

$$3x^2 + 3x - 2 = A(x + 1)(x - 1) + B(x - 1) +$$
$$C(x + 1)^2$$
$$3x^2 + 3x - 2 = A(x^2 - 1) + B(x - 1) +$$
$$C(x^2 + 2x + 1)$$
$$3x^2 + 3x - 2 = (A + C)x^2 + (B + 2C)x +$$
$$(-A - B + C)$$

If we equate the coefficients of x, we obtain

$$A + C = 3$$
$$B + 2C = 3$$
$$-A - B + C = -2.$$

Solving the system, we get $A = 2$, $B = 1$, and $C = 1$. The answer is $\dfrac{2}{x + 1} + \dfrac{1}{(x + 1)^2} + \dfrac{1}{x - 1}$.

27. Multiplying the equation by $(x + 1)(x^2 + 4)$, we get

$$x^2 - x - 7 = A(x^2 + 4) + (Bx + C)(x + 1)$$
$$x^2 - x - 7 = (A + B)x^2 + (B + C)x + (4A + C).$$

Equating the coefficients, we have

$$A + B = 1$$
$$B + C = -1$$
$$4A + C = -7.$$

Multiply $A + B = 1$ by -1 and add to $B + C = -1$.

$$\begin{array}{rcl} -A - B &=& -1 \\ B + C &=& -1 \\ \hline -A + C &=& -2 \end{array}$$

Multiply $4A + C = -7$ by -1 and add to $-A + C = -2$.

$$\begin{array}{rcl} -A + C &=& -2 \\ -4A - C &=& 7 \\ \hline -5A &=& 5 \end{array}$$

Using $A = -1$ in $A + B = 1$, $B = 2$.
Using $B = 2$ in $B + C = -1$, $C = -3$.
So $A = -1$, $B = 2$, and $C = -3$.

29. Note, $\dfrac{5x^2 + 5x}{(x + 2)(x^2 + 1)} = \dfrac{A}{x + 2} + \dfrac{Bx + C}{x^2 + 1}$.

$$5x^2 + 5x = A(x^2 + 1) + (Bx + C)(x + 2)$$
$$5x^2 + 5x = (A + B)x^2 + (2B + C)x + (A + 2C)$$

If we equate the coefficients of x, we obtain

$$A + B = 5$$
$$2B + C = 5$$
$$A + 2C = 0$$

Solving the system, we obtain $A = 2$, $B = 3$, and $C = -1$. The answer is $\dfrac{2}{x + 2} + \dfrac{3x - 1}{x^2 + 1}$.

31. Note,

$$\dfrac{x^2 - 2}{(x + 1)(x^2 + x + 1)} = \dfrac{A}{x + 1} + \dfrac{Bx + C}{x^2 + x + 1}.$$

$$x^2 - 2 = A(x^2 + x + 1) + (Bx + C)(x + 1)$$
$$x^2 - 2 = (A + B)x^2 + (A + B + C)x + (A + C)$$

If we equate the coefficients of x, we obtain

$$
\begin{aligned}
A + B &= 1 \\
A + B + C &= 0 \\
A + C &= -2.
\end{aligned}
$$

Solving the system, we obtain $A = -1$, $B = 2$, and $C = -1$. The answer is

$$\frac{-1}{x+1} + \frac{2x-1}{x^2+x+1}.$$

33. $\dfrac{-2x-7}{(x+2)^2} = \dfrac{A}{x+2} + \dfrac{B}{(x+2)^2}$

$$
\begin{aligned}
-2x - 7 &= A(x+2) + B \\
-2x - 7 &= Ax + (2A+B) \\
A = -2 \quad \text{and} \quad 2A+B &= -7
\end{aligned}
$$

Using $A = -2$ in $2A+B = -7$, we get $B = -3$. The answer is

$$\frac{-3}{(x+2)^2} + \frac{-2}{x+2}.$$

35. Note that $x^3 + x^2 + x + 1 =$
$x^2(x+1) + (x+1) = (x^2+1)(x+1)$.
Then we obtain

$$\frac{6x^2 - x + 1}{(x^2+1)(x+1)} = \frac{A}{x+1} + \frac{Bx+C}{x^2+1}$$

$$
\begin{aligned}
6x^2 - x + 1 &= A(x^2+1) + (Bx+C)(x+1) \\
6x^2 - x + 1 &= (A+B)x^2 + (B+C)x + (A+C).
\end{aligned}
$$

Equating the coefficients, we get

$$
\begin{aligned}
A + B &= 6 \\
B + C &= -1 \\
A + C &= 1.
\end{aligned}
$$

Multiply $A + B = 6$ by -1 and add to $B + C = -1$.

$$
\begin{array}{rcl}
-A - B &=& -6 \\
B + C &=& -1 \\
\hline
-A + C &=& -7
\end{array}
$$

Adding $-A+C = -7$ and $A+C = 1$, $2C = -6$. Using $C = -3$ in $B + C = -1$ and $A + C = 1$, we obtain $B = 2$ and $A = 4$.
The answer is $\dfrac{4}{x+1} + \dfrac{2x-3}{x^2+1}$.

37. Note,

$$\frac{3x^3 - x^2 + 19x - 9}{(x^2+9)^2} = \frac{Ax+B}{x^2+9} + \frac{Cx+D}{(x^2+9)^2}.$$

So $3x^3 - x^2 + 19x - 9 =$
$(Ax + B)(x^2+9) + (Cx + D) =$
$Ax^3 + Bx^2 + (9A+C)x + (9B+D)$.

Then $A = 3$ and $B = -1$. Since $9A + C = 19$ and $9B + D = -9$, we get $C = -8$ and $D = 0$.
The answer is $\dfrac{-8x}{(x^2+9)^2} + \dfrac{3x-1}{x^2+9}$.

39. Observe that

$$\frac{3x^2 + 17x + 14}{(x-2)(x^2+2x+4)} = \frac{A}{x-2} + \frac{Bx+C}{x^2+2x+4}.$$

Then $3x^2 + 17x + 14 =$
$A(x^2+2x+4) + (Bx+C)(x-2) =$
$(A+B)x^2 + (2A-2B+C)x + (4A-2C)$

Equating the coefficients, we obtain

$$
\begin{aligned}
A + B &= 3 \\
2A - 2B + C &= 17 \\
4A - 2C &= 14.
\end{aligned}
$$

Multiply $A + B = 3$ by 2 and add to $2A - 2B + C = 17$.

$$
\begin{array}{rcl}
2A + 2B &=& 6 \\
2A - 2B + C &=& 17 \\
\hline
4A + C &=& 23
\end{array}
$$

Multiplying $4A - 2C = 14$ by -1 and adding to $4A + C = 23$, $3C = 9$. So $C = 3$ and from $4A - 2C = 14$, $A = 5$. Using these values in $2A - 2B + C = 17$, we get

$B = -2$. The answer is $\dfrac{5}{x-2} + \dfrac{-2x+3}{x^2+2x+4}$.

41. Divide $2x^3 + x^2 + 3x - 2$ by $x^2 - 1$ by long division.

$$
\begin{array}{r}
2x + 1 \\
x^2 - 1 \enclose{longdiv}{2x^3 + x^2 + 3x - 2} \\
\underline{2x^3 + 0x^2 - 2x} \\
x^2 + 5x - 2 \\
\underline{x^2 + 0x - 1} \\
5x - 1
\end{array}
$$

Then $\dfrac{2x^3 + x^2 + 3x - 2}{x^2 - 1} = 2x + 1 + \dfrac{5x-1}{x^2-1}$.

Decompose $\dfrac{5x-1}{x^2-1} = \dfrac{A}{x-1} + \dfrac{B}{x+1}$.

$$5x - 1 = A(x+1) + B(x-1)$$
$$5x - 1 = (A+B)x + (A-B)$$

So $A + B = 5$ and $A - B = -1$.

Adding $A + B = 5$ and $A - B = -1$, $2A = 4$. Using $A = 2$ in $A + B = 5$, we find $B = 3$.

The answer is $2x + 1 + \dfrac{2}{x-1} + \dfrac{3}{x+1}$.

43.

Since $\dfrac{3x^3 - 2x^2 + x - 2}{(x^2 + x + 1)^2} =$

$\dfrac{Ax + B}{x^2 + x + 1} + \dfrac{Cx + D}{(x^2 + x + 1)^2}$, we get

$3x^3 - 2x^2 + x - 2 = (Ax + B)(x^2 + x + 1) + (Cx + D)$
$\qquad = Ax^3 + (A+B)x^2 + (A+B+C)x + (B+D).$

Equating the coefficients, we find $A = 3$. Since $A + B = -2$, we get $B = -5$. From $A + B + C = 1$ and $B + D = -2$, we have $C = 3$ and $D = 3$.

The answer is $\dfrac{3x - 5}{x^2 + x + 1} + \dfrac{3x + 3}{(x^2 + x + 1)^2}$.

45.

Since $\dfrac{3x^3 + 4x^2 - 12x + 16}{(x - 2)(x + 2)(x^2 + 4)} =$

$\dfrac{A}{x - 2} + \dfrac{B}{x + 2} + \dfrac{Cx + D}{x^2 + 4}$, we obtain

$3x^3 + 4x^2 - 12x + 16 =$
$\quad = A(x + 2)(x^2 + 4) + B(x - 2)(x^2 + 4) +$
$\qquad (Cx + D)(x^2 - 4)$
$\quad = (A + B + C)x^3 + (2A - 2B + D)x^2 +$
$\qquad (4A + 4B - 4C)x + (8A - 8B - 4D).$

Equating the coefficients, we get

$$
\begin{aligned}
A + B + C &= 3 \\
2A - 2B + D &= 4 \\
4A + 4B - 4C &= -12 \\
8A - 8B - 4D &= 16.
\end{aligned}
$$

Multiply first equation by -4 and add to the third. Multiply second equation by -4 and add to the fourth. Also multiply first equation

by 2 and add to the second.

$$
\begin{aligned}
-4A - 4B - 4C &= -12 \\
4A + 4B - 4C &= -12 \\
\hline
-8C &= -24
\end{aligned}
$$

$$
\begin{aligned}
-8A + 8B - 4D &= -16 \\
8A - 8B - 4D &= 16 \\
\hline
-8D &= 0
\end{aligned}
$$

$$
\begin{aligned}
2A + 2B + 2C &= 6 \\
2A - 2B + D &= 4 \\
\hline
4A + 2C + D &= 10
\end{aligned}
$$

So $C = 3$ and $D = 0$. From $4A + 2C + D = 10$, we find $A = 1$ and from $2A - 2B + D = 4$, we get $B = -1$.

The answer is $\dfrac{1}{x - 2} + \dfrac{-1}{x + 2} + \dfrac{3x}{x^2 + 4}$.

47.

$\dfrac{5x^3 + x^2 + x - 3}{x^3(x - 1)} = \dfrac{A}{x} + \dfrac{B}{x^2} + \dfrac{C}{x^3} + \dfrac{D}{x - 1}$,

$5x^3 + x^2 + x - 3 =$
$\quad = Ax^2(x - 1) + Bx(x - 1) + C(x - 1) + Dx^3$
$\quad = (A + D)x^3 + (-A + B)x^2 + (-B + C)x - C$

Equating the coefficients, we get $C = 3$. From $-B + C = 1$, we find $B = 2$. From $-A + B = 1$, we obtain $A = 1$. From $A + D = 5$, we have $D = 4$.

The answer is $\dfrac{1}{x} + \dfrac{2}{x^2} + \dfrac{3}{x^3} + \dfrac{4}{x - 1}$.

49. Note,

$\dfrac{6x^2 - 28x + 33}{(x - 2)^2(x - 3)} = \dfrac{A}{x - 2} + \dfrac{B}{(x - 2)^2} + \dfrac{C}{x - 3}$.

Then $6x^2 - 28x + 33 =$
$\quad = A(x - 2)(x - 3) + B(x - 3) + C(x - 2)^2$
$\quad = (A + C)x^2 + (-5A + B - 4C)x +$
$\qquad (6A - 3B + 4C).$

Equating the coefficients, we obtain

$$
\begin{aligned}
A + C &= 6 \\
-5A + B - 4C &= -28 \\
6A - 3B + 4C &= 33.
\end{aligned}
$$

Multiply second equation by 3 and add to the third.

$$
\begin{aligned}
-15A + 3B - 12C &= -84 \\
6A - 3B + 4C &= 33 \\
\hline
-9A - 8C &= -51
\end{aligned}
$$

Multiply $A + C = 6$ by 8 and add to
$-9A - 8C = -51$

$$\begin{array}{rcl} -9A - 8C & = & -51 \\ 8A + 8C & = & 48 \\ \hline -A & = & -3 \end{array}$$

Using $A = 3$ in $A + C = 6$, we find $C = 3$.
From $6A - 3B + 4C = 33$, we obtain $B = -1$.

The answer is $\dfrac{3}{x - 2} + \dfrac{-1}{(x - 2)^2} + \dfrac{3}{x - 3}$.

51. Use synthetic division to factor the denominator.

$$\begin{array}{r|rrrr} -5 & 1 & 4 & -11 & -30 \\ & & -5 & 5 & 30 \\ \hline & 1 & -1 & -6 & 0 \end{array}$$

$$\begin{aligned} x^3 + 4x^2 - 11x - 30 & = (x + 5)(x^2 - x - 6) \\ & = (x + 5)(x + 2)(x - 3) \end{aligned}$$

Decomposing, $\dfrac{9x^2 + 21x - 24}{(x + 5)(x + 2)(x - 3)} =$

$$= \dfrac{A}{x + 5} + \dfrac{B}{x + 2} + \dfrac{C}{x - 3}.$$

Then $9x^2 + 21x - 24 =$
$= A(x + 2)(x - 3) + B(x + 5)(x - 3) +$
$\quad C(x + 5)(x + 2)$

and substituting $x = -2, 3, -5$, we obtain

$$\begin{aligned} -30 & = -15B \\ 2 & = B \end{aligned}$$

$$\begin{aligned} 120 & = 40C \\ 3 & = C \end{aligned}$$

$$\begin{aligned} 96 & = -24A \\ 4 & = A. \end{aligned}$$

The answer is $\dfrac{4}{x + 5} + \dfrac{2}{x + 2} + \dfrac{3}{x - 3}$.

53. Note that $x^3 - 3x^2 + 3x - 1 = (x - 1)^3$.

Then $\dfrac{x^2 - 2}{(x - 1)^3} = \dfrac{A}{x - 1} + \dfrac{B}{(x - 1)^2} + \dfrac{C}{(x - 1)^3}$.

$$\begin{aligned} x^2 - 2 & = A(x - 1)^2 + B(x - 1) + C \\ & = Ax^2 + (-2A + B)x + (A - B + C) \end{aligned}$$

Then $A = 1$. Since $-2A + B = 0$, $B = 2$.
Since $A - B + C = -2$, we find $1 - 2 + C = -2$
and $C = -1$.

The answer is $\dfrac{1}{x - 1} + \dfrac{2}{(x - 1)^2} + \dfrac{-1}{(x - 1)^3}$.

55.

$$\begin{aligned} \dfrac{x}{(ax + b)^2} & = \dfrac{A}{ax + b} + \dfrac{B}{(ax + b)^2} \\ x & = A(ax + b) + B \\ & = aAx + (bA + B) \end{aligned}$$

So $aA = 1$ and $A = 1/a$. Since $bA + B = 0$,

we obtain $\dfrac{b}{a} + B = 0$ and $B = -b/a$.

The answer is $\dfrac{-b/a}{(ax + b)^2} + \dfrac{1/a}{ax + b}$.

57. Since $\dfrac{x + c}{x(ax + b)} = \dfrac{A}{x} + \dfrac{B}{ax + b}$, we have

$$\begin{aligned} x + c & = A(ax + b) + Bx \\ & = (aA + B)x + bA. \end{aligned}$$

So $bA = c$ and $A = \dfrac{c}{b}$. Since $aA + B = 1$,

we have $\dfrac{ac}{b} + B = 1$ and $B = 1 - \dfrac{ac}{b}$.

Answer is $\dfrac{c/b}{x} + \dfrac{1 - ac/b}{ax + b}$.

59. Since $\dfrac{1}{x^2(ax + b)} = \dfrac{A}{x} + \dfrac{B}{x^2} + \dfrac{C}{ax + b}$, we get

$1 = Ax(ax + b) + B(ax + b) + Cx^2$
$1 = (aA + C)x^2 + (bA + aB)x + bB.$

So $bB = 1$ and $B = \dfrac{1}{b}$. Since $bA + aB = 0$,

we obtain $bA + \dfrac{a}{b} = 0$ and $A = -\dfrac{a}{b^2}$.

Since $aA + C = 0$, $-\dfrac{a^2}{b^2} + C = 0$ and $C = \dfrac{a^2}{b^2}$.

The answer is

$$\dfrac{-a/b^2}{x} + \dfrac{1/b}{x^2} + \dfrac{a^2/b^2}{ax + b}.$$

For Thought

1. False, since $3 > 1 + 2$ is false.

2. False   **3.**   True

4. False, because $x^2 + y^2 > 5$ is the region outside of a circle of radius $\sqrt{5}$.

5. True, since $(-2, 1)$ satisfies both equations in system (a).

6. True

7. False, $(-2, 0)$ does not satisfy $y < x + 2$.

8. False, $(-1, 2)$ lies on the line $y - 3x = 5$.

9. True

10. True

5.5 Exercises

1. c   **3.**   d

5. $y < 2x$

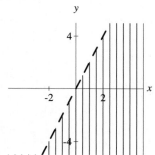

7. $x + y > 3$

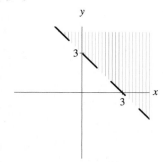

9. $2x - y \leq 4$

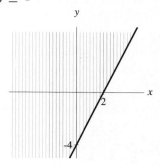

11. $y < -3x - 4$

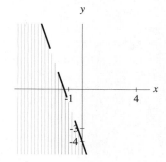

13. $x - 3 \geq 0$

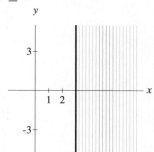

15. $20x - 30y \leq 6000$

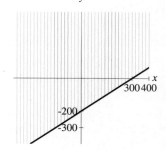

17. $y < 3$

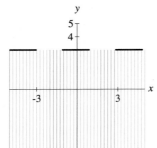

19. $y > -x^2$

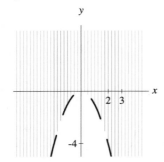

21. $x^2 + y^2 \geq 1$

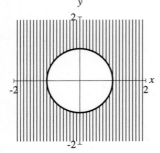

23. $x > |y|$

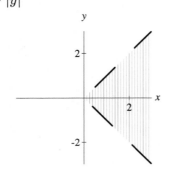

25. $x \geq y^2$

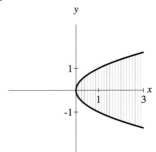

27. $y \geq x^3$

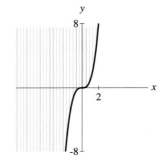

29. $y > 2^x$

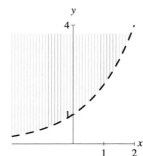

31. $y > x - 4,\ y < -x - 2$

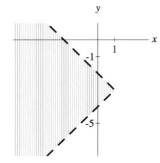

33. $3x - 4y \le 12$, $x + y \ge -3$

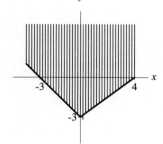

35. $3x - y < 4$, $y < 3x + 5$

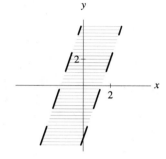

37. No solution since the graphs of $y + x < 0$ and $y > 3 - x$ do not overlap.

39. $x + y < 5$, $y \ge 2$

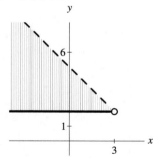

41. $y < x - 3$, $x \le 4$

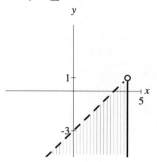

43. $y > x^2 - 3$, $y < x + 1$

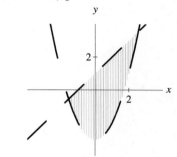

45. $x^2 + y^2 \ge 4$, $x^2 + y^2 \le 16$

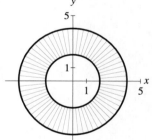

47. $(x - 3)^2 + y^2 \le 25$, $(x + 3)^2 + y^2 \le 25$

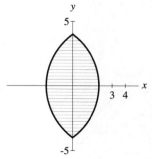

49. $x^2 + y^2 > 4$, $|x| \le 4$

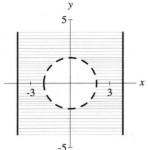

57. $x \ge 0$, $y \ge 0$, $x + y \ge 4$, $y \ge -2x + 6$

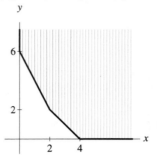

51. $y > |2x| - 4$, $y \le \sqrt{4 - x^2}$

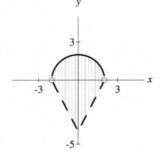

59. $x^2 + y^2 \ge 9$, $x^2 + y^2 \le 25$, $y \ge |x|$

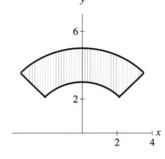

53. $|x - 1| < 2$, $|y - 1| < 4$

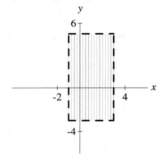

61. $y > (x - 1)^3$, $y > 1$, $x + y > -2$

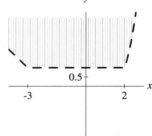

55. $x \ge 0$, $y \ge 0$, $x + y \le 4$

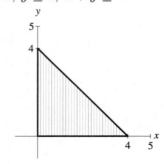

63. $y > 2^x$, $y < 6 - x^2$, $x + y > 0$

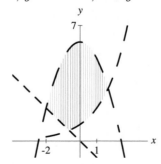

65. $x \geq 0$, $y \geq 0$, $y \leq -\dfrac{2}{3}x + 5$, $y \leq -3x + 12$

67. $x \geq 0$, $y \geq 0$, $y \geq -\dfrac{1}{2}x + 3$, $y \geq -\dfrac{3}{2}x + 5$

69. The system is

$$|x| < 2$$
$$|y| < 2.$$

71. Since a circle of radius 9 with center at the origin is given by $x^2 + y^2 = 81$, the system is

$$\begin{aligned} x^2 + y^2 &< 81 \\ x &> 0 \\ y &> 0. \end{aligned}$$

73. $(-1.17,\ 1.84)$ is a solution.

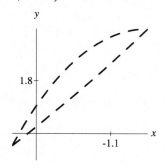

75. $(150,\ 22.4)$ is a solution.

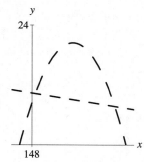

77. Let w and h be the width and height, respectively. Then $50 + 2w + 2h \leq 130$. The system is

$$\begin{aligned} w + h &\leq 40 \\ w, h &\geq 0. \end{aligned}$$

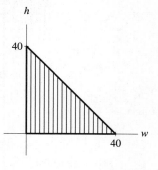

79. Let x and y be the number of mid-size and full-size cars, respectively. Divide $10,000x + 15,000y \leq 1,500,000$ by 1000. The system is

$$\begin{aligned} x + y &\leq 110 \\ x + 1.5y &\leq 150 \\ x &\geq 0 \\ y &\geq 0. \end{aligned}$$

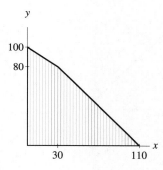

81. Let x and y be the number of \$50 tickets and \$100 tickets, respectively. Simplifying $y \leq 0.2(x + y)$, we get $0.8y \leq 0.2x$. Then $y \leq \dfrac{1}{4}x$. The system is

$$\begin{aligned} y &\leq \frac{1}{4}x \\ x + y &\leq 500 \\ x \geq 0, y &\geq 0. \end{aligned}$$

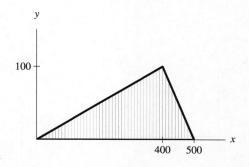

For Thought

1. False, $x \geq 0$ include points on the x-axis and the first and fourth quadrants.

2. False, $y \geq 2$ include points on or above the line $y = 2$.   **3.**   False

4. False, since x-intercept is $(6, 0)$ and y-intercept is $(0, 4)$.   **5.**   True   **6.**   True   **7.**   False

8. True, since $R(1, 3) = 30(1) + 15(3) = 75$.

9. False, since $C(0, 5) = 7(0) + 9(5) + 3 = 48$.

10. True

5.6 Exercises

1. Vertices are $(0, 0), (0, 4), (4, 0)$

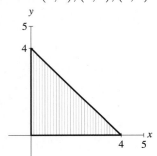

3. Vertices are $(0, 0), (1, 3), (1, 0), (0, 3)$

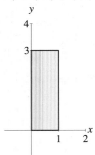

5. Vertices are $(0, 0), (2, 2), (0, 4), (3, 0)$

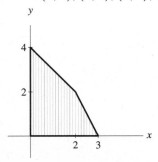

7. Vertices are $(3, 0), (1, 2), (0, 4)$

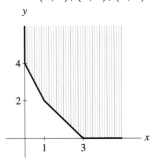

9. Vertices are $(1, 3), (4, 0), (0, 6)$

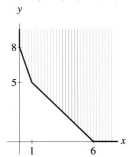

11. Vertices are $(1, 5), (6, 0), (0, 8)$

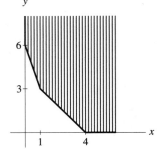

13. The values of $T(x, y) = 2x + 3y$ at the vertices are $T(0, 0) = 0$, $T(0, 4) = 12$, $T(3, 3) = 15$, and $T(5, 0) = 10$. The maximum value is 15.

15. The values of $H(x, y) = 2x + 2y$ at the vertices are $T(0, 6) = 12$, $T(2, 2) = 8$, and $T(5, 0) = 10$. The minimum value is 8.

17. The values of $P(x,y) = 5x + 9y$ at the vertices are $P(0,0) = 0$, $P(6,0) = 30$, $P(0,3) = 27$. Maximum value is 30.

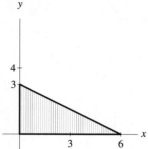

19. The values of $C(x,y) = 3x + 2y$ at the vertices are $C(0,4) = 8$ and $C(4,0) = 12$. The minimum value is 8.

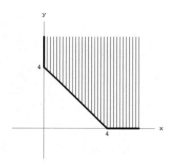

21. The values of $C(x,y) = 10x + 20y$ at the vertices are $C(0,8) = 160$, $C(5,3) = 110$, and $C(10,0) = 100$. Minimum value is 100.

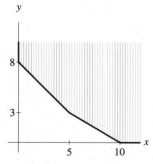

23. Let x and y be the number of bird houses and mailboxes, respectively.

$$\text{Maximize} \quad 12x + 20y$$
$$\text{subject to} \quad 3x + 4y \leq 48$$
$$x + 2y \leq 20$$
$$x,y \geq 0$$

The values of $R(x,y) = 12x + 20y$ at the vertices are $R(0,0) = 0$, $R(0,10) = 200$, $R(8,6) =$

216, and $R(16,0) = 192$. To maximize revenue, they must sell 8 bird houses and 6 mailboxes.

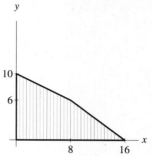

25. Let x and y be the number of bird houses and mailboxes, respectively. The values of $R(x,y) = 18x + 20y$ at the vertices are $R(0,0) = 0$, $R(0,10) = 200$, $R(8,6) = 264$, and $R(16,0) = 288$. To maximize revenue, they must sell 16 bird houses and 0 mailboxes.

27. Let x and y be the number of small and large truck loads, respectively.

$$\text{Minimize} \quad 70x + 60y$$
$$\text{subject to} \quad 12x + 20y \geq 120$$
$$x + y \geq 8$$
$$x,y \geq 0$$

The values of $C(x,y) = 70x + 60y$ at the vertices are $C(0,8) = 480$, $C(10,0) = 700$, and $C(5,3) = 530$, To minimize costs, they must make 8 large truck loads and 0 small truck loads.

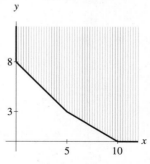

29. Let x and y be the number of small and large truck loads, respectively. The values of $C(x,y) = 70x + 75y$ at the vertices are $C(0,8) = 600$, $C(10,0) = 700$, and $C(5,3) = 575$.

To minimize costs, they must make 5 small truck loads and 3 large truck loads.

Review Exercises

1. The solution set is $\{(3,5)\}$.

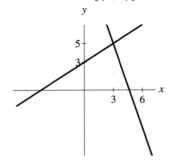

3. The solution set is $\{(-1,3)\}$

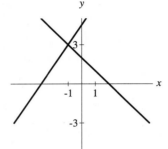

5. Substitute $y = x$ into $3x - 5y = 19$.

$$3x - 5x = 19$$
$$-2x = 19$$
$$x = -19/2$$

Independent and the solution set is $\{(-19/2, -19/2)\}$.

7. Multiply $4x - 3y = 6$ by 2 and $3x + 2y = 9$ by 3, then add the equations.

$$
\begin{aligned}
8x - 6y &= 12 \\
9x + 6y &= 27 \\
\hline
17x &= 39
\end{aligned}
$$

Substitute $x = 39/17$ into $4x - 3y = 6$.

$$\frac{156}{17} - 3y = 6$$
$$\frac{54}{17} = 3y$$
$$\frac{18}{17} = y$$

Independent and the solution set is $\{(39/17, 18/17)\}$.

9. Substitute $y = -3x + 1$ into $6x + 2y = 2$.

$$
\begin{aligned}
6x + 2(-3x + 1) &= 2 \\
6x - 6x + 2 &= 2 \\
2 &= 2
\end{aligned}
$$

Dependent and the solution set is $\{(x,y) \mid y = -3x + 1\}$.

11. Multiply $3x - 4y = 12$ by 2 and add to the second equation.

$$
\begin{aligned}
6x - 8y &= 24 \\
-6x + 8y &= 9 \\
\hline
0 &= 33
\end{aligned}
$$

Inconsistent and the solution set is $\emptyset$.

13. Add the first and second equations. Multiply $2x + y + z = 1$ by -3 and add to the third equation.

$$
\begin{aligned}
x + y - z &= 8 \\
2x + y + z &= 1 \\
\hline
3x + 2y &= 9
\end{aligned}
$$

$$
\begin{aligned}
-6x - 3y - 3z &= -3 \\
x + 2y + 3z &= -5 \\
\hline
-5x - y &= -8
\end{aligned}
$$

Multiply $-5x - y = -8$ by 2 and add to $3x + 2y = 9$.

$$
\begin{aligned}
-10x - 2y &= -16 \\
3x + 2y &= 9 \\
\hline
-7x &= -7
\end{aligned}
$$

Using $x = 1$ in $3x + 2y = 9$, $3 + 2y = 9$ or $y = 3$. From $x + y - z = 8$, $1 + 3 - z = 8$ or $z = -4$. The Solution set is $\{(1, 3, -4)\}$.

15. Multiply first equation by -2 and add to the second equation. Multiply first equation by -2 and add to the third one.

$$
\begin{aligned}
-2x - 2y - 2z &= -2 \\
2x - y + 2z &= 2 \\
\hline
-3y &= 0 \\
y &= 0
\end{aligned}
$$

$$
\begin{aligned}
-2x - 2y - 2z &= -2 \\
2x + 2y + 2z &= 2 \\
\hline
0 &= 0
\end{aligned}
$$

Using $y = 0$ in $x + y + z = 1$, $x + z = 1$ and $z = 1 - x$. The solution set is $\{(x, 0, 1 - x) \mid x \text{ is any real number}\}$.

17. Multiply first equation by -1 and add to the third equation.

$$
\begin{array}{rcl}
-x - y - z & = & -1 \\
x + y + z & = & 4 \\
\hline
0 & = & 3
\end{array}
$$

Inconsistent and the solution set is $\emptyset$.

19. Substitute $x = y^2$ into $x^2 + y^2 = 4$ and use the quadratic formula.

$$
\begin{array}{rcl}
y^4 + y^2 & = & 4 \\
y^4 + y^2 - 4 & = & 0 \\
y^2 & = & \dfrac{-1 + \sqrt{17}}{2} \\
y & = & \pm\sqrt{\dfrac{-1 + \sqrt{17}}{2}}
\end{array}
$$

Thus, $x = y^2 = \dfrac{-1 + \sqrt{17}}{2}$.

The solution set is

$$
\left\{ \left(\frac{-1 + \sqrt{17}}{2}, \pm\sqrt{\frac{-1 + \sqrt{17}}{2}} \right) \right\}.
$$

21. Substitute $y = x^2$ into $y = |x|$.

$$
\begin{array}{rcl}
x^2 & = & \sqrt{x^2} \\
x^4 & = & x^2 \\
x^2(x^2 - 1) & = & 0 \\
x & = & 0, \pm 1
\end{array}
$$

Using $x = 0, 1, -1$ in $y = x^2$, we get $y = 0, 1, 1$, respectively. The solution set is $\{(0, 0), (1, 1), (-1, 1)\}$.

23. Note, $\dfrac{7x - 7}{(x - 3)(x + 4)} = \dfrac{A}{x - 3} + \dfrac{B}{x + 4}$.
Then

$$
\begin{array}{rcl}
7x - 7 & = & A(x + 4) + B(x - 3) \\
7x - 7 & = & (A + B)x + (4A - 3B).
\end{array}
$$

Equating the coefficients, we obtain

$$
\begin{array}{rcl}
A + B & = & 7 \\
4A - 3B & = & -7.
\end{array}
$$

The solution of this system is $A = 2$, $B = 5$.

The answer is $\dfrac{2}{x - 3} + \dfrac{5}{x + 4}$.

25. Factoring the denominator, we obtain

$$
\begin{array}{rcl}
x^3 - 3x^2 + 4x - 12 & = & x^2(x - 3) + 4(x - 3) \\
& = & (x^2 + 4)(x - 3),
\end{array}
$$

and so $\dfrac{7x^2 - 7x + 23}{(x - 3)(x^2 + 4)} = \dfrac{A}{x - 3} + \dfrac{Bx + C}{x^2 + 4}$.
Then $7x^2 - 7x + 23 =$
$A(x^2 + 4) + (Bx + C)(x - 3) =$
$= (A + B)x^2 + (-3B + C)x + (4A - 3C)$.
Equating the coefficients, we have

$$
\begin{array}{rcl}
A + B & = & 7 \\
-3B + C & = & -7 \\
4A - 3C & = & 23.
\end{array}
$$

The solution of this system is $A = 5$, $B = 2$, and $C = -1$. The answer is $\dfrac{5}{x - 3} + \dfrac{2x - 1}{x^2 + 4}$.

27. $x^2 + (y - 3)^2 < 9$

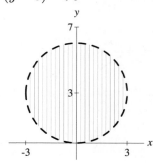

29. $x \leq (y - 1)^2$

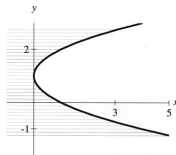

31. $2x - 3y \geq 6$, $x \leq 2$

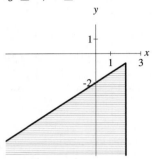

33. $y \geq 2x^2 - 6$, $x^2 + y^2 \leq 9$

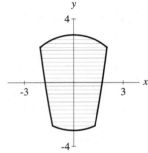

35. $x \geq 0$, $y \geq 1$, $x + 2y \leq 10$, $3x + 4y \leq 24$

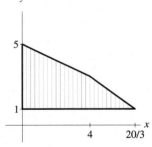

37. $x \geq 0$, $y \geq 0$, $x + 6y \geq 60$, $x + y \geq 35$

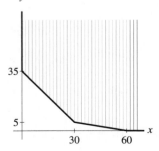

39. Substitute $(-2, 3)$ and $(4, -1)$ into $y = mx + b$.

$$
\begin{aligned}
-2m + b &= 3 \\
4m + b &= -1
\end{aligned}
$$

Multiply first equation by -1 and add to the second equation.

$$
\begin{aligned}
2m - b &= -3 \\
\underline{4m + b} &= \underline{-1} \\
6m &= -4
\end{aligned}
$$

Using $m = -2/3$ in $-2m + b = 3$, $4/3 + b = 3$ and $b = 5/3$. Equation of line is $y = -\dfrac{2}{3}x + \dfrac{5}{3}$.

41. Substitute $(1, 4)$, $(3, 20)$, and $(-2, 25)$ into $y = ax^2 + bx + c$.

$$
\begin{aligned}
a + b + c &= 4 \\
9a + 3b + c &= 20 \\
4a - 2b + c &= 25
\end{aligned}
$$

The solution of the above system is $a = 3$, $b = -4$, $c = 5$. The parabola is given by $y = 3x^2 - 4x + 5$.

43. Let x and y be the number of tacos and burritos, respectively.

$$
\begin{aligned}
x + 2y &= 181 \\
2x + 3y &= 300
\end{aligned}
$$

Solving the above system, we find $x = 57$ tacos and $y = 62$ burritos.

45. Let x, y and z be the selling price of a daisy, carnation, and a rose, respectively. Then

$$
\begin{aligned}
5x + 3y + 2z &= 3.05 \\
3x + y + 4z &= 2.75 \\
4x + 2y + z &= 2.10.
\end{aligned}
$$

Solving the above system, we find $x = 0.30$, $y = 0.25$, and $z = 0.40$. Esther's economy special sells for $x + y + z = \$0.95$.

47. The values of $C(x, y) = 0.42x + 0.84y$ at the vertices are $C(0, 35) = 29.4$, $C(30, 5) = 16.8$, and $C(60, 0) = 25.2$. The minimum value is 16.8.

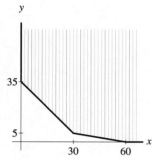

49. Let x and y be the number of barrels of oil obtained through the pipeline and barges, respectively.

$$\text{Minimize } 20x + 18y$$
$$\begin{array}{rcl} \text{subject to } x + y & \geq & 12{,}000{,}000 \\ x & \leq & 12{,}000{,}000 \\ x & \geq & 6{,}000{,}000 \\ y & \leq & 8{,}000{,}000 \\ x, y & \geq & 0 \end{array}$$

The values of $C(x, y) = 20x + 18y$ at the vertices are
$C(12 \text{ million}, 0) = 240$ million,
$C(12 \text{ million}, 8 \text{ million}) = 384$ million,
$C(6 \text{ million}, 8 \text{ million}) = 264$ million, and
$C(6 \text{ million}, 6 \text{ million}) = 228$ million.
To minimize cost, purchase 6 million barrels from each source.

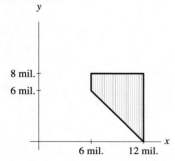

Chapter 5 Test

1. Solution set is $\{(-3, 4)\}$.

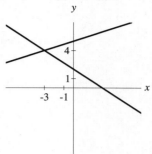

2. Substitute $y = 4 - 2x$ into $3x - 4y = 9$.

$$\begin{array}{rcl} 3x - 4(4 - 2x) & = & 9 \\ 3x - 16 + 8x & = & 9 \\ 11x & = & 25 \\ x & = & 25/11 \end{array}$$

Using $x = 25/11$ in $y = 4 - 2x$, we have $y = 4 - 50/11 = -6/11$. Solution set is $\{(25/11, -6/11)\}$.

3. Multiply $10x - 3y = 22$ by 2 and $7x + 2y = 40$ by 3. Then add the equations.

$$\begin{array}{rcl} 20x - 6y & = & 44 \\ 21x + 6y & = & 120 \\ \hline 41x & = & 164 \\ x & = & 4 \end{array}$$

Using $x = 4$ in $10x - 3y = 22$, we get $40 - 3y = 22$ and $y = 6$. The solution set is $\{(4, 6)\}$.

4. Substitute $x = 6 - y$ into $3x + 3y = 4$.

$$\begin{array}{rcl} 3(6 - y) + 3y & = & 4 \\ 18 - 3y + 3y & = & 4 \\ 18 & = & 4 \end{array}$$

The system is inconsistent.

5. Substitute $y = \dfrac{1}{2}x + 3$ into $x - 2y = -6$.

$$\begin{array}{rcl} x - 2\left(\dfrac{1}{2}x + 3\right) & = & -6 \\ x - x - 6 & = & -6 \\ -6 & = & -6 \end{array}$$

The system is dependent.

6. Substitute $y = 2x - 1$ into $y = 3x + 20$.

$$\begin{aligned} 2x - 1 &= 3x + 20 \\ -21 &= x \end{aligned}$$

Using $x = -21$ in $y = 2x - 1$, we get $y = -43$. The system is independent.

7. Substitute $y = -x + 2$ into $y = -x + 5$.

$$\begin{aligned} -x + 2 &= -x + 5 \\ 2 &= 5 \end{aligned}$$

The system is inconsistent.

8. Add the two equations.

$$\begin{aligned} 2x - y + z &= 4 \\ -x + 2y - z &= 6 \\ \hline x + y &= 10 \end{aligned}$$

Using $y = 10 - x$ in $2x - y + z = 4$, we find $2x - (10 - x) + z = 4$ and $z = 14 - 3x$. The solution set is
$\{(x, 10 - x, 14 - 3x) \mid x \text{ is any real number}\}$.

9. Add the first two equations. Also, multiply second equation by 3 and add to the third.

$$\begin{aligned} x - 2y - z &= 2 \\ 2x + 3y + z &= -1 \\ \hline 3x + y &= 1 \end{aligned}$$

$$\begin{aligned} 6x + 9y + 3z &= -3 \\ 3x - y - 3z &= -4 \\ \hline 9x + 8y &= -7 \end{aligned}$$

Multiply $3x + y = 1$ by -3 and add to $9x + 8y = -7$.

$$\begin{aligned} -9x - 3y &= -3 \\ 9x + 8y &= -7 \\ \hline 5y &= -10 \end{aligned}$$

Using $y = -2$ in $3x + y = 1$, we get $3x - 2 = 1$ or $x = 1$. From $x - 2y - z = 2$, we have $1 + 4 - z = 2$ or $z = 3$. The solution set is $\{(1, -2, 3)\}$.

10. Add the second and third equations.

$$\begin{aligned} x + y - z &= 4 \\ -x - y + z &= 2 \\ \hline 0 &= 6 \end{aligned}$$

Inconsistent and the solution set is $\emptyset$.

11. Multiply $x^2 + y^2 = 16$ by -1 and add to $x^2 - 4y^2 = 16$.

$$\begin{aligned} -x^2 - y^2 &= -16 \\ x^2 - 4y^2 &= 16 \\ \hline -5y^2 &= 0 \\ y &= 0 \end{aligned}$$

Using $y = 0$ in $x^2 + y^2 = 16$, we find $x^2 = 16$ and $x = \pm 4$. Solution set is $\{(4, 0), (-4, 0)\}$.

12. Substitute $y = x^2 - 5x$ into $x + y = -2$.

$$\begin{aligned} x + (x^2 - 5x) &= -2 \\ x^2 - 4x &= -2 \\ x^2 - 4x + 4 &= -2 + 4 \\ (x - 2)^2 &= 2 \\ x &= 2 \pm \sqrt{2} \end{aligned}$$

Using $x = 2 + \sqrt{2}$ and $x = 2 - \sqrt{2}$ in $y = -2 - x$, we have $y = -4 - \sqrt{2}$ and $y = -4 + \sqrt{2}$, respectively. The solution set is
$$\{(2 + \sqrt{2}, -4 - \sqrt{2}), (2 - \sqrt{2}, -4 + \sqrt{2})\}.$$

13.

$$\begin{aligned} \frac{2x + 10}{(x - 4)(x + 2)} &= \frac{A}{x - 4} + \frac{B}{x + 2} \\ 2x + 10 &= A(x + 2) + B(x - 4) \\ 2x + 10 &= (A + B)x + (2A - 4B) \end{aligned}$$

Equating the coefficients, we obtain

$$\begin{aligned} A + B &= 2 \\ 2A - 4B &= 10. \end{aligned}$$

The solution of this system is $A = 3, B = -1$.
The answer is $\dfrac{3}{x - 4} + \dfrac{-1}{x + 2}$.

14. Note, $\dfrac{4x^2 + x - 2}{x^2(x - 1)} = \dfrac{A}{x} + \dfrac{B}{x^2} + \dfrac{C}{x - 1}$. Then

$$\begin{aligned} 4x^2 + x - 2 &= Ax(x - 1) + B(x - 1) + Cx^2 \\ &= (A + C)x^2 + (-A + B)x - B. \end{aligned}$$

Equating the coefficients, we have

$$\begin{aligned} A + C &= 4 \\ -A + B &= 1 \\ -B &= -2. \end{aligned}$$

The solution of this system is $B = 2$, $A = 1$, and $C = 3$. The answer is $\dfrac{1}{x} + \dfrac{2}{x^2} + \dfrac{3}{x-1}$.

15. $2x - y < 8$

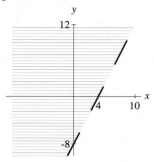

16. $x + y \leq 5$, $x - y < 0$

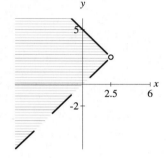

17. $x^2 + y^2 \leq 9$, $y \leq 1 - x^2$

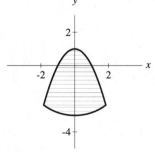

18. Let x and y be the number of male and female students, respectively. Then

$$\frac{1}{3}x + \frac{1}{4}y = 15$$
$$x + y = 52.$$

Multiply the first equation by 12 and multiply the second equation by -3. Then add the resulting equations.

$$\begin{array}{rcr} 4x + 3y &=& 180 \\ -3x - 3y &=& -156 \\ \hline x &=& 24 \end{array}$$

The solution of this system is $x = 24$ males and $y = 28$ females.

19. Let x and y be the number of TV commercials and newspaper ads, respectively. The linear program is given below.

$$\begin{array}{rrcr} \text{Maximize } 14{,}000x + 6000y & & & \\ \text{subject to } 9000x + 3000y &\leq& 99{,}000 \\ x + y &\leq& 23 \\ x, y &\geq& 0 \end{array}$$

The values of $N(x, y) = 14{,}000x + 6000y$ at the vertices are $N(0,0) = 0$, $N(0,23) = 138{,}000$, $N(5,18) = 178{,}000$, and $N(11,0) = 154{,}000$. To obtain maximum audience exposure, the hospital must have 5 TV commercials and 18 newspaper ads.

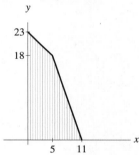

Tying It All Together

1. Multiply the equation by $24(x + 5)$.

$$\begin{array}{rcl} 24(x - 2) &=& 11(x + 5) \\ 24x - 48 &=& 11x + 55 \\ 13x &=& 103 \end{array}$$

The solution set is $\{103/13\}$.

2. Multiply the equation by $24x(x + 5)$.

$$\begin{array}{rcl} 24(x + 5) + 24x(x - 2) &=& 11x(x + 5) \\ 24x + 120 + 24x^2 - 48x &=& 11x^2 + 55x \\ 13x^2 - 79x + 120 &=& 0 \\ (13x - 40)(x - 3) &=& 0 \end{array}$$

The solution set is $\{40/13, 3\}$.

3.

$$\begin{array}{rcl} 5 - 3x - 6 - 2x + 4 &=& 7 \\ 3 - 5x &=& 7 \\ -5x &=& 4 \end{array}$$

The solution set is $\{-4/5\}$.

4. We will solve an equivalent statement without absolute values.

$$\begin{array}{ccc} 3-2x=5 & \text{or} & 3-2x=-5 \\ -2x=2 & \text{or} & -2x=-8 \\ x=-1 & \text{or} & x=4 \end{array}$$

The solution set is $\{4,-1\}$.

5. Square both sides of the equation.

$$\begin{aligned} 3-2x &= 25 \\ -2x &= 22 \end{aligned}$$

The solution set is $\{-11\}$.

6. Isolate x^2 on one side and take the square roots.

$$\begin{aligned} 3x^2 &= 4 \\ x^2 &= \frac{4}{3} \\ x &= \pm\frac{2}{\sqrt{3}} \\ x &= \pm\frac{2}{\sqrt{3}}\cdot\frac{\sqrt{3}}{\sqrt{3}} \end{aligned}$$

The solution set is $\left\{\pm\frac{2\sqrt{3}}{3}\right\}$.

7. Multiply equation by x^2.

$$\begin{aligned} (x-2)^2 &= x^2 \\ x^2-4x+4 &= x^2 \\ -4x &= -4 \end{aligned}$$

The solution set is $\{1\}$.

8. Since $2^{x-1}=9$, we obtain $x-1=\log_2(9)$ by using the definition of a logarithm. The solution set is $\{1+\log_2(9)\}$.

9. Write left-hand side as a single logarithm.

$$\begin{aligned} \log((x+1)(x+4)) &= 1 \\ x^2+5x+4 &= 10^1 \\ x^2+5x-6 &= 0 \\ (x+6)(x-1) &= 0 \\ x &= -6, 1 \end{aligned}$$

But $\log(x+1)$ is undefined when $x=-6$. The solution set is $\{1\}$.

10. Raise both sides of equation to the power $-3/2$.

$$\begin{aligned} x^{-2/3} &= \frac{1}{4} \\ x &= \pm\left(\frac{1}{4}\right)^{-3/2} \\ x &= \pm(4)^{3/2} \\ x &= \pm(4^{1/2})^3 \\ x &= \pm(2)^3 \end{aligned}$$

The solution set is $\{\pm 8\}$.

11. Use the quadratic formula.

$$\begin{aligned} x^2-3x-6 &= 0 \\ x &= \frac{3\pm\sqrt{33}}{2} \end{aligned}$$

The solution set is $\left\{\frac{3\pm\sqrt{33}}{2}\right\}$.

12. By using the square root property, we obtain

$$\begin{aligned} (x-3)^2 &= \frac{1}{2} \\ x-3 &= \pm\frac{\sqrt{2}}{2} \\ x &= \frac{6}{2}\pm\frac{\sqrt{2}}{2}. \end{aligned}$$

The solution set is $\left\{\frac{6\pm\sqrt{2}}{2}\right\}$.

13. Since $3-2x>0$, we get $3>2x$ and $x<3/2$. The solution set is $(-\infty,3/2)$ and its graph is

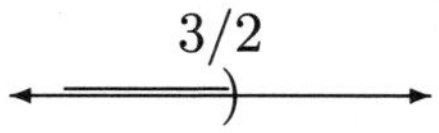

14. Since we must exclude $x=\frac{3}{2}$, the solution set is $(-\infty,1.5)\cup(1.5,\infty)$ and the graph is

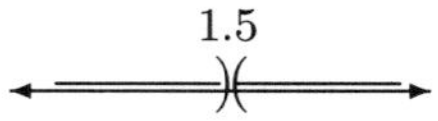

15. The sign graph of $(x-3)(x+3)\geq 0$ is shown below.

- - - - - - - - 0 + + + +

- - - - 0 + + + + + + + +

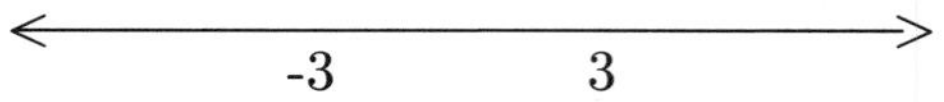

So the solution set is $(-\infty,-3]\cup[3,\infty)$ and the graph is

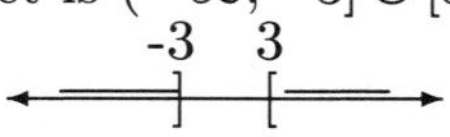

16. Note that $x^2 + 2x - 8 \le 27$ is equivalent to $x^2 + 2x - 35 \le 0$. The sign graph of $(x + 7)(x - 5) \le 0$ is

```
- - - - - - - - - 0 + + + +
- - - - 0 + + + + + + + +
```
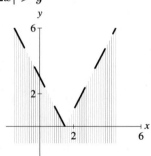

$$-7 \qquad 5$$

So the solution set is $[-7, 5]$ and the graph is

-7   5

17. $3 - 2x > y$

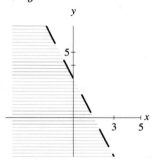

18. $|3 - 2x| > y$

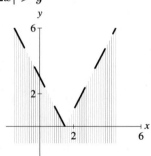

19. $x^2 \ge 9$

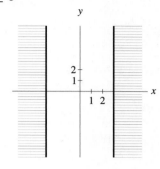

20. $(x - 2)(x + 4) \le y$

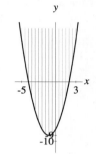

For Thought

1. False, the augmented matrix is a 2×3 matrix.

2. False, the required matrix is $\begin{bmatrix} 1 & -1 & | & 4 \\ 3 & 1 & | & 5 \end{bmatrix}$.

3. True   **4.**   True

5. True, since the row operation done is $R_1 + R_2 \to R_2$.

6. True, since the row operation done is $-R_1 + R_2 \to R_2$.

7. False, since it corresponds to

$$x = 2$$
$$y = 7.$$

8. True, since $0 \cdot x + 0 \cdot y = 7$ has no solution.

9. False, the system is dependent.   **10.**   True

6.1 Exercises

1. 1×3 **3.** 1×1 **5.** 3×2

7.
$$\begin{bmatrix} 1 & -2 & | & 4 \\ 3 & 2 & | & -5 \end{bmatrix}$$

9.
$$\begin{bmatrix} 1 & -1 & -1 & | & 4 \\ 1 & 3 & -1 & | & 1 \\ 0 & 2 & -5 & | & -6 \end{bmatrix}$$

11.
$$\begin{bmatrix} 1 & 3 & -1 & | & 5 \\ 1 & 0 & 1 & | & 0 \end{bmatrix}$$

13.
$$3x + 4y = -2$$
$$3x - 5y = 0$$

15.
$$5x = 6$$
$$-4x + 2z = -1$$
$$4x + 4y = 7$$

17.
$$x - y + 2z = 1$$
$$y + 4z = 3$$

19. Interchange R_1 and R_2
$$\begin{bmatrix} 1 & 2 & | & 0 \\ -2 & 4 & | & 1 \end{bmatrix}$$

21. Multiply $\dfrac{1}{2}$ to R_1
$$\begin{bmatrix} 1 & 4 & | & 1 \\ 0 & 3 & | & 6 \end{bmatrix}$$

23. Multiply 3 to R_1 then add the product to R_2. This is the new R_2.

$$\begin{bmatrix} 1 & -2 & | & 1 \\ -3 & 5 & | & 0 \end{bmatrix} =$$
$$\begin{bmatrix} 1 & -2 & | & 1 \\ 3 \cdot 1 + (-3) & 3 \cdot (-2) + 5 & | & 3 \cdot 1 + 0 \end{bmatrix} =$$
$$\begin{bmatrix} 1 & -2 & | & 1 \\ 0 & -1 & | & 3 \end{bmatrix}$$

25. From the augmented matrix
$$\begin{bmatrix} 2 & 4 & | & 14 \\ 5 & 4 & | & 5 \end{bmatrix}$$
the system is
$$2x + 4y = 14$$
$$5x + 4y = 5.$$

From the final augmented matrix
$$\begin{bmatrix} 1 & 0 & | & -3 \\ 0 & 1 & | & 5 \end{bmatrix}$$
we obtain that the solution set is $\{(-3, 5)\}$.

The row operations are $\dfrac{1}{2} R_1 \to R_1$,

$-5R_1 + R_2 \to R_2$, $-\dfrac{1}{6} R_2 \to R_2$,

and $-2R_2 + R_1 \to R_1$.

27. On $\begin{bmatrix} 1 & 1 & | & 5 \\ -2 & 1 & | & -1 \end{bmatrix}$ use $2R_1 + R_2 \to R_2$ to get

$\begin{bmatrix} 1 & 1 & | & 5 \\ 0 & 3 & | & 9 \end{bmatrix}$, use $\frac{1}{3}R_2 \to R_2$ to get

$\begin{bmatrix} 1 & 1 & | & 5 \\ 0 & 1 & | & 3 \end{bmatrix}$, use $-R_2 + R_1 \to R_1$ to get

$\begin{bmatrix} 1 & 0 & | & 2 \\ 0 & 1 & | & 3 \end{bmatrix}$, solution set is $\{(2, 3)\}$,

and the system is independent.

29. On $\begin{bmatrix} 2 & 2 & | & 8 \\ -3 & -1 & | & -6 \end{bmatrix}$ use $\frac{1}{2}R_1 \to R_1$ to get

$\begin{bmatrix} 1 & 1 & | & 4 \\ -3 & -1 & | & -6 \end{bmatrix}$, use $3R_1 + R_2 \to R_2$ to get

$\begin{bmatrix} 1 & 1 & | & 4 \\ 0 & 2 & | & 6 \end{bmatrix}$, use $\frac{1}{2}R_2 \to R_2$ to get

$\begin{bmatrix} 1 & 1 & | & 4 \\ 0 & 1 & | & 3 \end{bmatrix}$, use $-1R_2 + R_1 \to R_1$ to get

$\begin{bmatrix} 1 & 0 & | & 1 \\ 0 & 1 & | & 3 \end{bmatrix}$, solution set is $\{(1, 3)\}$,

and the system is independent.

31. On $\begin{bmatrix} 2 & -1 & | & 3 \\ 3 & 2 & | & 15 \end{bmatrix}$ use $-R_1 + R_2 \to R_1$ to get

$\begin{bmatrix} 1 & 3 & | & 12 \\ 3 & 2 & | & 15 \end{bmatrix}$, use $-3R_1 + R_2 \to R_2$ to get

$\begin{bmatrix} 1 & 3 & | & 12 \\ 0 & -7 & | & -21 \end{bmatrix}$, use $-\frac{1}{7}R_2 \to R_2$ to get

$\begin{bmatrix} 1 & 3 & | & 12 \\ 0 & 1 & | & 3 \end{bmatrix}$, use $-3R_2 + R_1 \to R_1$ to get

$\begin{bmatrix} 1 & 0 & | & 3 \\ 0 & 1 & | & 3 \end{bmatrix}$, solution set is $\{(3, 3)\}$,

and the system is independent.

33. On $\begin{bmatrix} 0.4 & -0.2 & | & 0 \\ 1 & 1.5 & | & 2 \end{bmatrix}$ use $5R_1 \to R_1$ and $2R_2 \to R_2$ to get

$\begin{bmatrix} 2 & -1 & | & 0 \\ 2 & 3 & | & 4 \end{bmatrix}$, use $R_1 + (-R_2) \to R_2$ to get

$\begin{bmatrix} 2 & -1 & | & 0 \\ 0 & -4 & | & -4 \end{bmatrix}$, use $-\frac{1}{4}R_2 \to R_2$ to get

$\begin{bmatrix} 2 & -1 & | & 0 \\ 0 & 1 & | & 1 \end{bmatrix}$, use $R_1 + R_2 \to R_1$ to get

$\begin{bmatrix} 2 & 0 & | & 1 \\ 0 & 1 & | & 1 \end{bmatrix}$, use $\frac{1}{2}R_1 \to R_1$ to get

$\begin{bmatrix} 1 & 0 & | & 0.5 \\ 0 & 1 & | & 1 \end{bmatrix}$, solution set is $\{(0.5, 1)\}$,

and the system is independent.

35. On $\begin{bmatrix} 3 & -5 & | & 7 \\ -3 & 5 & | & 4 \end{bmatrix}$ use $R_1 + R_2 \to R_2$ to get

$\begin{bmatrix} 3 & -5 & | & 7 \\ 0 & 0 & | & 11 \end{bmatrix}$, inconsistent system, and the solution set is $\emptyset$.

37. On $\begin{bmatrix} 0.5 & 1.5 & | & 2 \\ 3 & 9 & | & 12 \end{bmatrix}$ use $2R_1 \to R_1$ to get

$\begin{bmatrix} 1 & 3 & | & 4 \\ 3 & 9 & | & 12 \end{bmatrix}$, use $-3R_1 + R_2 \to R_2$ to get

$\begin{bmatrix} 1 & 3 & | & 4 \\ 0 & 0 & | & 0 \end{bmatrix}$, dependent system, and solution set is $\{(u, v) \mid u + 3v = 4\}$.

39. Rewrite system as

$$2x + y = 4$$
$$x - y = 8.$$

On $\begin{bmatrix} 2 & 1 & | & 4 \\ 1 & -1 & | & 8 \end{bmatrix}$ use $R_1 + (-2R_2) \to R_2$ to get

$\begin{bmatrix} 2 & 1 & | & 4 \\ 0 & 3 & | & -12 \end{bmatrix}$, use $\frac{1}{3}R_2 \to R_2$ to get

$\begin{bmatrix} 2 & 1 & | & 4 \\ 0 & 1 & | & -4 \end{bmatrix}$, use $-1R_2 + R_1 \to R_1$ to get

$\begin{bmatrix} 2 & 0 & | & 8 \\ 0 & 1 & | & -4 \end{bmatrix}$, use $\frac{1}{2}R_1 \to R_1$ to get

$\begin{bmatrix} 1 & 0 & | & 4 \\ 0 & 1 & | & -4 \end{bmatrix}$, solution set is $\{(4, -4)\}$,

and the system is independent.

41.

On $\begin{bmatrix} 1 & 1 & 1 & | & 6 \\ 1 & -1 & -1 & | & 0 \\ 0 & 2 & -1 & | & 3 \end{bmatrix}$ use $-1R_2 + R_1 \to R_2$ to get

$\begin{bmatrix} 1 & 1 & 1 & | & 6 \\ 0 & 2 & 2 & | & 6 \\ 0 & 2 & -1 & | & 3 \end{bmatrix}$, use $\frac{1}{2}R_2 \to R_2$ to get

$\begin{bmatrix} 1 & 1 & 1 & | & 6 \\ 0 & 1 & 1 & | & 3 \\ 0 & 2 & -1 & | & 3 \end{bmatrix}$, use $-1R_2 + R_1 \to R_1$ to get

$\begin{bmatrix} 1 & 0 & 0 & | & 3 \\ 0 & 1 & 1 & | & 3 \\ 0 & 2 & -1 & | & 3 \end{bmatrix}$, use $-2R_2 + R_3 \to R_3$ to get

$\begin{bmatrix} 1 & 0 & 0 & | & 3 \\ 0 & 1 & 1 & | & 3 \\ 0 & 0 & -3 & | & -3 \end{bmatrix}$, use $-\frac{1}{3}R_3 \to R_3$ to get

$\begin{bmatrix} 1 & 0 & 0 & | & 3 \\ 0 & 1 & 1 & | & 3 \\ 0 & 0 & 1 & | & 1 \end{bmatrix}$, use $-1R_3 + R_2 \to R_2$ to get

$\begin{bmatrix} 1 & 0 & 0 & | & 3 \\ 0 & 1 & 0 & | & 2 \\ 0 & 0 & 1 & | & 1 \end{bmatrix}$, solution set is $\{(3,2,1)\}$,

and the system is independent.

43.

Rewrite system as

$$2x + y - z = 2$$
$$x + 2y - z = 2$$
$$x - y + 2z = 2.$$

On $\begin{bmatrix} 2 & 1 & -1 & | & 2 \\ 1 & 2 & -1 & | & 2 \\ 1 & -1 & 2 & | & 2 \end{bmatrix}$, use $-1R_2 + R_1 \to R_1$

and $-1R_3 + R_2 \to R_3$ to get

$\begin{bmatrix} 1 & -1 & 0 & | & 0 \\ 1 & 2 & -1 & | & 2 \\ 0 & 3 & -3 & | & 0 \end{bmatrix}$, use $-1R_2 + R_1 \to R_2$

and $\frac{1}{3}R_3 \to R_3$ to get

$\begin{bmatrix} 1 & -1 & 0 & | & 0 \\ 0 & -3 & 1 & | & -2 \\ 0 & 1 & -1 & | & 0 \end{bmatrix}$, use $R_1 + R_3 \to R_1$

and $R_3 + R_2 \to R_2$ to get

$\begin{bmatrix} 1 & 0 & -1 & | & 0 \\ 0 & -2 & 0 & | & -2 \\ 0 & 1 & -1 & | & 0 \end{bmatrix}$, use $-\frac{1}{2}R_2 \to R_2$ to get

$\begin{bmatrix} 1 & 0 & -1 & | & 0 \\ 0 & 1 & 0 & | & 1 \\ 0 & 1 & -1 & | & 0 \end{bmatrix}$, use $-1R_3 + R_2 \to R_3$ to get

$\begin{bmatrix} 1 & 0 & -1 & | & 0 \\ 0 & 1 & 0 & | & 1 \\ 0 & 0 & 1 & | & 1 \end{bmatrix}$, use $R_1 + R_3 \to R_1$ to get

$\begin{bmatrix} 1 & 0 & 0 & | & 1 \\ 0 & 1 & 0 & | & 1 \\ 0 & 0 & 1 & | & 1 \end{bmatrix}$, solution set is $\{(1,1,1)\}$,

and the system is independent.

45.

Interchange rows of $\begin{bmatrix} 2 & -2 & 1 & | & -2 \\ 1 & 1 & -3 & | & 3 \\ 1 & -3 & 1 & | & -5 \end{bmatrix}$ to get

$\begin{bmatrix} 1 & 1 & -3 & | & 3 \\ 1 & -3 & 1 & | & -5 \\ 2 & -2 & 1 & | & -2 \end{bmatrix}$, use $-1R_2 + R_1 \to R_2$

and $-1R_3 + R_1 + R_2 \to R_3$ to get

$\begin{bmatrix} 1 & 1 & -3 & | & 3 \\ 0 & 4 & -4 & | & 8 \\ 0 & 0 & -3 & | & 0 \end{bmatrix}$, use $\frac{1}{4}R_2 \to R_2$

and $-\frac{1}{3}R_3 \to R_3$ to get

$\begin{bmatrix} 1 & 1 & -3 & | & 3 \\ 0 & 1 & -1 & | & 2 \\ 0 & 0 & 1 & | & 0 \end{bmatrix}$, use $R_2 + R_3 \to R_2$ to get

$\begin{bmatrix} 1 & 1 & -3 & | & 3 \\ 0 & 1 & 0 & | & 2 \\ 0 & 0 & 1 & | & 0 \end{bmatrix}$, use $-1R_2 + R_1 \to R_1$ to get

$\begin{bmatrix} 1 & 0 & -3 & | & 1 \\ 0 & 1 & 0 & | & 2 \\ 0 & 0 & 1 & | & 0 \end{bmatrix}$, use $3R_3 + R_1 \to R_1$ to get

$\begin{bmatrix} 1 & 0 & 0 & | & 1 \\ 0 & 1 & 0 & | & 2 \\ 0 & 0 & 1 & | & 0 \end{bmatrix}$, solution set is $\{(1,2,0)\}$,

and the system is independent.

47.

Rewrite system as

$$\begin{aligned} x - 3y + z &= 0 \\ x - y - 3z &= 4 \\ x + y + 2z &= -1. \end{aligned}$$

On $\begin{bmatrix} 1 & -3 & 1 & 0 \\ 1 & -1 & -3 & 4 \\ 1 & 1 & 2 & -1 \end{bmatrix}$, use

$-1R_2 + R_1 \to R_2$ and $-1R_3 + R_1 \to R_3$ to get

$\begin{bmatrix} 1 & -3 & 1 & 0 \\ 0 & -2 & 4 & -4 \\ 0 & -4 & -1 & 1 \end{bmatrix}$, use $-\dfrac{1}{2}R_2 \to R_2$

and $-1R_3 \to R_3$ to get

$\begin{bmatrix} 1 & -3 & 1 & 0 \\ 0 & 1 & -2 & 2 \\ 0 & 4 & 1 & -1 \end{bmatrix}$, use $-1R_3 + R_1 \to R_1$

to get

$\begin{bmatrix} 1 & -7 & 0 & 1 \\ 0 & 1 & -2 & 2 \\ 0 & 4 & 1 & -1 \end{bmatrix}$, use $-4R_2 + R_3 \to R_3$

to get

$\begin{bmatrix} 1 & -7 & 0 & 1 \\ 0 & 1 & -2 & 2 \\ 0 & 0 & 9 & -9 \end{bmatrix}$, use $\dfrac{1}{9}R_3 \to R_3$ to get

$\begin{bmatrix} 1 & -7 & 0 & 1 \\ 0 & 1 & -2 & 2 \\ 0 & 0 & 1 & -1 \end{bmatrix}$, use $2R_3 + R_2 \to R_2$

to get

$\begin{bmatrix} 1 & -7 & 0 & 1 \\ 0 & 1 & 0 & 0 \\ 0 & 0 & 1 & -1 \end{bmatrix}$, use $R_1 + 7R_2 \to R_1$ to

get

$\begin{bmatrix} 1 & 0 & 0 & 1 \\ 0 & 1 & 0 & 0 \\ 0 & 0 & 1 & -1 \end{bmatrix}$, solution set is $\{(1, 0, -1)\}$,

and the system is independent.

49.

On $\begin{bmatrix} 1 & -2 & 3 & 1 \\ 2 & -4 & 6 & 2 \\ -3 & 6 & -9 & -3 \end{bmatrix}$, use $\dfrac{1}{2}R_2 \to R_2$

and $-\dfrac{1}{3}R_3 \to R_3$ to get

$\begin{bmatrix} 1 & -2 & 3 & 1 \\ 1 & -2 & 3 & 1 \\ 1 & -2 & 3 & 1 \end{bmatrix}$, use $-1R_1 + R_2 \to R_2$

and $-1R_1 + R_3 \to R_3$ to get

$\begin{bmatrix} 1 & -2 & 3 & 1 \\ 0 & 0 & 0 & 0 \\ 0 & 0 & 0 & 0 \end{bmatrix}$, dependent system,

and solution set is $\{(x, y, z) \mid x - 2y + 3z = 1\}$.

51.

On $\begin{bmatrix} 1 & -1 & 1 & 2 \\ 2 & 1 & -1 & 1 \\ 2 & -2 & 2 & 5 \end{bmatrix}$, use $-\dfrac{1}{2}R_3 + R_1 \to R_3$

to get $\begin{bmatrix} 1 & -1 & 1 & 2 \\ 2 & 1 & -1 & 1 \\ 0 & 0 & 0 & -1/2 \end{bmatrix}$, inconsistent,

and the solution set is $\emptyset$.

53.

On $\begin{bmatrix} 1 & 1 & -1 & 3 \\ 3 & 1 & 1 & 7 \\ 1 & -1 & 3 & 1 \end{bmatrix}$, use $-1R_3 + R_1 \to R_3$

and $3R_1 + (-1R_2) \to R_2$ to get

$\begin{bmatrix} 1 & 1 & -1 & 3 \\ 0 & 2 & -4 & 2 \\ 0 & 2 & -4 & 2 \end{bmatrix}$, use $-1R_3 + R_2 \to R_3$ to

get

$\begin{bmatrix} 1 & 1 & -1 & 3 \\ 0 & 2 & -4 & 2 \\ 0 & 0 & 0 & 0 \end{bmatrix}$, use $\dfrac{1}{2}R_2 \to R_2$ to get

$\begin{bmatrix} 1 & 1 & -1 & 3 \\ 0 & 1 & -2 & 1 \\ 0 & 0 & 0 & 0 \end{bmatrix}$, use $-1R_2 + R_1 \to R_1$ to

get

$\begin{bmatrix} 1 & 0 & 1 & 2 \\ 0 & 1 & -2 & 1 \\ 0 & 0 & 0 & 0 \end{bmatrix}$. Substitute $z = 2 - x$

into $y = 1 + 2z$. Then $y = 1 + 2(2 - x) = 5 - 2x$.
The solution set is

$$\{(x, 5 - 2x, 2 - x) \mid x \text{ is any real number}\}$$

and the system is dependent.

55.

On $\begin{bmatrix} 2 & -1 & 3 & | & 1 \\ 1 & 1 & -1 & | & 4 \end{bmatrix}$, use $-2R_2 + R_1 \rightarrow R_2$

to get

$\begin{bmatrix} 2 & -1 & 3 & | & 1 \\ 0 & -3 & 5 & | & -7 \end{bmatrix}$, use $-3R_1 + R_2 \rightarrow R_1$

to get

$\begin{bmatrix} -6 & 0 & -4 & | & -10 \\ 0 & -3 & 5 & | & -7 \end{bmatrix}$, use $-\frac{1}{6}R_1 \rightarrow R_1$

and $-\frac{1}{3}R_2 \rightarrow R_2$ to get

$\begin{bmatrix} 1 & 0 & 2/3 & | & 5/3 \\ 0 & 1 & -5/3 & | & 7/3 \end{bmatrix}$. Note $x = \dfrac{5 - 2z}{3}$ and

$y = \dfrac{7 + 5z}{3}$. Solving for z, we get $z = \dfrac{5 - 3x}{2}$.

Then $y = \dfrac{7 + 5 \cdot \frac{5-3x}{2}}{3} = \dfrac{7 + 5 \cdot \frac{5-3x}{2}}{3} \cdot \dfrac{2}{2} =$

$\dfrac{39 - 15x}{6} = \dfrac{13 - 5x}{2}$. The solution set is

$$\left\{ \left(x, \frac{13 - 5x}{2}, \frac{5 - 3x}{2} \right) \mid x \text{ is any real number} \right\}$$

and the system is dependent.

57.

On $\begin{bmatrix} 1 & -1 & 1 & -1 & | & 2 \\ -1 & 2 & -1 & -1 & | & -1 \\ 2 & -1 & -1 & 1 & | & 4 \\ 1 & 3 & -2 & -3 & | & 6 \end{bmatrix}$,

use $2R_2 + R_3 \rightarrow R_3$ and $R_2 + R_4 \rightarrow R_4$ to get

$\begin{bmatrix} 1 & -1 & 1 & -1 & | & 2 \\ 0 & 1 & 0 & -2 & | & 1 \\ 0 & 3 & -3 & -1 & | & 2 \\ 0 & 5 & -3 & -4 & | & 5 \end{bmatrix}$, use

$-3R_2 + R_3 \rightarrow R_3$ and $-5R_2 + R_4 \rightarrow R_4$ to get

$\begin{bmatrix} 1 & -1 & 1 & -1 & | & 2 \\ 0 & 1 & 0 & -2 & | & 1 \\ 0 & 0 & -3 & 5 & | & -1 \\ 0 & 0 & -3 & 6 & | & 0 \end{bmatrix}$, use

$R_1 + R_2 \rightarrow R_1$ and $-\frac{1}{3}R_4 \rightarrow R_4$ to get

$\begin{bmatrix} 1 & 0 & 1 & -3 & | & 3 \\ 0 & 1 & 0 & -2 & | & 1 \\ 0 & 0 & -3 & 5 & | & -1 \\ 0 & 0 & 1 & -2 & | & 0 \end{bmatrix}$, use $R_3 \rightarrow R_4$

and $R_4 \rightarrow R_3$ to get

$\begin{bmatrix} 1 & 0 & 1 & -3 & | & 3 \\ 0 & 1 & 0 & -2 & | & 1 \\ 0 & 0 & 1 & -2 & | & 0 \\ 0 & 0 & -3 & 5 & | & -1 \end{bmatrix}$, use

$-3R_3 + (-1R_4) \rightarrow R_4$ to get

$\begin{bmatrix} 1 & 0 & 1 & -3 & | & 3 \\ 0 & 1 & 0 & -2 & | & 1 \\ 0 & 0 & 1 & -2 & | & 0 \\ 0 & 0 & 0 & 1 & | & 1 \end{bmatrix}$, use $-1R_3 + R_1 \rightarrow R_1$

to get

$\begin{bmatrix} 1 & 0 & 0 & -1 & | & 3 \\ 0 & 1 & 0 & -2 & | & 1 \\ 0 & 0 & 1 & -2 & | & 0 \\ 0 & 0 & 0 & 1 & | & 1 \end{bmatrix}$, use $2R_4 + R_3 \rightarrow R_3$ to

get

$\begin{bmatrix} 1 & 0 & 0 & -1 & | & 3 \\ 0 & 1 & 0 & -2 & | & 1 \\ 0 & 0 & 1 & 0 & | & 2 \\ 0 & 0 & 0 & 1 & | & 1 \end{bmatrix}$, use $2R_4 + R_2 \rightarrow R_2$ to

get

$\begin{bmatrix} 1 & 0 & 0 & -1 & | & 3 \\ 0 & 1 & 0 & 0 & | & 3 \\ 0 & 0 & 1 & 0 & | & 2 \\ 0 & 0 & 0 & 1 & | & 1 \end{bmatrix}$, use $R_4 + R_1 \rightarrow R_1$ to

get

$\begin{bmatrix} 1 & 0 & 0 & 0 & | & 4 \\ 0 & 1 & 0 & 0 & | & 3 \\ 0 & 0 & 1 & 0 & | & 2 \\ 0 & 0 & 0 & 1 & | & 1 \end{bmatrix}$, the solution set is

$\{(4, 3, 2, 1)\}$, and the system is independent.

59. Let x and y be the number of hours Mike worked at Burgers and the Soap Opera, respectively. The augmented matrix is

$A = \begin{bmatrix} 1 & 1 & | & 60 \\ 8 & 9 & | & 502 \end{bmatrix}$. On A use
$-8R_1 + R_2 \rightarrow R_2$ to get

$\begin{bmatrix} 1 & 1 & | & 60 \\ 0 & 1 & | & 22 \end{bmatrix}$, use $-R_2 + R_1 \rightarrow R_1$ to get

$\begin{bmatrix} 1 & 0 & | & 38 \\ 0 & 1 & | & 22 \end{bmatrix}$. Mike worked $x = 38$ hours

at Burgers and $y = 22$ hours at Soap Opera.

61. Let $x, y,$ and z be the amounts invested in a mutual fund, in treasury bills, and in bonds, respectively. Augmented matrix is

$$A = \begin{bmatrix} 1 & 1 & 1 & | & 40,000 \\ 0.08 & 0.09 & 0.12 & | & 3,660 \\ 1 & -1 & -1 & | & 0 \end{bmatrix}.$$

On A use $100R_2 \to R_2$ to get

$$\begin{bmatrix} 1 & 1 & 1 & | & 40,000 \\ 8 & 9 & 12 & | & 366,000 \\ 1 & -1 & -1 & | & 0 \end{bmatrix}, \text{ use}$$

$-8R_1 + R_2 \to R_2$ to get

$$\begin{bmatrix} 1 & 1 & 1 & | & 40,000 \\ 0 & 1 & 4 & | & 46,000 \\ 1 & -1 & -1 & | & 0 \end{bmatrix}, \text{ use}$$

$-1R_3 + R_1 \to R_3$ and $-1R_2 + R_1 \to R_1$ to get

$$\begin{bmatrix} 1 & 0 & -3 & | & -6,000 \\ 0 & 1 & 4 & | & 46,000 \\ 0 & 2 & 2 & | & 40,000 \end{bmatrix}, \text{ use } \frac{1}{2}R_3 \to R_3 \text{ to}$$

get

$$\begin{bmatrix} 1 & 0 & -3 & | & -6,000 \\ 0 & 1 & 4 & | & 46,000 \\ 0 & 1 & 1 & | & 20,000 \end{bmatrix}, \text{ use}$$

$-1R_3 + R_2 \to R_3$ to get

$$\begin{bmatrix} 1 & 0 & -3 & | & -6,000 \\ 0 & 1 & 4 & | & 46,000 \\ 0 & 0 & 3 & | & 26,000 \end{bmatrix}, \text{ use}$$

$-1R_3 + R_2 \to R_2$ and $\frac{1}{3}R_3 \to R_3$ to get

$$\begin{bmatrix} 1 & 0 & -3 & | & -6,000 \\ 0 & 1 & 1 & | & 20,000 \\ 0 & 0 & 1 & | & 8,666.67 \end{bmatrix}, \text{ use } 3R_3 + R_1 \to R_1$$

and $-1R_3 + R_2 \to R_2$ to get

$$\begin{bmatrix} 1 & 0 & 0 & | & 20,000 \\ 0 & 1 & 0 & | & 11,333.33 \\ 0 & 0 & 1 & | & 8,666.67 \end{bmatrix}. \text{ Investments were}$$

$x = \$20,000$ in a mutual fund, $y = \$11,333.33$ in treasury bills, and $z = \$8,666.67$ in bonds.

63. The augmented matrix is

$$A = \begin{bmatrix} -1 & -1 & 1 & | & 4 \\ 1 & 1 & 1 & | & 2 \\ 8 & 2 & 1 & | & 7 \end{bmatrix}.$$

On A use $R_1 \to R_2$ and $R_2 \to R_1$ to get

$$\begin{bmatrix} 1 & 1 & 1 & | & 2 \\ -1 & -1 & 1 & | & 4 \\ 8 & 2 & 1 & | & 7 \end{bmatrix}, \text{ use } 8R_1 + (-1R_3) \to R_3$$

and $-\frac{8}{3}R_2 + \left(-\frac{1}{3}R_3\right) \to R_2$ to get

$$\begin{bmatrix} 1 & 1 & 1 & | & 2 \\ 0 & 2 & -3 & | & -13 \\ 0 & 6 & 7 & | & 9 \end{bmatrix}, \text{ use } -3R_2 + R_3 \to R_3$$

to get

$$\begin{bmatrix} 1 & 1 & 1 & | & 2 \\ 0 & 2 & -3 & | & -13 \\ 0 & 0 & 16 & | & 48 \end{bmatrix}, \text{ use } \frac{1}{2}R_2 \to R_2$$

and $\frac{1}{16}R_3 \to R_3$ to get

$$\begin{bmatrix} 1 & 1 & 1 & | & 2 \\ 0 & 1 & -3/2 & | & -13/2 \\ 0 & 0 & 1 & | & 3 \end{bmatrix}, \text{ use}$$

$\frac{3}{2}R_3 + R_2 \to R_2$ to get

$$\begin{bmatrix} 1 & 1 & 1 & | & 2 \\ 0 & 1 & 0 & | & -2 \\ 0 & 0 & 1 & | & 3 \end{bmatrix}, \text{ use } -1R_3 + R_1 \to R_1 \text{ to}$$

get

$$\begin{bmatrix} 1 & 1 & 0 & | & -1 \\ 0 & 1 & 0 & | & -2 \\ 0 & 0 & 1 & | & 3 \end{bmatrix}, \text{ use } -1R_2 + R_1 \to R_1 \text{ to}$$

get

$$\begin{bmatrix} 1 & 0 & 0 & | & 1 \\ 0 & 1 & 0 & | & -2 \\ 0 & 0 & 1 & | & 3 \end{bmatrix}. \text{ Then } a = 1, b = -2,$$

$c = 3,$ and the equation is $y = x^3 - 2x + 3.$

65. Since the number of cars entering M.L. King Dr. and Washington St. is 750 and $x + y$ is the number of cars leaving the intersection of M.L. King Dr. and Washington St. then $x + y = 750.$

On the intersection of M.L. King Dr. and JFK Blvd., the number of cars entering this intersection is $450 + x$ and the number of cars leaving is $700 + z$. So $450 + x = 700 + z.$

Simplifying, one gets $y = 750 - x$ and $z = x - 250$; and since y and z are nonnegative,

$250 \le x \le 750$. The values of x, y, and z that realizes this traffic flow must satisfy

$$
\begin{aligned}
y &= 750 - x \\
z &= x - 250 \\
250 \le \quad x &\le 750
\end{aligned}
$$

If $z = 50$, then $50 = x - 250$ or $x = 300$, and $y = 750 - 300 = 450$.

For Thought

1. True

2. False, since the orders of matrices A and C are different.

3. False, $A + B = \begin{bmatrix} 2 \\ 6 \end{bmatrix}$.

4. True, $C + D = \begin{bmatrix} 1-3 & 1+5 \\ 3+1 & 3-2 \end{bmatrix} =$

$= \begin{bmatrix} -2 & 6 \\ 4 & 1 \end{bmatrix}$.

5. True, $A - B = \begin{bmatrix} 1-1 \\ 3-3 \end{bmatrix} = \begin{bmatrix} 0 \\ 0 \end{bmatrix}$.

6. False, $3B = 3 \begin{bmatrix} 1 \\ 3 \end{bmatrix} = \begin{bmatrix} 3 \\ 9 \end{bmatrix}$.

7. False, $-A = -\begin{bmatrix} 1 \\ 3 \end{bmatrix} = \begin{bmatrix} -1 \\ -3 \end{bmatrix}$.

8. False, matrices of different orders cannot be added.   **9.**   False, matrices of different orders cannot be subtracted.

10. False, $C + 2D = \begin{bmatrix} 1 & 1 \\ 3 & 3 \end{bmatrix} + \begin{bmatrix} -6 & 10 \\ 2 & -4 \end{bmatrix} =$

$= \begin{bmatrix} -5 & 11 \\ 5 & -1 \end{bmatrix}$.

6.2 Exercises

1. $x = 2, \; y = 5$

3. Since $2x = 6$ and $4y = 16$, $x = 3$ and $y = 4$. Also $3z = z + y$ and so $z = y/2 = 4/2 = 2$. Then $x = 3$, $y = 4$, and $z = 2$.

5. $\begin{bmatrix} 3+2 \\ 5+1 \end{bmatrix} = \begin{bmatrix} 5 \\ 6 \end{bmatrix}$

7. $\begin{bmatrix} -0.5+2 & -0.03+1 \\ 2-0.05 & -0.33+1 \end{bmatrix} = \begin{bmatrix} 1.5 & 0.97 \\ 1.95 & 0.67 \end{bmatrix}$

9. $\begin{bmatrix} 2+1 & -3-1 & 4+1 \\ 4+0 & -6+1 & 8-1 \\ 6+0 & -3+0 & 1+1 \end{bmatrix} = \begin{bmatrix} 3 & -4 & 5 \\ 4 & -5 & 7 \\ 6 & -3 & 2 \end{bmatrix}$

11. $-A = -\begin{bmatrix} 1 & -4 \\ -5 & 6 \end{bmatrix} = \begin{bmatrix} -1 & 4 \\ 5 & -6 \end{bmatrix}$

and $A + (-A) = \begin{bmatrix} 0 & 0 \\ 0 & 0 \end{bmatrix}$

13. $-A = -\begin{bmatrix} 3 & 0 & -1 \\ 8 & -2 & 1 \\ -3 & 6 & 3 \end{bmatrix} =$

$\begin{bmatrix} -3 & 0 & 1 \\ -8 & 2 & -1 \\ 3 & -6 & -3 \end{bmatrix}$ and

$A + (-A) = \begin{bmatrix} 0 & 0 & 0 \\ 0 & 0 & 0 \\ 0 & 0 & 0 \end{bmatrix}$

15. $B - A = \begin{bmatrix} -1+4 & -2-1 \\ 7-3 & 4-0 \end{bmatrix} = \begin{bmatrix} 3 & -3 \\ 4 & 4 \end{bmatrix}$

17. $B - C = \begin{bmatrix} -1+3 & -2+4 \\ 7-2 & 4+5 \end{bmatrix} = \begin{bmatrix} 2 & 2 \\ 5 & 9 \end{bmatrix}$

19. $B - E$ is undefined since B and E have different sizes

21. $3A = 3 \begin{bmatrix} -4 & 1 \\ 3 & 0 \end{bmatrix} = \begin{bmatrix} -12 & 3 \\ 9 & 0 \end{bmatrix}$

23. $-1D = -1 \begin{bmatrix} -4 \\ 5 \end{bmatrix} = \begin{bmatrix} 4 \\ -5 \end{bmatrix}$

25.

$$3A + 3C = \begin{bmatrix} -12 & 3 \\ 9 & 0 \end{bmatrix} + \begin{bmatrix} -9 & -12 \\ 6 & -15 \end{bmatrix} =$$

$$\begin{bmatrix} -21 & -9 \\ 15 & -15 \end{bmatrix}$$

27.

$$2A - B = \begin{bmatrix} -8 & 2 \\ 6 & 0 \end{bmatrix} - \begin{bmatrix} -1 & -2 \\ 7 & 4 \end{bmatrix} =$$

$$\begin{bmatrix} -7 & 4 \\ -1 & -4 \end{bmatrix}$$

29.

$$2D - 3E = \begin{bmatrix} -8 \\ 10 \end{bmatrix} - \begin{bmatrix} -3 \\ 6 \end{bmatrix} = \begin{bmatrix} -5 \\ 4 \end{bmatrix}$$

31. $D+A$ is undefined since D and A have different sizes

33.

$$(A + B) + C = \begin{bmatrix} -5 & -1 \\ 10 & 4 \end{bmatrix} + \begin{bmatrix} -3 & -4 \\ 2 & -5 \end{bmatrix} =$$

$$\begin{bmatrix} -8 & -5 \\ 12 & -1 \end{bmatrix}$$

35. $\begin{bmatrix} 0.2 + 0.2 & 0.1 + 0.05 \\ 0.4 + 0.3 & 0.3 + 0.8 \end{bmatrix} = \begin{bmatrix} 0.4 & 0.15 \\ 0.7 & 1.1 \end{bmatrix}$

37. $\begin{bmatrix} 1/2 & 3/2 \\ 3 & -12 \end{bmatrix}$

39. $\begin{bmatrix} -2 & 4 \\ 6 & 8 \end{bmatrix} - \begin{bmatrix} -12 & 4 \\ 8 & -8 \end{bmatrix} = \begin{bmatrix} 10 & 0 \\ -2 & 16 \end{bmatrix}$

41. Undefined since we cannot add matrices with different sizes

43. $\begin{bmatrix} -1 & 13 \\ -9 & 3 \\ 6 & -2 \end{bmatrix}$

45. $\begin{bmatrix} 3\sqrt{2} & 2 & 3\sqrt{3} \end{bmatrix}$

47.

$$\begin{bmatrix} 2a \\ 2b \end{bmatrix} + \begin{bmatrix} 6a \\ 12b \end{bmatrix} + \begin{bmatrix} 5a \\ -15b \end{bmatrix} = \begin{bmatrix} 13a \\ -b \end{bmatrix}$$

49.

$$\begin{bmatrix} -0.4x - 0.6x & 0.4y - 0.9y \\ 0.8x - 1.5x & 3.2y + 0.3y \end{bmatrix} =$$

$$\begin{bmatrix} -x & -0.5y \\ -0.7x & 3.5y \end{bmatrix}$$

51.

$$\begin{bmatrix} 2x & 2y & 2z \\ -2x & 4y & 6z \\ 2x & -2y & -6z \end{bmatrix} - \begin{bmatrix} -x & 0 & 3z \\ 4x & y & -z \\ 2x & 5y & z \end{bmatrix} =$$

$$\begin{bmatrix} 3x & 2y & -z \\ -6x & 3y & 7z \\ 0 & -7y & -7z \end{bmatrix}$$

53. Equate the corresponding entries.

$$x + y = 5$$
$$x - y = 1$$

Adding the two equations, we get $2x = 6$. Substitute $x = 3$ into $x+y = 5$. Then $3+y = 5$ and $y = 2$. Solution set is $\{(3, 2)\}$.

55. Equate the corresponding entries.

$$2x + 3y = 7$$
$$x - 4y = -13$$

Multiply second equation by -2 and add to the first one.

$$\begin{array}{rl} 2x + 3y & = 7 \\ -2x + 8y & = 26 \\ \hline 11y & = 33 \\ y & = 3 \end{array}$$

Substitute $y = 3$ into $x - 4y = -13$. Then $x - 12 = -13$ and $x = -1$. The solution set is $\{(-1, 3)\}$.

57. Equate the corresponding entries.

$$x + y + z = 8$$
$$x - y - z = -7$$
$$x - y + z = 2$$

Adding the first and second equations, $2x = 1$ and $x = 0.5$. Multiply second equation by -1 and add to the third.

$$\begin{array}{rl} -x + y + z & = 7 \\ x - y + z & = 2 \\ \hline 2z & = 9 \\ z & = 4.5 \end{array}$$

Substitute $x = 0.5$ and $z = 4.5$ into $x + y + z = 8$. Then $y + 5 = 8$ and $y = 3$. The solution set is $\{(0.5, 3, 4.5)\}$.

59. The matrices for January, February and March are, respectively,

$$J = \begin{bmatrix} 120 \\ 30 \\ 40 \end{bmatrix}, F = \begin{bmatrix} 130 \\ 70 \\ 50 \end{bmatrix}, \text{ and}$$

$$M = \begin{bmatrix} 140 \\ 60 \\ 45 \end{bmatrix}. \text{ The sum}$$

$$J + F + M = \begin{bmatrix} \$390 \\ \$160 \\ \$135 \end{bmatrix} \text{ represents the}$$

total expenses on food, clothing and utilities for the three months.

61. The supply matrix for the first week is

$$S = \begin{bmatrix} 40 & 80 \\ 30 & 90 \\ 80 & 200 \end{bmatrix}. \text{ Next week's supply matrix}$$

is $S + 0.5S = \begin{bmatrix} 40 & 80 \\ 30 & 90 \\ 80 & 200 \end{bmatrix} + \begin{bmatrix} 20 & 40 \\ 15 & 45 \\ 40 & 100 \end{bmatrix}$

$$= \begin{bmatrix} 60 & 120 \\ 45 & 135 \\ 120 & 300 \end{bmatrix}.$$

63. yes, yes

65. yes, yes

67. yes, yes

69. $\begin{bmatrix} 0 & 0 \\ 0 & 0 \end{bmatrix}$

For Thought

1. True   **2.** True   **3.** False, they cannot be multiplied since the number of columns of A is not the same as the number of rows of C.

4. False, the order of CA is 2×1.   **5.** True

6. True, $BC = [7 \cdot 2 + 9 \cdot 4 \quad 7 \cdot 3 + 9 \cdot 5] =$
$= [14 + 36 \quad 21 + 45] = [50 \quad 66].$

7. True, $AB = \begin{bmatrix} 1 \\ 6 \end{bmatrix} [7 \quad 9] = \begin{bmatrix} 1 \cdot 7 & 1 \cdot 9 \\ 6 \cdot 7 & 6 \cdot 9 \end{bmatrix} =$

$$= \begin{bmatrix} 7 & 9 \\ 42 & 54 \end{bmatrix}.$$

8. True, $\begin{bmatrix} 2 & 3 \\ 4 & 5 \end{bmatrix} \begin{bmatrix} 2 & -1 \\ 0 & 3 \end{bmatrix} =$

$$\begin{bmatrix} 4 + 0 & -2 + 9 \\ 8 + 0 & -4 + 15 \end{bmatrix} = \begin{bmatrix} 4 & 7 \\ 8 & 11 \end{bmatrix}.$$

9. True, $BA = [7 \quad 9] \begin{bmatrix} 1 \\ 6 \end{bmatrix} = [7 + 54] = [61].$

10. False, since $EC = \begin{bmatrix} 2 & -1 \\ 0 & 3 \end{bmatrix} \begin{bmatrix} 2 & 3 \\ 4 & 5 \end{bmatrix} =$

$$= \begin{bmatrix} 4 - 4 & 6 - 5 \\ 0 + 12 & 0 + 15 \end{bmatrix} = \begin{bmatrix} 0 & 1 \\ 12 & 15 \end{bmatrix} \text{ and from}$$

Exercise 8 one sees $EC \neq CE$.

6.3 Exercises

1. 3×5 **3.** 1×1 **5.** 5×5

7. 3×3

9. undefined

11. $[-3(4) + 2(1)] = [-10]$

13. $\begin{bmatrix} 1(1) + 3(3) \\ 2(1) + (-4)(3) \end{bmatrix} = \begin{bmatrix} 10 \\ -10 \end{bmatrix}$

15. $\begin{bmatrix} 5(1) + 1(3) & 5(2) + 1(1) \\ 2(1) + 1(3) & 2(2) + 1(1) \end{bmatrix} = \begin{bmatrix} 8 & 11 \\ 5 & 5 \end{bmatrix}$

17. $\begin{bmatrix} 3(5) & 3(6) \\ 1(5) & 1(6) \end{bmatrix} = \begin{bmatrix} 15 & 18 \\ 5 & 6 \end{bmatrix}$

19. $AB = \begin{bmatrix} 1(1) + 3(-1) & 1(0) + 3(1) & 1(1) + 3(0) \\ 2(1) + 4(-1) & 2(0) + 4(1) & 2(1) + 4(0) \\ 5(1) + 6(-1) & 5(0) + 6(1) & 5(1) + 6(0) \end{bmatrix}$

$$= \begin{bmatrix} -2 & 3 & 1 \\ -2 & 4 & 2 \\ -1 & 6 & 5 \end{bmatrix} \text{ and}$$

$$BA = \begin{bmatrix} 1(1) + 0(2) + 1(5) & 1(3) + 0(4) + 1(6) \\ -1(1) + 1(2) + 0(5) & -1(3) + 1(4) + 0(6) \end{bmatrix}$$

$$= \begin{bmatrix} 6 & 9 \\ 1 & 1 \end{bmatrix}$$

21. $AB =$

$$\begin{bmatrix} 1(1) + 2(0) + 3(0) & 1(1) + 2(1) + 3(0) & 1(1) + 2(1) + 3(\\ 2(1) + 1(0) + 3(0) & 2(1) + 1(1) + 3(0) & 2(1) + 1(1) + 3(\\ 3(1) + 2(0) + 1(0) & 3(1) + 2(1) + 1(0) & 3(1) + 2(1) + 1(\end{bmatrix}$$

$$= \begin{bmatrix} 1 & 3 & 6 \\ 2 & 3 & 6 \\ 3 & 5 & 6 \end{bmatrix} \text{ and}$$

$BA =$

$$\begin{bmatrix} 1(1)+1(2)+1(3) & 1(2)+1(1)+1(2) & 1(3)+1(3)+1(1) \\ 0(1)+1(2)+1(3) & 0(2)+1(1)+1(2) & 0(3)+1(3)+1(1) \\ 0(1)+0(2)+1(3) & 0(2)+0(1)+1(2) & 0(3)+0(3)+1(1) \end{bmatrix}$$

$$= \begin{bmatrix} 6 & 5 & 7 \\ 5 & 3 & 4 \\ 3 & 2 & 1 \end{bmatrix}$$

23.

$$AB = \begin{bmatrix} 2\cdot 2 & 2\cdot 3 & 2\cdot 4 \\ -3\cdot 2 & -3\cdot 3 & -3\cdot 4 \\ 1\cdot 2 & 1\cdot 3 & 1\cdot 4 \end{bmatrix} =$$

$$\begin{bmatrix} 4 & 6 & 8 \\ -6 & -9 & -12 \\ 2 & 3 & 4 \end{bmatrix}$$

25. $\begin{bmatrix} 2+0+0 & 2+3+0 & 2+3+4 \end{bmatrix} =$

$\begin{bmatrix} 2 & 5 & 9 \end{bmatrix}$

27. undefined

29.

$$EC = \begin{bmatrix} 2+4+1 & 3+5+0 \\ 0+4+1 & 0+5+0 \\ 0+0+1 & 0+0+0 \end{bmatrix} = \begin{bmatrix} 7 & 8 \\ 5 & 5 \\ 1 & 0 \end{bmatrix}$$

31.

$$DC = \begin{bmatrix} 4-4+1 & 6-5+0 \\ 0+12+2 & 0+15+0 \end{bmatrix} =$$

$$\begin{bmatrix} 1 & 1 \\ 14 & 15 \end{bmatrix}$$

33. undefined

35.

$$EA = \begin{bmatrix} 2-3+1 \\ 0-3+1 \\ 0+0+1 \end{bmatrix} = \begin{bmatrix} 0 \\ -2 \\ 1 \end{bmatrix}$$

37. undefined

39. We will use the answer in Exercise 23.

$AB + 2E =$

$$\begin{bmatrix} 4 & 6 & 8 \\ -6 & -9 & -12 \\ 2 & 3 & 4 \end{bmatrix} + \begin{bmatrix} 2 & 2 & 2 \\ 0 & 2 & 2 \\ 0 & 0 & 2 \end{bmatrix}$$

$$= \begin{bmatrix} 6 & 8 & 10 \\ -6 & -7 & -10 \\ 2 & 3 & 6 \end{bmatrix}$$

41. $A^2 = \begin{bmatrix} 1 & 0 \\ 1 & 1 \end{bmatrix} \begin{bmatrix} 1 & 0 \\ 1 & 1 \end{bmatrix} =$

$$\begin{bmatrix} 1(1)+0(1) & 1(0)+0(1) \\ 1(1)+1(1) & 1(0)+1(1) \end{bmatrix} = \begin{bmatrix} 1 & 0 \\ 2 & 1 \end{bmatrix}$$

43. We will use the answer in Exercise 42.

$A^4 = A^3 A = \begin{bmatrix} 1 & 0 \\ 3 & 1 \end{bmatrix} \begin{bmatrix} 1 & 0 \\ 1 & 1 \end{bmatrix} =$

$$\begin{bmatrix} 1(1)+0(1) & 1(0)+0(1) \\ 3(1)+1(1) & 3(0)+1(1) \end{bmatrix} = \begin{bmatrix} 1 & 0 \\ 4 & 1 \end{bmatrix}$$

45.

$$\begin{bmatrix} 2\cdot 1+0\cdot 0 & 2\cdot 1+0\cdot 1 \\ 3\cdot 1+1\cdot 0 & 3\cdot 1+1\cdot 1 \end{bmatrix} = \begin{bmatrix} 2 & 2 \\ 3 & 4 \end{bmatrix}$$

47.

$$\begin{bmatrix} 7\cdot 3+4\cdot(-5) & 7\cdot(-4)+4\cdot 7 \\ 5\cdot 3+3\cdot(-5) & 5\cdot(-4)+3\cdot 7 \end{bmatrix} =$$

$$\begin{bmatrix} 1 & 0 \\ 0 & 1 \end{bmatrix}$$

49.

$$\begin{bmatrix} -0.5\cdot 1+4\cdot 0 & -0.5\cdot 0+4\cdot 1 \\ 9\cdot 1+0.7\cdot 0 & 9\cdot 0+0.7\cdot 1 \end{bmatrix} =$$

$$\begin{bmatrix} -0.5 & 4 \\ 9 & 0.7 \end{bmatrix}$$

51. $[-2a+6a \quad -6b+3b] = [4a \quad -3b]$

53.

$$\begin{bmatrix} -2a+0\cdot 1 & 5a+0\cdot 4 & 3a+0\cdot 6 \\ 0\cdot(-2)+b & 0\cdot 5+4b & 0\cdot 3+6b \end{bmatrix} =$$

$$\begin{bmatrix} -2a & 5a & 3a \\ b & 4b & 6b \end{bmatrix}$$

55. $[1\cdot 1+2\cdot 0+3\cdot 1 \quad 1\cdot 0+2\cdot 1+3\cdot 0 \quad 1\cdot 1+2\cdot 1+3\cdot 1]$
$= [4 \quad 2 \quad 6]$

57. $[-1 \cdot (-5) + 0 \cdot 1 + 3 \cdot 4] = [17]$

59.
$$\begin{bmatrix} x^2 & xy \\ xy & y^2 \end{bmatrix}$$

61.
$$\begin{bmatrix} -1 \cdot \sqrt{2} + 2 \cdot 0 + 3\sqrt{2} \\ 3 \cdot \sqrt{2} + 4 \cdot 0 + 4\sqrt{2} \end{bmatrix} = \begin{bmatrix} 2\sqrt{2} \\ 7\sqrt{2} \end{bmatrix}$$

63.
$$\begin{bmatrix} (1/2) \cdot (-8) + (1/3) \cdot (-5) & 6 + (1/3) \cdot 15 \\ (1/4) \cdot (-8) + (1/5) \cdot (-5) & 3 + (1/5) \cdot 15 \end{bmatrix}$$
$$= \begin{bmatrix} -4 - (5/3) & 6 + 5 \\ -2 - 1 & 3 + 3 \end{bmatrix} = \begin{bmatrix} -17/3 & 11 \\ -3 & 6 \end{bmatrix}$$

65. undefined

67.
$$\begin{bmatrix} 9+0-7 & 8+0-8 & 10+0-4 \\ 0+3+0 & 0+5+0 & 0+2+0 \\ 9+3+7 & 8+5+8 & 10+2+4 \end{bmatrix} =$$
$$\begin{bmatrix} 2 & 0 & 6 \\ 3 & 5 & 2 \\ 19 & 21 & 16 \end{bmatrix}$$

69.
$$\begin{bmatrix} 1-0.6-0.6 & 1.5+1.2-1 \\ 0.8+0.4+1.8 & 1.2-0.8+3 \\ 0.4+0.6-2.4 & 0.6-1.2-4 \end{bmatrix} =$$
$$\begin{bmatrix} -0.2 & 1.7 \\ 3 & 3.4 \\ -1.4 & -4.6 \end{bmatrix}$$

71. System of equations is
$$\begin{aligned} 2x - 3y &= 0 \\ x + 2y &= 7. \end{aligned}$$

Multiply second equation by -2 and add to the first one.
$$\begin{aligned} 2x - 3y &= 0 \\ -2x - 4y &= -14 \\ \hline -7y &= -14 \end{aligned}$$

Substitute $y = 2$ into $x + 2y = 7$. Then $x + 4 = 7$ and $x = 3$. Solution set is $\{(3,2)\}$.

73. System of equations is
$$\begin{aligned} 2x + 3y &= 5 \\ 4x + 6y &= 9 \end{aligned}$$

Multiply first equation by -2 and add to the second one.
$$\begin{aligned} -4x - 6y &= -10 \\ 4x + 6y &= 9 \\ \hline 0 &= -1 \end{aligned}$$

Inconsistent and the solution set is $\emptyset$.

75. System of equations is
$$\begin{aligned} x + y + z &= 4 \\ y + z &= 5 \\ z &= 6. \end{aligned}$$

Substitute $z = 6$ into $y + z = 5$ to get $y + 6 = 5$ and $y = -1$. From $x + y + z = 4$, we have $x - 1 + 6 = 4$ and $x = -1$. Solution set is $\{(-1, -1, 6)\}$.

77.
$$\begin{bmatrix} 2 & 3 \\ 4 & -1 \end{bmatrix} \begin{bmatrix} x \\ y \end{bmatrix} = \begin{bmatrix} 9 \\ 6 \end{bmatrix}$$

79.
$$\begin{bmatrix} 1 & 2 & -1 \\ 3 & -1 & 3 \\ 2 & 1 & -4 \end{bmatrix} \begin{bmatrix} x \\ y \\ z \end{bmatrix} = \begin{bmatrix} 3 \\ 1 \\ 0 \end{bmatrix}$$

81.
$$A = \begin{bmatrix} \$24,000 & \$40,000 \\ \$38,000 & \$70,000 \end{bmatrix}, Q = \begin{bmatrix} 4 \\ 7 \end{bmatrix}, \text{ and}$$

matrix product $AQ = \begin{bmatrix} \$376,000 \\ \$642,000 \end{bmatrix}$

represents the costs for labor and material for building 4 economy houses and 7 deluxe models.

83. False

85. True

87. True

For Thought

1. True, $AB = \begin{bmatrix} 10-9 & -6+6 \\ 15-15 & -9+10 \end{bmatrix} =$

$= \begin{bmatrix} 1 & 0 \\ 0 & 1 \end{bmatrix} = \begin{bmatrix} 10-9 & 15-15 \\ -6+6 & -9+10 \end{bmatrix} = BA.$

2. True, $AB = BA = I$ by Exercise 1.

3. True, by Exercise 2 the inverse of B is A.

4. False, AC is undefined, although CA is defined.　　**5.**　True

6. False, a non-square matrix has no inverse.

7. False, the coefficient matrix is $\begin{bmatrix} 2 & 3 \\ 3 & 1 \end{bmatrix}$.

8. True, since $A^{-1} = B$ then $A^{-1}D =$

$= \begin{bmatrix} 5 & -3 \\ -3 & 2 \end{bmatrix}\begin{bmatrix} 11 \\ 19 \end{bmatrix} = \begin{bmatrix} -2 \\ 5 \end{bmatrix}.$

9. False, $(-2, 5)$ does not satisfy $3x + y = 19$.

10. False, the solution is $\begin{bmatrix} 2 & 3 \\ 3 & 1 \end{bmatrix}^{-1}\begin{bmatrix} 3 \\ -7 \end{bmatrix}.$

6.4 Exercises

1.

$AI = \begin{bmatrix} 1 & 3 \\ 4 & 6 \end{bmatrix}\begin{bmatrix} 1 & 0 \\ 0 & 1 \end{bmatrix}$

$= \begin{bmatrix} 1(1) + 3(0) & 1(0) + 3(1) \\ 4(1) + 6(0) & 4(0) + 6(1) \end{bmatrix}$

$= \begin{bmatrix} 1 & 3 \\ 4 & 6 \end{bmatrix}$

$= A$

and

$IA = \begin{bmatrix} 1 & 0 \\ 0 & 1 \end{bmatrix}\begin{bmatrix} 1 & 3 \\ 4 & 6 \end{bmatrix}$

$= \begin{bmatrix} 1(1) + 0(4) & 1(3) + 0(6) \\ 0(1) + 1(4) & 0(3) + 1(6) \end{bmatrix}$

$= \begin{bmatrix} 1 & 3 \\ 4 & 6 \end{bmatrix}$

$= A$

3. $AI = \begin{bmatrix} 3 & 2 & 1 \\ 5 & 6 & 2 \\ 7 & 8 & 3 \end{bmatrix}\begin{bmatrix} 1 & 0 & 0 \\ 0 & 1 & 0 \\ 0 & 0 & 1 \end{bmatrix} =$

$\begin{bmatrix} 3(1) + 2(0) + 1(0) & 3(0) + 2(1) + 1(0) & 3(0) + 2(0) + 1(1) \\ 5(1) + 6(0) + 2(0) & 5(0) + 6(1) + 2(0) & 5(0) + 6(0) + 2(1) \\ 7(1) + 8(0) + 3(0) & 7(0) + 8(1) + 3(0) & 7(0) + 8(0) + 3(1) \end{bmatrix}$

$= \begin{bmatrix} 3 & 2 & 1 \\ 5 & 6 & 2 \\ 7 & 8 & 3 \end{bmatrix} = A$

and

$IA = \begin{bmatrix} 1 & 0 & 0 \\ 0 & 1 & 0 \\ 0 & 0 & 1 \end{bmatrix}\begin{bmatrix} 3 & 2 & 1 \\ 5 & 6 & 2 \\ 7 & 8 & 3 \end{bmatrix} =$

$\begin{bmatrix} 1(3) + 0(5) + 0(7) & 1(2) + 0(6) + 0(8) & 1(1) + 0(2) + 0(3) \\ 0(3) + 1(5) + 0(7) & 0(2) + 1(6) + 0(8) & 0(1) + 1(2) + 0(3) \\ 0(3) + 0(5) + 1(7) & 0(2) + 0(6) + 1(8) & 0(1) + 0(2) + 1(3) \end{bmatrix}$

$= \begin{bmatrix} 3 & 2 & 1 \\ 5 & 6 & 2 \\ 7 & 8 & 3 \end{bmatrix} = A$

5.

$I\begin{bmatrix} -3 & 5 \\ 12 & 6 \end{bmatrix} = \begin{bmatrix} -3 & 5 \\ 12 & 6 \end{bmatrix}$

7.

$\begin{bmatrix} -8+9 & -6+6 \\ 12-12 & 9-8 \end{bmatrix} = \begin{bmatrix} 1 & 0 \\ 0 & 1 \end{bmatrix}$

9.

$\begin{bmatrix} 5-4 & -4+4 \\ 5-5 & -4+5 \end{bmatrix} = \begin{bmatrix} 1 & 0 \\ 0 & 1 \end{bmatrix}$

11.

$\begin{bmatrix} 3 & 5 & 1 \\ 4 & 5 & 7 \\ 4 & 9 & 2 \end{bmatrix}I = \begin{bmatrix} 3 & 5 & 1 \\ 4 & 5 & 7 \\ 4 & 9 & 2 \end{bmatrix}$

13.

$\begin{bmatrix} 0+0+1 & 1+0-1 & -3+0+3 \\ 0+0+0 & 1+0+0 & -3+3+0 \\ 0+0+0 & 0+0+0 & 0+1+0 \end{bmatrix} =$

$\begin{bmatrix} 1 & 0 & 0 \\ 0 & 1 & 0 \\ 0 & 0 & 1 \end{bmatrix}$

15.

$$\begin{bmatrix} 0.5+0.5+0 & -0.5+0.5+0 & 0.5-0.5+0 \\ 0+0.5-0.5 & 0+0.5+0.5 & 0-0.5+0.5 \\ 0.5+0-0.5 & -0.5+0+0.5 & 0.5+0+0.5 \end{bmatrix} =$$

$$\begin{bmatrix} 1 & 0 & 0 \\ 0 & 1 & 0 \\ 0 & 0 & 1 \end{bmatrix}$$

17.

Yes, since $\begin{bmatrix} 3 & 1 \\ 11 & 4 \end{bmatrix}\begin{bmatrix} 4 & -1 \\ -11 & 3 \end{bmatrix} =$

$$\begin{bmatrix} 12-11 & -3+3 \\ 44-44 & -11+12 \end{bmatrix} = \begin{bmatrix} 1 & 0 \\ 0 & 1 \end{bmatrix} \text{ and}$$

similarly $\begin{bmatrix} 4 & -1 \\ -11 & 3 \end{bmatrix}\begin{bmatrix} 3 & 1 \\ 11 & 4 \end{bmatrix} = I.$

19.

No, since $\begin{bmatrix} 1/2 & -1 \\ 3 & -12 \end{bmatrix}\begin{bmatrix} 4 & 2 \\ 1 & 1 \end{bmatrix} =$

$$\begin{bmatrix} 2-1 & 1-1 \\ 12-12 & 6-12 \end{bmatrix} = \begin{bmatrix} 1 & 0 \\ 0 & -6 \end{bmatrix} \neq I.$$

21. No, since only square matrices may have inverses.

23.

On $\begin{bmatrix} 1 & 4 & | & 1 & 0 \\ 0 & 2 & | & 0 & 1 \end{bmatrix}$, use $-2R_2 + R_1 \to R_1$ to get

$$\begin{bmatrix} 1 & 0 & | & 1 & -2 \\ 0 & 2 & | & 0 & 1 \end{bmatrix}, \text{ use } \frac{1}{2}R_2 \to R_2 \text{ to get}$$

$$\begin{bmatrix} 1 & 0 & | & 1 & -2 \\ 0 & 1 & | & 0 & 1/2 \end{bmatrix}. \text{ Then } A^{-1} = \begin{bmatrix} 1 & -2 \\ 0 & 1/2 \end{bmatrix}.$$

25.

On $\begin{bmatrix} 1 & 6 & | & 1 & 0 \\ 1 & 9 & | & 0 & 1 \end{bmatrix}$, use $-1R_1 + R_2 \to R_2$ to get

$$\begin{bmatrix} 1 & 6 & | & 1 & 0 \\ 0 & 3 & | & -1 & 1 \end{bmatrix}, \text{ use } -2R_2 + R_1 \to R_1 \text{ to get}$$

$$\begin{bmatrix} 1 & 0 & | & 3 & -2 \\ 0 & 3 & | & -1 & 1 \end{bmatrix}, \text{ use } \frac{1}{3}R_2 \to R_2 \text{ to get}$$

$$\begin{bmatrix} 1 & 0 & | & 3 & -2 \\ 0 & 1 & | & -1/3 & 1/3 \end{bmatrix}.$$

Thus, $A^{-1} = \begin{bmatrix} 3 & -2 \\ -1/3 & 1/3 \end{bmatrix}.$

27.

On $\begin{bmatrix} -2 & -3 & | & 1 & 0 \\ 3 & 4 & | & 0 & 1 \end{bmatrix}$, use $R_2 + R_1 \to R_1$ to get

$$\begin{bmatrix} 1 & 1 & | & 1 & 1 \\ 3 & 4 & | & 0 & 1 \end{bmatrix}, \text{ use } -3R_1 + R_2 \to R_2 \text{ to get}$$

$$\begin{bmatrix} 1 & 1 & | & 1 & 1 \\ 0 & 1 & | & -3 & -2 \end{bmatrix}, \text{ use } -1R_2 + R_1 \to R_1 \text{ to get}$$

$$\begin{bmatrix} 1 & 0 & | & 4 & 3 \\ 0 & 1 & | & -3 & -2 \end{bmatrix}. \text{ So } A^{-1} = \begin{bmatrix} 4 & 3 \\ -3 & -2 \end{bmatrix}.$$

29.

On $\begin{bmatrix} 1 & -5 & | & 1 & 0 \\ -1 & 3 & | & 0 & 1 \end{bmatrix}$, use $R_1 + R_2 \to R_2$ to get

$$\begin{bmatrix} 1 & -5 & | & 1 & 0 \\ 0 & -2 & | & 1 & 1 \end{bmatrix}, \text{ use } -\frac{1}{2}R_2 \to R_2 \text{ to get}$$

$$\begin{bmatrix} 1 & -5 & | & 1 & 0 \\ 0 & 1 & | & -1/2 & -1/2 \end{bmatrix}, \text{ use}$$

$5R_2 + R_1 \to R_1$ to get

$$\begin{bmatrix} 1 & 0 & | & -3/2 & -5/2 \\ 0 & 1 & | & -1/2 & -1/2 \end{bmatrix}.$$

Then $A^{-1} = \begin{bmatrix} -3/2 & -5/2 \\ -1/2 & -1/2 \end{bmatrix}.$

31.

On $\begin{bmatrix} -1 & 5 & | & 1 & 0 \\ 2 & -10 & | & 0 & 1 \end{bmatrix}$, use $R_1 + R_2 \to R_1$

and $\frac{1}{2}R_2 \to R_2$ to get

$$\begin{bmatrix} 1 & -5 & | & 1 & 1 \\ 1 & -5 & | & 0 & 1/2 \end{bmatrix}, \text{ use } -1R_2 + R_1 \to R_1 \text{ to get}$$

$$\begin{bmatrix} 0 & 0 & | & 1 & 1/2 \\ 1 & -5 & | & 0 & 1/2 \end{bmatrix}. \text{ So } A \text{ has no inverse.}$$

33.

On $\begin{bmatrix} 1 & 1 & 0 & | & 1 & 0 & 0 \\ 0 & -1 & -1 & | & 0 & 1 & 0 \\ 1 & 0 & -1 & | & 0 & 0 & 1 \end{bmatrix}$, use

$R_2 + R_1 \to R_1$ and $-1R_3 + R_1 \to R_3$ to get

$$\begin{bmatrix} 1 & 0 & -1 & | & 1 & 1 & 0 \\ 0 & -1 & -1 & | & 0 & 1 & 0 \\ 0 & 1 & 1 & | & 1 & 0 & -1 \end{bmatrix}, \text{ use}$$

$R_2 + R_3 \to R_2$ to get

$$\left[\begin{array}{ccc|ccc} 1 & 0 & -1 & 1 & 1 & 0 \\ 0 & 0 & 0 & 1 & 1 & -1 \\ 0 & 1 & 1 & 1 & 0 & 1 \end{array}\right]. \text{ So } A^{-1} \text{ does not}$$

exist.

35.

On $\left[\begin{array}{ccc|ccc} 1 & 1 & 1 & 1 & 0 & 0 \\ 1 & -1 & -1 & 0 & 1 & 0 \\ 1 & -1 & 1 & 0 & 0 & 1 \end{array}\right]$, use

$R_1 + R_2 \to R_2$ and $R_2 + (-1R_3) \to R_3$ to get

$$\left[\begin{array}{ccc|ccc} 1 & 1 & 1 & 1 & 0 & 0 \\ 2 & 0 & 0 & 1 & 1 & 0 \\ 0 & 0 & -2 & 0 & 1 & -1 \end{array}\right], \text{ use}$$

$-2R_1 + R_2 \to R_2$ and $-\dfrac{1}{2}R_3 \to R_3$ to get

$$\left[\begin{array}{ccc|ccc} 1 & 1 & 1 & 1 & 0 & 0 \\ 0 & -2 & -2 & -1 & 1 & 0 \\ 0 & 0 & 1 & 0 & -1/2 & 1/2 \end{array}\right], \text{ use}$$

$-\dfrac{1}{2}R_2 \to R_2$ to get

$$\left[\begin{array}{ccc|ccc} 1 & 1 & 1 & 1 & 0 & 0 \\ 0 & 1 & 1 & 1/2 & -1/2 & 0 \\ 0 & 0 & 1 & 0 & -1/2 & 1/2 \end{array}\right], \text{ use}$$

$-1R_2 + R_1 \to R_1$ to get

$$\left[\begin{array}{ccc|ccc} 1 & 0 & 0 & 1/2 & 1/2 & 0 \\ 0 & 1 & 1 & 1/2 & -1/2 & 0 \\ 0 & 0 & 1 & 0 & -1/2 & 1/2 \end{array}\right], \text{ use}$$

$-1R_3 + R_2 \to R_2$ to get

$$\left[\begin{array}{ccc|ccc} 1 & 0 & 0 & 1/2 & 1/2 & 0 \\ 0 & 1 & 0 & 1/2 & 0 & -1/2 \\ 0 & 0 & 1 & 0 & -1/2 & 1/2 \end{array}\right].$$

So $A^{-1} = \left[\begin{array}{ccc} 1/2 & 1/2 & 0 \\ 1/2 & 0 & -1/2 \\ 0 & -1/2 & 1/2 \end{array}\right].$

37.

On $\left[\begin{array}{ccc|ccc} 0 & 2 & 0 & 1 & 0 & 0 \\ 3 & 3 & 2 & 0 & 1 & 0 \\ 2 & 5 & 1 & 0 & 0 & 1 \end{array}\right]$, use

$-1R_3 + R_2 \to R_3$ to get

$$\left[\begin{array}{ccc|ccc} 0 & 2 & 0 & 1 & 0 & 0 \\ 3 & 3 & 2 & 0 & 1 & 0 \\ 1 & -2 & 1 & 0 & 1 & -1 \end{array}\right], \text{ use}$$

$R_3 \to R_1$ and $R_1 \to R_3$ to get

$$\left[\begin{array}{ccc|ccc} 1 & -2 & 1 & 0 & 1 & -1 \\ 3 & 3 & 2 & 0 & 1 & 0 \\ 0 & 2 & 0 & 1 & 0 & 0 \end{array}\right], \text{ use}$$

$R_3 + R_1 \to R_1$ and $-1R_3 + R_2 \to R_2$ to get

$$\left[\begin{array}{ccc|ccc} 1 & 0 & 1 & 1 & 1 & -1 \\ 3 & 1 & 2 & -1 & 1 & 0 \\ 0 & 2 & 0 & 1 & 0 & 0 \end{array}\right], \text{ use}$$

$-3R_1 + R_2 \to R_2$ to get

$$\left[\begin{array}{ccc|ccc} 1 & 0 & 1 & 1 & 1 & -1 \\ 0 & 1 & -1 & -4 & -2 & 3 \\ 0 & 2 & 0 & 1 & 0 & 0 \end{array}\right], \text{ use}$$

$-2R_2 + R_3 \to R_3$ to get

$$\left[\begin{array}{ccc|ccc} 1 & 0 & 1 & 1 & 1 & -1 \\ 0 & 1 & -1 & -4 & -2 & 3 \\ 0 & 0 & 2 & 9 & 4 & -6 \end{array}\right], \text{ use}$$

$\dfrac{1}{2}R_3 \to R_3$ to get

$$\left[\begin{array}{ccc|ccc} 1 & 0 & 1 & 1 & 1 & -1 \\ 0 & 1 & -1 & -4 & -2 & 3 \\ 0 & 0 & 1 & 9/2 & 2 & -3 \end{array}\right], \text{ use}$$

$R_2 + R_3 \to R_2$ to get

$$\left[\begin{array}{ccc|ccc} 1 & 0 & 1 & 1 & 1 & -1 \\ 0 & 1 & 0 & 1/2 & 0 & 0 \\ 0 & 0 & 1 & 9/2 & 2 & -3 \end{array}\right], \text{ use}$$

$-1R_3 + R_1 \to R_1$ to get

$$\left[\begin{array}{ccc|ccc} 1 & 0 & 0 & -7/2 & -1 & 2 \\ 0 & 1 & 0 & 1/2 & 0 & 0 \\ 0 & 0 & 1 & 9/2 & 2 & -3 \end{array}\right].$$

So $A^{-1} = \left[\begin{array}{ccc} -7/2 & -1 & 2 \\ 1/2 & 0 & 0 \\ 9/2 & 2 & -3 \end{array}\right].$

39.

On $\left[\begin{array}{ccc|ccc} 1 & 0 & 1 & 1 & 0 & 0 \\ 0 & 2 & 2 & 0 & 1 & 0 \\ 2 & 1 & 0 & 0 & 0 & 1 \end{array}\right]$, use

$2R_1 + (-1R_3) \to R_3$ and $\dfrac{1}{2}R_2 \to R_2$ to get

$$\left[\begin{array}{ccc|ccc} 1 & 0 & 1 & 1 & 0 & 0 \\ 0 & 1 & 1 & 0 & 1/2 & 0 \\ 0 & -1 & 2 & 2 & 0 & -1 \end{array}\right], \text{ use}$$

$R_2 + R_3 \to R_3$ to get

$$\left[\begin{array}{ccc|ccc} 1 & 0 & 1 & 1 & 0 & 0 \\ 0 & 1 & 1 & 0 & 1/2 & 0 \\ 0 & 0 & 3 & 2 & 1/2 & -1 \end{array}\right], \text{ use}$$

$\frac{1}{3}R_3 \to R_3$ to get

$$\left[\begin{array}{ccc|ccc} 1 & 0 & 1 & 1 & 0 & 0 \\ 0 & 1 & 1 & 0 & 1/2 & 0 \\ 0 & 0 & 1 & 2/3 & 1/6 & -1/3 \end{array}\right], \text{ use}$$

$-1R_3 + R_2 \to R_2$ and $-1R_3 + R_1 \to R_1$ to get

$$\left[\begin{array}{ccc|ccc} 1 & 0 & 0 & 1/3 & -1/6 & 1/3 \\ 0 & 1 & 0 & -2/3 & 1/3 & 1/3 \\ 0 & 0 & 1 & 2/3 & 1/6 & -1/3 \end{array}\right].$$

Then $A^{-1} = \left[\begin{array}{ccc} 1/3 & -1/6 & 1/3 \\ -2/3 & 1/3 & 1/3 \\ 2/3 & 1/6 & -1/3 \end{array}\right].$

41.

On $\left[\begin{array}{ccc|ccc} 0 & 4 & 2 & 1 & 0 & 0 \\ 0 & 3 & 2 & 0 & 1 & 0 \\ 1 & -1 & 1 & 0 & 0 & 1 \end{array}\right]$, use

$R_1 \to R_3$ and $R_3 \to R_1$ to get

$$\left[\begin{array}{ccc|ccc} 1 & -1 & 1 & 0 & 0 & 1 \\ 0 & 3 & 2 & 0 & 1 & 0 \\ 0 & 4 & 2 & 1 & 0 & 0 \end{array}\right], \text{ use}$$

$-1R_2 + R_3 \to R_2$ to get

$$\left[\begin{array}{ccc|ccc} 1 & -1 & 1 & 0 & 0 & 1 \\ 0 & 1 & 0 & 1 & -1 & 0 \\ 0 & 4 & 2 & 1 & 0 & 0 \end{array}\right], \text{ use}$$

$R_1 + R_2 \to R_1$ to get

$$\left[\begin{array}{ccc|ccc} 1 & 0 & 1 & 1 & -1 & 1 \\ 0 & 1 & 0 & 1 & -1 & 0 \\ 0 & 4 & 2 & 1 & 0 & 0 \end{array}\right], \text{ use}$$

$-4R_2 + R_3 \to R_3$ to get

$$\left[\begin{array}{ccc|ccc} 1 & 0 & 1 & 1 & -1 & 1 \\ 0 & 1 & 0 & 1 & -1 & 0 \\ 0 & 0 & 2 & -3 & 4 & 0 \end{array}\right], \text{ use}$$

$\frac{1}{2}R_3 \to R_3$ to get

$$\left[\begin{array}{ccc|ccc} 1 & 0 & 1 & 1 & -1 & 1 \\ 0 & 1 & 0 & 1 & -1 & 0 \\ 0 & 0 & 1 & -3/2 & 2 & 0 \end{array}\right], \text{ use}$$

$-1R_3 + R_1 \to R_1$ to get

$$\left[\begin{array}{ccc|ccc} 1 & 0 & 0 & 5/2 & -3 & 1 \\ 0 & 1 & 0 & 1 & -1 & 0 \\ 0 & 0 & 1 & -3/2 & 2 & 0 \end{array}\right].$$

Then $A^{-1} = \left[\begin{array}{ccc} 5/2 & -3 & 1 \\ 1 & -1 & 0 \\ -3/2 & 2 & 0 \end{array}\right].$

43.

On $\left[\begin{array}{cccc|cccc} 1 & 2 & 3 & 4 & 1 & 0 & 0 & 0 \\ 0 & 1 & 2 & 3 & 0 & 1 & 0 & 0 \\ 0 & 0 & 1 & 2 & 0 & 0 & 1 & 0 \\ 0 & 0 & 0 & 1 & 0 & 0 & 0 & 1 \end{array}\right]$, use

$-2R_2 + R_1 \to R_1$ to get

$$\left[\begin{array}{cccc|cccc} 1 & 0 & -1 & -2 & 1 & -2 & 0 & 0 \\ 0 & 1 & 2 & 3 & 0 & 1 & 0 & 0 \\ 0 & 0 & 1 & 2 & 0 & 0 & 1 & 0 \\ 0 & 0 & 0 & 1 & 0 & 0 & 0 & 1 \end{array}\right], \text{ use}$$

$-2R_4 + R_3 \to R_3$ to get

$$\left[\begin{array}{cccc|cccc} 1 & 0 & -1 & -2 & 1 & -2 & 0 & 0 \\ 0 & 1 & 2 & 3 & 0 & 1 & 0 & 0 \\ 0 & 0 & 1 & 0 & 0 & 0 & 1 & -2 \\ 0 & 0 & 0 & 1 & 0 & 0 & 0 & 1 \end{array}\right], \text{ use}$$

$-2R_3 + R_2 \to R_2$ and $-2R_4 + R_3 \to R_3$ to get

$$\left[\begin{array}{cccc|cccc} 1 & 0 & -1 & -2 & 1 & -2 & 0 & 0 \\ 0 & 1 & 0 & 3 & 0 & 1 & -2 & 4 \\ 0 & 0 & 1 & -2 & 0 & 0 & 1 & -4 \\ 0 & 0 & 0 & 1 & 0 & 0 & 0 & 1 \end{array}\right], \text{ use}$$

$R_1 + R_3 \to R_1$ to get

$$\left[\begin{array}{cccc|cccc} 1 & 0 & 0 & -4 & 1 & -2 & 1 & -4 \\ 0 & 1 & 0 & 3 & 0 & 1 & -2 & 4 \\ 0 & 0 & 1 & -2 & 0 & 0 & 1 & -4 \\ 0 & 0 & 0 & 1 & 0 & 0 & 0 & 1 \end{array}\right], \text{ use}$$

$2R_4 + R_3 \to R_3$, $-3R_4 + R_2 \to R_2$ and $4R_4 + R_1 \to R_1$ to get

$$\left[\begin{array}{cccc|cccc} 1 & 0 & 0 & 0 & 1 & -2 & 1 & 0 \\ 0 & 1 & 0 & 0 & 0 & 1 & -2 & 1 \\ 0 & 0 & 1 & 0 & 0 & 0 & 1 & -2 \\ 0 & 0 & 0 & 1 & 0 & 0 & 0 & 1 \end{array}\right].$$

So $A^{-1} = \left[\begin{array}{cccc} 1 & -2 & 1 & 0 \\ 0 & 1 & -2 & 1 \\ 0 & 0 & 1 & -2 \\ 0 & 0 & 0 & 1 \end{array}\right].$

45.

Since the coefficient matrix is $A = \left[\begin{array}{cc} 1 & 6 \\ 1 & 9 \end{array}\right]$,

$A^{-1}\left[\begin{array}{c} -3 \\ -6 \end{array}\right] = \left[\begin{array}{cc} 3 & -2 \\ -1/3 & 1/3 \end{array}\right]\left[\begin{array}{c} -3 \\ -6 \end{array}\right] =$

$= \left[\begin{array}{c} 3 \\ -1 \end{array}\right].$ The solution set is $\{(3, -1)\}$.

47.

Since the coefficient matrix is $A = \begin{bmatrix} 1 & 6 \\ 1 & 9 \end{bmatrix}$,

$$A^{-1}\begin{bmatrix} 4 \\ 5 \end{bmatrix} = \begin{bmatrix} 3 & -2 \\ -1/3 & 1/3 \end{bmatrix}\begin{bmatrix} 4 \\ 5 \end{bmatrix} =$$

$$= \begin{bmatrix} 2 \\ 1/3 \end{bmatrix}. \text{ The solution set is } \{(2, 1/3)\}.$$

49.

Since the coefficient matrix is

$A = \begin{bmatrix} -2 & -3 \\ 3 & 4 \end{bmatrix}$, we get

$$A^{-1}\begin{bmatrix} 1 \\ -1 \end{bmatrix} = \begin{bmatrix} 4 & 3 \\ -3 & -2 \end{bmatrix}\begin{bmatrix} 1 \\ -1 \end{bmatrix} =$$

$$= \begin{bmatrix} 1 \\ -1 \end{bmatrix}. \text{ The solution set is } \{(1, -1)\}.$$

51.

Since the coefficient matrix is

$A = \begin{bmatrix} 1 & -5 \\ -1 & 3 \end{bmatrix}$, we obtain

$$A^{-1}\begin{bmatrix} -5 \\ 1 \end{bmatrix} = \begin{bmatrix} -3/2 & -5/2 \\ -1/2 & -1/2 \end{bmatrix}\begin{bmatrix} -5 \\ 1 \end{bmatrix} =$$

$$= \begin{bmatrix} 5 \\ 2 \end{bmatrix}. \text{ The solution set is } \{(5, 2)\}.$$

53.

Since coefficient matrix is

$A = \begin{bmatrix} 1 & 1 & 1 \\ 1 & -1 & -1 \\ 1 & -1 & 1 \end{bmatrix}$, we find $A^{-1}\begin{bmatrix} 3 \\ -1 \\ 5 \end{bmatrix} =$

$$= \begin{bmatrix} 1/2 & 1/2 & 0 \\ 1/2 & 0 & -1/2 \\ 0 & -1/2 & 1/2 \end{bmatrix}\begin{bmatrix} 3 \\ -1 \\ 5 \end{bmatrix} = \begin{bmatrix} 1 \\ -1 \\ 3 \end{bmatrix}.$$

The solution set is $\{(1, -1, 3)\}$.

55.

Since the coefficient matrix is

$A = \begin{bmatrix} 0 & 2 & 0 \\ 3 & 3 & 2 \\ 2 & 5 & 1 \end{bmatrix}$, we get $A^{-1}\begin{bmatrix} 6 \\ 16 \\ 19 \end{bmatrix} =$

$$= \begin{bmatrix} -7/2 & -1 & 2 \\ 1/2 & 0 & 0 \\ 9/2 & 2 & -3 \end{bmatrix}\begin{bmatrix} 6 \\ 16 \\ 19 \end{bmatrix} = \begin{bmatrix} 1 \\ 3 \\ 2 \end{bmatrix}.$$

The solution set is $\{(1, 3, 2)\}$.

57. The coefficient matrix of

$$0.3x + 0.1y = 3$$
$$2x + 4y = 7$$

is $A = \begin{bmatrix} 0.3 & 0.1 \\ 2 & 4 \end{bmatrix}$. Note $A^{-1}\begin{bmatrix} 3 \\ 7 \end{bmatrix} =$

$$\begin{bmatrix} 4 & -0.1 \\ -2 & 0.3 \end{bmatrix}\begin{bmatrix} 3 \\ 7 \end{bmatrix} = \begin{bmatrix} 11.3 \\ -3.9 \end{bmatrix}.$$

The solution set is $\{(11.3, -3.9)\}$.

59. Use the Gauss-Jordan method. On the

augmented matrix $\begin{bmatrix} 1 & -1 & 1 & | & 5 \\ 2 & -1 & 3 & | & 1 \\ 0 & 1 & 1 & | & -9 \end{bmatrix}$,

use $-1R_1 + R_2 \rightarrow R_1$ to get

$\begin{bmatrix} 1 & 0 & 2 & | & -4 \\ 2 & -1 & 3 & | & 1 \\ 0 & 1 & 1 & | & -9 \end{bmatrix}$, use

$-2R_1 + R_2 \rightarrow R_2$ to get

$\begin{bmatrix} 1 & 0 & 2 & | & -4 \\ 0 & -1 & -1 & | & 9 \\ 0 & 1 & 1 & | & -9 \end{bmatrix}$, use

$R_2 + R_3 \rightarrow R_3$ and $-1R_2 \rightarrow R_2$ to get

$\begin{bmatrix} 1 & 0 & 2 & | & -4 \\ 0 & 1 & 1 & | & -9 \\ 0 & 0 & 0 & | & 0 \end{bmatrix}$. Since $y + z = -9$

and $x + 2z = -4$, the solution set is

$\{(-2z - 4, -z - 9, z) \mid z \text{ is any real number }\}$.

61.

Note coefficient matrix is $A = \begin{bmatrix} 1 & 1 & 1 \\ 2 & 4 & 1 \\ 1 & 3 & 6 \end{bmatrix}$

and $A^{-1}\begin{bmatrix} 1 \\ 2 \\ 3 \end{bmatrix} =$

$$\begin{bmatrix} 7/4 & -1/4 & -1/4 \\ -11/12 & 5/12 & 1/12 \\ 1/6 & -1/6 & 1/6 \end{bmatrix}\begin{bmatrix} 1 \\ 2 \\ 3 \end{bmatrix} = \begin{bmatrix} 1/2 \\ 1/6 \\ 1/3 \end{bmatrix}.$$

The solution set is $\{(1/2, 1/6, 1/3)\}$.

63.

Note $A^{-1} = \begin{bmatrix} -55/6 & 5/2 & 5/3 \\ 35/12 & -5/4 & 5/6 \\ 40/3 & 0 & -10/3 \end{bmatrix}$

and $A^{-1} \begin{bmatrix} 27 \\ 9 \\ 16 \end{bmatrix} = \begin{bmatrix} -165 \\ 97.5 \\ 240 \end{bmatrix}$.

The solution set is $\{(-165, 97.5, 240)\}$.

65.

We get $A^{-1} = \begin{bmatrix} 68/133 & -36/133 & 8/19 \\ 10/19 & -12/19 & 6/19 \\ 127/133 & -122/133 & 6/19 \end{bmatrix}$

and $A^{-1} \begin{bmatrix} 16 \\ 24 \\ -8 \end{bmatrix} \approx \begin{bmatrix} -1.6842 \\ -9.2632 \\ -9.2632 \end{bmatrix}$.

Solution set is approximately

$$\{(-1.6842, -9.2632, -9.2632)\}.$$

67.

Since $AA^{-1} = \begin{bmatrix} a & 7 \\ 3 & b \end{bmatrix} \begin{bmatrix} -b & 7 \\ 3 & -a \end{bmatrix} =$

$= \begin{bmatrix} 21 - ab & 0 \\ 0 & 21 - ab \end{bmatrix}$, $21 - ab = 1$ and

$ab = 20$. List of permissible pairs (a, b):

a	1	2	4	5	10	20
b	20	10	5	4	2	1

Matrices are $\begin{bmatrix} 1 & 7 \\ 3 & 20 \end{bmatrix}$, $\begin{bmatrix} 2 & 7 \\ 3 & 10 \end{bmatrix}$,

$\begin{bmatrix} 4 & 7 \\ 3 & 5 \end{bmatrix}$, $\begin{bmatrix} 5 & 7 \\ 3 & 4 \end{bmatrix}$,

$\begin{bmatrix} 10 & 7 \\ 3 & 2 \end{bmatrix}$, and $\begin{bmatrix} 20 & 7 \\ 3 & 1 \end{bmatrix}$.

69. Let x and y be the costs of a dozen eggs and a magazine before taxes. Then

$$0.08x + 0.05y = 0.15$$
$$x + y = 2.70 - 0.15.$$

If $A = \begin{bmatrix} .08 & .05 \\ 1 & 1 \end{bmatrix}$, then $A^{-1} \begin{bmatrix} .15 \\ 2.55 \end{bmatrix} =$

$= \begin{bmatrix} 100/3 & -5/3 \\ -100/3 & 8/3 \end{bmatrix} \begin{bmatrix} .15 \\ 2.55 \end{bmatrix} = \begin{bmatrix} .75 \\ 1.80 \end{bmatrix}$.

Eggs costs $0.75 a dozen, the magazine costs $1.80.

71. Let x and y be the costs of one load of plywood and a load insulation, respectively. So

$$4x + 6y = 2500$$
$$3x + 5y = 1950.$$

If $A = \begin{bmatrix} 4 & 6 \\ 3 & 5 \end{bmatrix}$ then $A^{-1} \begin{bmatrix} 2500 \\ 1950 \end{bmatrix} =$

$= \begin{bmatrix} 5/2 & -3 \\ -3/2 & 2 \end{bmatrix} \begin{bmatrix} 2500 \\ 1950 \end{bmatrix} = \begin{bmatrix} 400 \\ 150 \end{bmatrix}$.

One load of plywood costs $400 and a load of insulation costs $150 .

73.

One computes $A^{-1} = \begin{bmatrix} 2 & -1 \\ -5 & 3 \end{bmatrix}$. To decode the message, we find

$\begin{bmatrix} 2 & -1 \\ -5 & 3 \end{bmatrix} \begin{bmatrix} 36 \\ 65 \end{bmatrix} = \begin{bmatrix} 7 \\ 15 \end{bmatrix} = \begin{bmatrix} g \\ o \end{bmatrix}$,

$\begin{bmatrix} 2 & -1 \\ -5 & 3 \end{bmatrix} \begin{bmatrix} 49 \\ 83 \end{bmatrix} = \begin{bmatrix} 15 \\ 4 \end{bmatrix} = \begin{bmatrix} o \\ d \end{bmatrix}$,

$\begin{bmatrix} 2 & -1 \\ -5 & 3 \end{bmatrix} \begin{bmatrix} 12 \\ 24 \end{bmatrix} = \begin{bmatrix} 0 \\ 12 \end{bmatrix} = \begin{bmatrix} space \\ l \end{bmatrix}$,

$\begin{bmatrix} 2 & -1 \\ -5 & 3 \end{bmatrix} \begin{bmatrix} 66 \\ 111 \end{bmatrix} = \begin{bmatrix} 21 \\ 3 \end{bmatrix} = \begin{bmatrix} u \\ c \end{bmatrix}$,

$\begin{bmatrix} 2 & -1 \\ -5 & 3 \end{bmatrix} \begin{bmatrix} 33 \\ 55 \end{bmatrix} = \begin{bmatrix} 11 \\ 0 \end{bmatrix} = \begin{bmatrix} k \\ space \end{bmatrix}$.

The message is 'good luck.'

75. Let x and y be the amounts invested in the Asset Manager Fund and Magellan Fund, respectively. Since $0.86(60,000) = 51,600$, we get

$$x + y = 60,000$$
$$0.76x + 0.90y = 51,600.$$

If $A = \begin{bmatrix} 1 & 1 \\ 0.76 & 0.90 \end{bmatrix}$, then

$A^{-1} = \begin{bmatrix} 45/7 & -50/7 \\ -38/7 & 50/7 \end{bmatrix}$ and

$A^{-1} \begin{bmatrix} 60,000 \\ 51,600 \end{bmatrix} = \begin{bmatrix} 17,142.86 \\ 42,857.14 \end{bmatrix}$. In the

Asset Manager Fund the amount invested was $17,142.86; in the Magellan Fund the amount invested was $42,857.14.

77. Let $x, y,$ and z be the prices of an animal totem, a trade-bead necklace, and a tribal mask, respectively. Then we obtain the system

$$24x + 33y + 12z = 202.23$$
$$19x + 40y + 22z = 209.38$$
$$30x + 9y + 19z = 167.66.$$

The inverse of the coefficient matrix A is

$$A^{-1} = \frac{1}{11,007} \begin{bmatrix} 562 & -519 & 246 \\ 299 & 96 & -300 \\ 1029 & 774 & 333 \end{bmatrix}.$$

Since $A^{-1} \begin{bmatrix} 202.23 \\ 209.38 \\ 167.66 \end{bmatrix} \approx \begin{bmatrix} \$4.20 \\ \$2.75 \\ \$0.89 \end{bmatrix}$, an

animal totem costs \$4.20, a necklace costs \$2.75, and a tribal mask costs \$0.89.

For Thought

1. False, $|A| = 12 - (-5) = 17$.

2. True, $|A| \neq 0$.

3. True, $|B| = 4 \cdot 5 - (-2)(-10) = 0$.

4. False, $|B| = 0$.

5. True, since the determinant of the coefficient matrix A is nonzero.

6. True, in general $|LM| = |L||M|$ for any square matrices L and M of the same size.

7. False, the system is not linear.   **8.**   True

9. False, $\begin{vmatrix} 2 & 0.1 \\ 100 & 5 \end{vmatrix} = 2 \cdot 5 - 100(0.1) = 0.$

10. False, because a 2×2 matrix is not equal to the number 27.

6.5 Exercises

1.

$$\begin{vmatrix} 1 & 3 \\ 0 & 2 \end{vmatrix} = 1 \cdot 2 - 0 \cdot 3 = 2$$

3.

$$\begin{vmatrix} 3 & 4 \\ 2 & 9 \end{vmatrix} = 3(9) - 2(4) = 19$$

5.

$$\begin{vmatrix} -0.3 & -0.5 \\ -0.7 & 0.2 \end{vmatrix} = (-0.3)(0.2) - (-0.7)(-0.5)$$
$$= -0.41$$

7.

$$\begin{vmatrix} 1/8 & -3/8 \\ 2 & -1/4 \end{vmatrix} = (1/8)(-1/4) - (2)(-3/8)$$
$$= -1/32 + 3/4 = 23/32$$

9.

$$\begin{vmatrix} 0.02 & 0.4 \\ 1 & 20 \end{vmatrix} = (.02)(20) - (.4)(1) = 0$$

11.

$$\begin{vmatrix} 3 & -5 \\ -9 & 15 \end{vmatrix} = (3)(15) - (-9)(-5) = 0$$

13.

Since $\begin{vmatrix} a & 2 \\ 3 & 4 \end{vmatrix} = 4a - 6 = 10$, we find

$4a = 16$ or $a = 4$.

15.

Since $\begin{vmatrix} a & 8 \\ 2 & a \end{vmatrix} = a^2 - 16 = 0$, we find

$a^2 = 16$ or $a = \pm 4$.

17.

Note $D = \begin{vmatrix} 2 & 1 \\ 1 & 2 \end{vmatrix} = 3$, $D_x = \begin{vmatrix} 5 & 1 \\ 7 & 2 \end{vmatrix} = 3$,

and $D_y = \begin{vmatrix} 2 & 5 \\ 1 & 7 \end{vmatrix} = 9$.

Then $x = \dfrac{D_x}{D} = \dfrac{3}{3} = 1$ and $y = \dfrac{D_y}{D} = \dfrac{9}{3} = 3$.

Solution set is $\{(1,3)\}$.

19.

Note $D = \begin{vmatrix} 1 & -2 \\ 1 & 2 \end{vmatrix} = 4$, $D_x = \begin{vmatrix} 7 & -2 \\ -5 & 2 \end{vmatrix} =$

4, and $D_y = \begin{vmatrix} 1 & 7 \\ 1 & -5 \end{vmatrix} = -12$.

So $x = \dfrac{D_x}{D} = \dfrac{4}{4} = 1$ and $y = \dfrac{D_y}{D} = \dfrac{-12}{4} =$

-3. The solution set is $\{(1, -3)\}$.

21.

Note $D = \begin{vmatrix} 2 & -1 \\ 1 & 3 \end{vmatrix} = 7$, $D_x = \begin{vmatrix} -11 & -1 \\ 12 & 3 \end{vmatrix}$

$= -21$, and $D_y = \begin{vmatrix} 2 & -11 \\ 1 & 12 \end{vmatrix} = 35$. Then

$x = \dfrac{D_x}{D} = -\dfrac{21}{7} = -3$ and $y = \dfrac{D_y}{D} = \dfrac{35}{7} = 5$.

Solution set is $\{(-3, 5)\}$.

23. Rewrite system as

$$\begin{aligned} x - y &= 6 \\ x + y &= 5. \end{aligned}$$

Note $D = \begin{vmatrix} 1 & -1 \\ 1 & 1 \end{vmatrix} = 2$, $D_x = \begin{vmatrix} 6 & -1 \\ 5 & 1 \end{vmatrix}$

$= 11$, and $D_y = \begin{vmatrix} 1 & 6 \\ 1 & 5 \end{vmatrix} = -1$.

So $x = \dfrac{D_x}{D} = \dfrac{11}{2}$ and $y = \dfrac{D_y}{D} = -\dfrac{1}{2}$.

Solution set is $\{(11/2, -1/2)\}$.

25.

Note $D = \begin{vmatrix} 1/2 & -1/3 \\ 1/4 & 1/2 \end{vmatrix} = 1/3$,

$D_x = \begin{vmatrix} 4 & -1/3 \\ 6 & 1/2 \end{vmatrix} = 4$, and

$D_y = \begin{vmatrix} 1/2 & 4 \\ 1/4 & 6 \end{vmatrix} = 2$. So $x = \dfrac{D_x}{D}$

$= \dfrac{4}{1/3} = 12$ and $y = \dfrac{D_y}{D} = \dfrac{2}{1/3} = 6$.

Solution set is $\{(12, 6)\}$.

27.

Note $D = \begin{vmatrix} 0.2 & 0.12 \\ 1 & 1 \end{vmatrix} = 0.08$,

$D_x = \begin{vmatrix} 148 & 0.12 \\ 900 & 1 \end{vmatrix} = 40$, and

$D_y = \begin{vmatrix} 0.2 & 148 \\ 1 & 900 \end{vmatrix} = 32$. Then

$x = \dfrac{D_x}{D} = \dfrac{40}{0.08} = 500$ and

$y = \dfrac{D_y}{D} = \dfrac{32}{0.08} = 400$.

Solution set is $\{(500, 400)\}$.

29. Cramer's rule does not apply since

$D = \begin{vmatrix} 3 & 1 \\ -6 & -2 \end{vmatrix} = 0$. Dividing the second

equation by -2, one gets the first equation.

Solution set is $\{(x, y) \mid 3x + y = 6 \}$.

31. Cramer's Rule does not apply since the determinant D is zero.

Adding the two equations, one gets $0 = 19$. Inconsistent and the solution set is $\emptyset$.

33. We use Cramer's Rule on the system

$$\begin{aligned} x - y &= 3 \\ 3x - y &= -9. \end{aligned}$$

Note, $D = \begin{vmatrix} 1 & -1 \\ 3 & -1 \end{vmatrix} = 2$,

$D_x = \begin{vmatrix} 3 & -1 \\ -9 & -1 \end{vmatrix} = -12$, and

$D_y = \begin{vmatrix} 1 & 3 \\ 3 & -9 \end{vmatrix} = -18$.

So $x = \dfrac{D_x}{D} = \dfrac{-12}{2} = -6$ and

$y = \dfrac{D_y}{D} = \dfrac{-18}{2} = -9$.

Solution set is $\{(-6, -9)\}$.

35. Note, $D = \begin{vmatrix} \sqrt{2} & \sqrt{3} \\ 3\sqrt{2} & -2\sqrt{3} \end{vmatrix} = -5\sqrt{6}$,

$D_x = \begin{vmatrix} 4 & \sqrt{3} \\ -3 & -2\sqrt{3} \end{vmatrix} = -5\sqrt{3}$, and

$D_y = \begin{vmatrix} \sqrt{2} & 4 \\ 3\sqrt{2} & -3 \end{vmatrix} = -15\sqrt{2}$.

So $x = \dfrac{D_x}{D} = \dfrac{-5\sqrt{3}}{-5\sqrt{6}} = \dfrac{1}{\sqrt{2}} = \dfrac{\sqrt{2}}{2}$

and $y = \dfrac{D_y}{D} = \dfrac{-15\sqrt{2}}{-5\sqrt{6}} = \dfrac{3}{\sqrt{3}} = \sqrt{3}$.

The solution set is $\{(\sqrt{2}/2, \sqrt{3})\}$.

37. Multiply second equation by -1 and add to the first one.

$$
\begin{aligned}
x^2 + y^2 &= 25 \\
-x^2 + y &= -5 \\
\hline
y^2 + y &= 20 \\
y^2 + y - 20 &= 0 \\
(y+5)(y-4) &= 0
\end{aligned}
$$

If $y = -5$, then $x^2 = 0$ and $x = 0$.
If $y = 4$, then $x^2 = 9$ and $x = \pm 3$.
The solution set is $\{(0, -5), (\pm 3, 4)\}$.

39. Solving for x in the second equation, we find $x = 4 + 2y$. Substituting into the first equation, we obtain

$$
\begin{aligned}
4 + 2y - 2y &= y^2 \\
4 &= y^2 \\
\pm 2 &= y.
\end{aligned}
$$

Using $y = 2$ in $x = 4 + 2y$, we get $x = 8$.
Similarly, if $y = -2$ then $x = 0$.
Solution set is $\{(8, 2), (0, -2)\}$.

41.

Invertible, since $\begin{vmatrix} 4 & 0.5 \\ 2 & 3 \end{vmatrix} = 12 - 1 = 11 \neq 0$

43.

Not invertible, for $\begin{vmatrix} 3 & -4 \\ 9 & -12 \end{vmatrix} = -36 + 36 = 0$

45. Note, $D = \begin{vmatrix} 3.47 & 23.09 \\ 12.48 & 3.98 \end{vmatrix} = -274.3526$,

$D_x = \begin{vmatrix} 5978.95 & 23.09 \\ 2765.34 & 3.98 \end{vmatrix} = -40,055.4796$,

and

$D_y = \begin{vmatrix} 3.47 & 5978.95 \\ 12.48 & 2765.34 \end{vmatrix} = -65,021.5662$.

Then $x = \dfrac{D_x}{D} = 146$ and $y = \dfrac{D_y}{D} = 237$.
The solution set is $\{(146, 237)\}$.

47. Let x and y be the number of boys and girls, respectively. Then

$$
\begin{aligned}
0.44x + 0.35y &= 231 \\
x + y &= 615.
\end{aligned}
$$

Note, $D = \begin{vmatrix} 0.44 & 0.35 \\ 1 & 1 \end{vmatrix} = 0.09$,

$D_x = \begin{vmatrix} 231 & 0.35 \\ 615 & 1 \end{vmatrix} = 15.75$, and

$D_y = \begin{vmatrix} 0.44 & 231 \\ 1 & 615 \end{vmatrix} = 39.6$.

There were $x = \dfrac{D_x}{D} = \dfrac{15.75}{0.09} = 175$ boys

and $y = \dfrac{D_y}{D} = \dfrac{39.6}{0.09} = 440$ girls.

49. Let x and y be the measurements of the two acute angles. Then we obtain

$$
\begin{aligned}
x + y &= 90 \\
x - 2y &= 1.
\end{aligned}
$$

Note, $D = \begin{vmatrix} 1 & 1 \\ 1 & -2 \end{vmatrix} = -3$,

$D_x = \begin{vmatrix} 90 & 1 \\ 1 & -2 \end{vmatrix} = -181$, and

$D_y = \begin{vmatrix} 1 & 90 \\ 1 & 1 \end{vmatrix} = -89$.

The acute angles are $x = \dfrac{D_x}{D} = \dfrac{181}{3}$ degrees

and $y = \dfrac{D_y}{D} = \dfrac{89}{3}$ degrees.

51. Let x and y be the salaries of the president and vice-president, respectively. Then we have

$$
\begin{aligned}
x + y &= 400,000 \\
x - y &= 100,000.
\end{aligned}
$$

Note, $D = \begin{vmatrix} 1 & 1 \\ 1 & -1 \end{vmatrix} = -2$,

$D_x = \begin{vmatrix} 400,000 & 1 \\ 100,000 & -1 \end{vmatrix} = -500,000$, and

$D_y = \begin{vmatrix} 1 & 400,000 \\ 1 & 100,000 \end{vmatrix} = -300,000$.

Thus, $x = \dfrac{D_x}{D} = \dfrac{-500,000}{-2} = 250,000$ and

$y = \dfrac{D_y}{D} = \dfrac{-300,000}{-2} = 150,000$.

The president's salary is \$250,000 and the vice-president's salary is \$150,000.

53. Yes, $|MN| = |M||N|$ since $|M| = 2$, $|N| = 3$,

and $|MN| = \begin{vmatrix} 8 & 31 \\ 14 & 55 \end{vmatrix} = 6.$

55.

$\left| \begin{bmatrix} a & b \\ c & d \end{bmatrix} \begin{bmatrix} e & f \\ g & h \end{bmatrix} \right| =$

$\begin{vmatrix} ae + bg & af + bh \\ ce + dg & cf + dh \end{vmatrix} =$

$(ae + bg)(cf + dh) - (ce + dg)(af + bh) =$
$aecf + bgcf + aedh + bgdh - ceaf - dgaf -$
$cebh - dgbh = bgcf + aedh - dgaf - cebh =$
$ad(eh - gf) - bc(eh - gf) = (ad - bc)(eh - gf) =$

$= \begin{vmatrix} a & b \\ c & d \end{vmatrix} \begin{vmatrix} e & f \\ g & h \end{vmatrix}$

57.

No, since $|-2M| = \begin{vmatrix} -6 & -4 \\ -10 & -8 \end{vmatrix} = 8$

and $-2|M| = -4.$

For Thought

1. False, the sign array of A is used in evaluating $|A|$.

2. False, the last term should be $1 \cdot \begin{vmatrix} 3 & 4 \\ 0 & 0 \end{vmatrix}.$

3. True

4. True, $|A|$ was expanded about the third row.

5. False, $|A|$ can be expanded only about a row or column.   **6.** False, a minor is a 2×2 matrix only if it comes from a 3×3 matrix.

7. False, $x = \dfrac{D_x}{D}$.   **8.**   True

9. False, it can happen that $D = 0$ and there are infinitely many solutions.

10. False, Cramer's Rule applies only to a system of linear equations.

6.6 Exercises

1.

$\begin{vmatrix} 5 & -6 \\ 9 & -8 \end{vmatrix} = -40 + 54 = 14$

3.

$\begin{vmatrix} 4 & 5 \\ 7 & 9 \end{vmatrix} = 36 - 35 = 1$

5.

$\begin{vmatrix} 2 & 1 \\ 7 & -8 \end{vmatrix} = -16 - 7 = -23$

7.

$\begin{vmatrix} 2 & 1 \\ 4 & -6 \end{vmatrix} = -12 - 4 = -16$

9.

$1 \begin{vmatrix} 1 & -2 \\ -1 & 5 \end{vmatrix} - (-3) \begin{vmatrix} -4 & 0 \\ -1 & 5 \end{vmatrix} + 3 \begin{vmatrix} -4 & 0 \\ 1 & -2 \end{vmatrix}$

$= 1(3) - (-3)(-20) + 3(8) = -33$

11.

$3 \begin{vmatrix} 4 & -1 \\ 1 & -2 \end{vmatrix} - 0 \begin{vmatrix} -1 & 2 \\ 1 & -2 \end{vmatrix} + 5 \begin{vmatrix} -1 & 2 \\ 4 & -1 \end{vmatrix}$

$= 3(-7) - 0 + 5(-7) = -56$

13.

$-2 \begin{vmatrix} 0 & -1 \\ 2 & -7 \end{vmatrix} - (-3) \begin{vmatrix} 5 & 1 \\ 2 & -7 \end{vmatrix} + 0 \begin{vmatrix} 5 & 1 \\ 0 & -1 \end{vmatrix}$

$= -2(2) - (-3)(-37) + 0 = -115$

15.

$0.1 \begin{vmatrix} 20 & 6 \\ 90 & 8 \end{vmatrix} - 0.4 \begin{vmatrix} 30 & 1 \\ 90 & 8 \end{vmatrix} + 0.7 \begin{vmatrix} 30 & 1 \\ 20 & 6 \end{vmatrix}$

$= 0.1(-380) - 0.4(150) + 0.7(160) = 14$

17. Expanding about the second row,

$D = -(-2) \begin{vmatrix} 3 & 5 \\ 3 & -4 \end{vmatrix} = -(-2)(-27) = -54.$

19. Expanding about the first row,

$D = 1 \begin{vmatrix} 2 & 2 \\ 4 & 4 \end{vmatrix} - 1 \begin{vmatrix} 2 & 2 \\ 4 & 4 \end{vmatrix} + 1 \begin{vmatrix} 2 & 2 \\ 4 & 4 \end{vmatrix}$

$= 1(0) - 1(0) + 1(0) = 0.$

21. Expanding about the first row,

$D = -(-1) \begin{vmatrix} 3 & 6 \\ -2 & -5 \end{vmatrix} = -(-1)(-3) = -3.$

23. Expanding about the second column,

$D = -9 \begin{vmatrix} 2 & 1 \\ 4 & 6 \end{vmatrix} = -9(8) = -72.$

25. Expanding about the first row,

$$3\begin{vmatrix} -3 & 2 & 0 \\ 3 & 1 & 2 \\ -4 & 1 & 3 \end{vmatrix} + 0 + 1\begin{vmatrix} 2 & -3 & 0 \\ -2 & 3 & 2 \\ 2 & -4 & 3 \end{vmatrix} -$$

$$5\begin{vmatrix} 2 & -3 & 2 \\ -2 & 3 & 1 \\ 2 & -4 & 1 \end{vmatrix} =$$

$$= 3(-37) + 1(4) - 5(6) = -137.$$

27. Expand the determinant about the second row.

$$-(1)\begin{vmatrix} -3 & 4 & 6 \\ 3 & 1 & -3 \\ 0 & 2 & 1 \end{vmatrix} + (-5)\begin{vmatrix} 2 & 4 & 6 \\ 1 & 1 & -3 \\ -2 & 2 & 1 \end{vmatrix} =$$

$$= -(1)(3) + (-5)(58) = -293.$$

29.

One finds $D = \begin{vmatrix} 1 & 1 & 1 \\ 1 & -1 & 1 \\ 2 & 1 & 1 \end{vmatrix} = 2,$

$$D_x = \begin{vmatrix} 6 & 1 & 1 \\ 2 & -1 & 1 \\ 7 & 1 & 1 \end{vmatrix} = 2, \; D_y = \begin{vmatrix} 1 & 6 & 1 \\ 1 & 2 & 1 \\ 2 & 7 & 1 \end{vmatrix} =$$

4, and $D_z = \begin{vmatrix} 1 & 1 & 6 \\ 1 & -1 & 2 \\ 2 & 1 & 7 \end{vmatrix} = 6.$

Since $x = \dfrac{D_x}{D} = 2/2 = 1$, $y = \dfrac{D_y}{D} = 4/2 = 2,$

and $z = \dfrac{D_z}{D} = 6/2 = 3,$

the solution set is $\{(1, 2, 3)\}$.

31.

One finds $D = \begin{vmatrix} 1 & 2 & 0 \\ 1 & -3 & 1 \\ 2 & -1 & 0 \end{vmatrix} = 5,$

$$D_x = \begin{vmatrix} 8 & 2 & 0 \\ -2 & -3 & 1 \\ 1 & -1 & 0 \end{vmatrix} = 10,$$

$$D_y = \begin{vmatrix} 1 & 8 & 0 \\ 1 & -2 & 1 \\ 2 & 1 & 0 \end{vmatrix} = 15,$$

and $D_z = \begin{vmatrix} 1 & 2 & 8 \\ 1 & -3 & -2 \\ 2 & -1 & 1 \end{vmatrix} = 25.$

Since $x = \dfrac{D_x}{D} = 10/5 = 2,$

$y = \dfrac{D_y}{D} = 15/5 = 3,$

and $z = \dfrac{D_z}{D} = 25/5 = 5,$

the solution set is $\{(2, 3, 5)\}$.

33.

One finds $D = \begin{vmatrix} 2 & -3 & 1 \\ 1 & 4 & -1 \\ 3 & -1 & 2 \end{vmatrix} = 16,$

$$D_x = \begin{vmatrix} 1 & -3 & 1 \\ 0 & 4 & -1 \\ 0 & -1 & 2 \end{vmatrix} = 7,$$

$$D_y = \begin{vmatrix} 2 & 1 & 1 \\ 1 & 0 & -1 \\ 3 & 0 & 2 \end{vmatrix} = -5,$$

and $D_z = \begin{vmatrix} 2 & -3 & 1 \\ 1 & 4 & 0 \\ 3 & -1 & 0 \end{vmatrix} = -13.$

Then $x = \dfrac{D_x}{D} = 7/16,$

$y = \dfrac{D_y}{D} = -5/16$, and $z = \dfrac{D_z}{D} = -13/16.$

The solution set is $\{(7/16, -5/16, -13/16)\}$.

35.

One finds $D = \begin{vmatrix} 1 & 1 & 1 \\ 2 & -1 & 3 \\ 3 & 1 & -1 \end{vmatrix} = 14,$

$$D_x = \begin{vmatrix} 2 & 1 & 1 \\ 0 & -1 & 3 \\ 0 & 1 & -1 \end{vmatrix} = -4,$$

$$D_y = \begin{vmatrix} 1 & 2 & 1 \\ 2 & 0 & 3 \\ 3 & 0 & -1 \end{vmatrix} = 22,$$

and $D_z = \begin{vmatrix} 1 & 1 & 2 \\ 2 & -1 & 0 \\ 3 & 1 & 0 \end{vmatrix} = 10.$

Then $x = \dfrac{D_x}{D} = -4/14 = -2/7,$

$y = \dfrac{D_y}{D} = 22/14 = 11/7$, and

$z = \dfrac{D_z}{D} = 10/14 = 5/7.$

The solution set is $\{(-2/7, 11/7, 5/7)\}$.

37. This system is dependent since the sum of the first two equations is the third equation. Adding the first and third equations, we get $3x - 3z = 4$ or $z = x - 4/3$. Substituting into the first equation, we find

$$x + y - 2\left(x - \frac{4}{3}\right) = 1$$
$$y = 1 - \frac{8}{3} + x$$
$$y = x - \frac{5}{3}.$$

Solution set is

$$\left\{\left(x, x - \frac{5}{3}, x - \frac{4}{3}\right) \mid x \text{ is any real number}\right\}.$$

39. Multiply the first equation by -2 and add to the third equation.

$$\begin{array}{rcl} -2x + 2y - 2z & = & -10 \\ 2x - 2y + 2z & = & 16 \\ \hline 0 & = & 6 \end{array}$$

Inconsistent and the solution set is $\emptyset$.

41. Let $x, y,$ and z be the ages of Jackie, Rochelle, and Alisha, respectively. Since $\dfrac{x+y}{2} = 33,$ $\dfrac{y+z}{2} = 25,$ and $\dfrac{x+z}{2} = 19,$ we obtain

$$\begin{array}{rcl} x + y & = & 66 \\ y + z & = & 50 \\ x + z & = & 38. \end{array}$$

One finds $D = \begin{vmatrix} 1 & 1 & 0 \\ 0 & 1 & 1 \\ 1 & 0 & 1 \end{vmatrix} = 2,$

$D_x = \begin{vmatrix} 66 & 1 & 0 \\ 50 & 1 & 1 \\ 38 & 0 & 1 \end{vmatrix} = 54, D_y = \begin{vmatrix} 1 & 66 & 0 \\ 0 & 50 & 1 \\ 1 & 38 & 1 \end{vmatrix} = $

$78,$ and $D_z = \begin{vmatrix} 1 & 1 & 66 \\ 0 & 1 & 50 \\ 1 & 0 & 38 \end{vmatrix} = 22.$

So Jackie is $x = \dfrac{D_x}{D} = 54/2 = 27$ years old,

Rochelle is $y = \dfrac{D_y}{D} = 78/2 = 39$ years old,

and Alisha is $z = \dfrac{D_z}{D} = 22/2 = 11$ years old.

43. Let $x, y,$ and z be the scores in the first test, second test, and final exam, respectively. Then

$$\begin{array}{rcl} x + y + z & = & 180 \\ 0.2x + 0.2y + 0.6z & = & 76 \\ 0.1x + 0.2y + 0.7z & = & 83. \end{array}$$

One finds $D = \begin{vmatrix} 1 & 1 & 1 \\ 0.2 & 0.2 & 0.6 \\ 0.1 & 0.2 & 0.7 \end{vmatrix} = -0.04,$

$D_x = \begin{vmatrix} 180 & 1 & 1 \\ 76 & 0.2 & 0.6 \\ 83 & 0.2 & 0.7 \end{vmatrix} = -1.2,$

$D_y = \begin{vmatrix} 1 & 180 & 1 \\ 0.2 & 76 & 0.6 \\ 0.1 & 83 & 0.7 \end{vmatrix} = -2,$

and $D_z = \begin{vmatrix} 1 & 1 & 180 \\ 0.2 & 0.2 & 76 \\ 0.1 & 0.2 & 83 \end{vmatrix} = -4.$

Test scores are $x = \dfrac{D_x}{D} = 1.2/(0.04) = 30,$

$y = \dfrac{D_y}{D} = 2/(0.04) = 50,$ and

the final exam is $z = \dfrac{D_z}{D} = 4/(0.04) = 100.$

45. One finds $D = \begin{vmatrix} 0.2 & -0.3 & 1.2 \\ 0.25 & 0.35 & -0.9 \\ 2.4 & -1 & 1.25 \end{vmatrix} = -\dfrac{527}{800},$

$D_x = \begin{vmatrix} 13.11 & -0.3 & 1.2 \\ -1.575 & 0.35 & -0.9 \\ 42.02 & -1 & 1.25 \end{vmatrix} = -\dfrac{11,067}{1000},$

$D_y = \begin{vmatrix} 0.2 & 13.11 & 1.2 \\ 0.25 & -1.575 & -0.9 \\ 2.4 & 42.02 & 1.25 \end{vmatrix} = -\dfrac{64,821}{8000},$

and $D_z = \begin{vmatrix} 0.2 & -0.3 & 13.11 \\ 0.25 & 0.35 & -1.575 \\ 2.4 & -1 & 42.02 \end{vmatrix} = -\dfrac{3689}{500}.$

Note, $\dfrac{D_x}{D} = 16.8, \dfrac{D_y}{D} = 12.3,$ and $\dfrac{D_z}{D} = 11.2.$

The solution set is $\{(16.8, 12.3, 11.2)\}$.

47. Let x, y, and z be the prices per gallon of regular, plus, and supreme gasoline, respectively.

One finds

$$D = \begin{vmatrix} 1270 & 980 & 890 \\ 1450 & 1280 & 1050 \\ 1340 & 1190 & 1060 \end{vmatrix} = 18{,}038{,}000,$$

$$D_x = \begin{vmatrix} 5924.86 & 980 & 890 \\ 7138.22 & 1280 & 1050 \\ 6789.41 & 1190 & 1060 \end{vmatrix} = 32{,}450{,}362,$$

$$D_y = \begin{vmatrix} 1270 & 5924.86 & 890 \\ 1450 & 7138.12 & 1050 \\ 1340 & 6789.41 & 1060 \end{vmatrix} = 34{,}254{,}162,$$

$$D_z = \begin{vmatrix} 1270 & 980 & 5724.86 \\ 1450 & 1280 & 7138.22 \\ 1340 & 1190 & 6789.41 \end{vmatrix} = 36{,}057{,}962.$$

Regular gas costs $\dfrac{D_x}{D} \approx \$1.799$,

plus costs $\dfrac{D_y}{D} \approx \$1.899$, and

supreme costs $\dfrac{D_z}{D} \approx \$1.999$ per gallon.

49.

One finds $D = \begin{vmatrix} 1 & 1 & 1 & 1 \\ 2 & -1 & 1 & 3 \\ 1 & 2 & -1 & 2 \\ 1 & -1 & -1 & 4 \end{vmatrix} = 11,$

$$D_w = \begin{vmatrix} 4 & 1 & 1 & 1 \\ 13 & -1 & 1 & 3 \\ -2 & 2 & -1 & 2 \\ 8 & -1 & -1 & 4 \end{vmatrix} = 11,$$

$$D_x = \begin{vmatrix} 1 & 4 & 1 & 1 \\ 2 & 13 & 1 & 3 \\ 1 & -2 & -1 & 2 \\ 1 & 8 & -1 & 4 \end{vmatrix} = -22,$$

$$D_y = \begin{vmatrix} 1 & 1 & 4 & 1 \\ 2 & -1 & 13 & 3 \\ 1 & 2 & -2 & 2 \\ 1 & -1 & 8 & 4 \end{vmatrix} = 33,$$

and $D_z = \begin{vmatrix} 1 & 1 & 1 & 4 \\ 2 & -1 & 1 & 13 \\ 1 & 2 & -1 & -2 \\ 1 & -1 & -1 & 8 \end{vmatrix} = 22.$

Note, $\dfrac{D_w}{D} = 1$, $\dfrac{D_x}{D} = -2$, $\dfrac{D_y}{D} = 3$, and

$\dfrac{D_z}{D} = 2$. The solution set is $\{(1, -2, 3, 2)\}$.

51.

Note, $\begin{vmatrix} x & y & 1 \\ 3 & -5 & 1 \\ -2 & 6 & 1 \end{vmatrix} = 8 - 11x - 5y = 0.$

Since both points $(3, -5)$ and $(-2, 6)$ satisfies $8 - 11x - 5y = 0$, this is an equation of the line through the two points.

53. We will show it for one particular case and the other cases can be proved similarly.

Let us suppose $A = \begin{bmatrix} a & b & c \\ 0 & 0 & 0 \\ g & h & i \end{bmatrix}$.

Then $|A| = a \cdot 0 \cdot i + b \cdot 0 \cdot g + c \cdot 0 \cdot h - g \cdot 0 \cdot c - h \cdot 0 \cdot a - i \cdot 0 \cdot b = 0$

55. We will prove it for one particular case and the other cases can be shown similarly.

Let us suppose $A = \begin{bmatrix} a & b & c \\ d & e & f \\ g & h & i \end{bmatrix}$ and

$B = \begin{bmatrix} a & b & c \\ d & e & f \\ kg & kh & ki \end{bmatrix}$. Then $|B| =$

$aeki + bfkg + cdkh - kgec - khfa - kidb = k(aei + bfg + cdh - gec - hfa - idb) = k|A|.$

Review Exercises

1.

$$\begin{bmatrix} 2+3 & -3+7 \\ -2+1 & 4+2 \end{bmatrix} = \begin{bmatrix} 5 & 4 \\ -1 & 6 \end{bmatrix}$$

3.

$$\begin{bmatrix} 4 & -6 \\ -4 & 8 \end{bmatrix} - \begin{bmatrix} 3 & 7 \\ 1 & 2 \end{bmatrix} =$$

$$\begin{bmatrix} 1 & -13 \\ -5 & 6 \end{bmatrix}$$

5.

$$AB = \begin{bmatrix} 6-3 & 14-6 \\ -6+4 & -14+8 \end{bmatrix} =$$

$$\begin{bmatrix} 3 & 8 \\ -2 & -6 \end{bmatrix}$$

7. $D + E$ is undefined

9.

$$AC = \begin{bmatrix} -2 - 9 \\ 2 + 12 \end{bmatrix} = \begin{bmatrix} -11 \\ 14 \end{bmatrix}$$

11.

$$EF = \begin{bmatrix} 3 & 2 & -1 \\ -12 & -8 & 4 \\ 9 & 6 & -3 \end{bmatrix}$$

13. $FG = \begin{bmatrix} -3 + 2 + 2 & 2 - 3 & -1 \end{bmatrix} = \begin{bmatrix} 1 & -1 & -1 \end{bmatrix}$

15. GF is undefined

17.

On $\left[\begin{array}{cc|cc} 2 & -3 & 1 & 0 \\ -2 & 4 & 0 & 1 \end{array} \right]$, use $R_1 + R_2 \to R_2$

to get

$\left[\begin{array}{cc|cc} 2 & -3 & 1 & 0 \\ 0 & 1 & 1 & 1 \end{array} \right]$, use $3R_2 + R_1 \to R_1$ to get

$\left[\begin{array}{cc|cc} 2 & 0 & 4 & 3 \\ 0 & 1 & 1 & 1 \end{array} \right]$, use $\frac{1}{2}R_1 \to R_1$ to get

$\left[\begin{array}{cc|cc} 1 & 0 & 2 & 1.5 \\ 0 & 1 & 1 & 1 \end{array} \right]$. So $A^{-1} = \begin{bmatrix} 2 & 1.5 \\ 1 & 1 \end{bmatrix}$.

19.

On $\left[\begin{array}{ccc|ccc} -1 & 0 & 0 & 1 & 0 & 0 \\ 1 & 1 & 0 & 0 & 1 & 0 \\ -2 & 3 & 1 & 0 & 0 & 1 \end{array} \right]$,

use $R_1 + R_2 \to R_2$, $2R_2 + R_3 \to R_3$, and

$-1R_1 \to R_1$ to get

$\left[\begin{array}{ccc|ccc} 1 & 0 & 0 & -1 & 0 & 0 \\ 0 & 1 & 0 & 1 & 1 & 0 \\ 0 & 5 & 1 & 0 & 2 & 1 \end{array} \right]$,

use $-5R_2 + R_3 \to R_3$ to get

$\left[\begin{array}{ccc|ccc} 1 & 0 & 0 & -1 & 0 & 0 \\ 0 & 1 & 0 & 1 & 1 & 0 \\ 0 & 0 & 1 & -5 & -3 & 1 \end{array} \right]$.

So $G^{-1} = \begin{bmatrix} -1 & 0 & 0 \\ 1 & 1 & 0 \\ -5 & -3 & 1 \end{bmatrix}$.

21.

From Exercise 5, $(AB)^{-1} = \begin{bmatrix} 3 & 8 \\ -2 & -6 \end{bmatrix}^{-1}$.

On $\left[\begin{array}{cc|cc} 3 & 8 & 1 & 0 \\ -2 & -6 & 0 & 1 \end{array} \right]$, use $-\frac{1}{2}R_2 \to R_2$

to get

$\left[\begin{array}{cc|cc} 3 & 8 & 1 & 0 \\ 1 & 3 & 0 & -0.5 \end{array} \right]$, use $-3R_2 + R_1 \to R_1$

to get

$\left[\begin{array}{cc|cc} 0 & -1 & 1 & 1.5 \\ 1 & 3 & 0 & -0.5 \end{array} \right]$, use $3R_1 + R_2 \to R_2$

to get

$\left[\begin{array}{cc|cc} 0 & -1 & 1 & 1.5 \\ 1 & 0 & 3 & 4 \end{array} \right]$, use $R_1 \leftrightarrow R_2$ to get

$\left[\begin{array}{cc|cc} 1 & 0 & 3 & 4 \\ 0 & -1 & 1 & 1.5 \end{array} \right]$, use $-1R_2 \to R_2$ to get

$\left[\begin{array}{cc|cc} 1 & 0 & 3 & 4 \\ 0 & 1 & -1 & -1.5 \end{array} \right]$.

So $(AB)^{-1} = \begin{bmatrix} 3 & 4 \\ -1 & -1.5 \end{bmatrix}$.

23.

$$AA^{-1} = I = \begin{bmatrix} 1 & 0 \\ 0 & 1 \end{bmatrix}$$

25. $|A| = 8 - 6 = 2$

27. Expanding about the third column,

$$|G| = 1 \cdot \begin{vmatrix} -1 & 0 \\ 1 & 1 \end{vmatrix} = 1(-1) = -1.$$

29. Solution set is $\{(10/3, 17/3)\}$.

First solution is by Gaussian elimination.

On $\left[\begin{array}{cc|c} 1 & 1 & 9 \\ 2 & -1 & 1 \end{array} \right]$, use $-2R_1 + R_2 \to R_2$ to

get

$\left[\begin{array}{cc|c} 1 & 1 & 9 \\ 0 & -3 & -17 \end{array} \right]$, use $-\frac{1}{3}R_2 \to R_2$ to get

$\left[\begin{array}{cc|c} 1 & 1 & 9 \\ 0 & 1 & 17/3 \end{array} \right]$, use $-1R_2 + R_1 \to R_1$ to get

$\left[\begin{array}{cc|c} 1 & 0 & 10/3 \\ 0 & 1 & 17/3 \end{array} \right]$.

Secondly, by matrix inversion note $A^{-1} =$

$\begin{bmatrix} 1/3 & 1/3 \\ 2/3 & -1/3 \end{bmatrix}$ and $A^{-1} \begin{bmatrix} 9 \\ 1 \end{bmatrix} = \begin{bmatrix} 10/3 \\ 17/3 \end{bmatrix}$.

Thirdly, by Cramer's Rule note

$$D = \begin{vmatrix} 1 & 1 \\ 2 & -1 \end{vmatrix} = -3, \ D_x = \begin{vmatrix} 9 & 1 \\ 1 & -1 \end{vmatrix} = -10,$$

and $D_y = \begin{vmatrix} 1 & 9 \\ 2 & 1 \end{vmatrix} = -17.$

So $\dfrac{D_x}{D} = 10/3, \ \dfrac{D_y}{D} = 17/3.$

31. Solution set is $\{(-2, 3)\}$.

First solution is by Gaussian elimination. On

$\begin{bmatrix} 2 & 1 & | & -1 \\ 3 & 2 & | & 0 \end{bmatrix}$, use $-\dfrac{3}{2}R_1 + R_2 \to R_2$ to get

$\begin{bmatrix} 2 & 1 & | & -1 \\ 0 & 1/2 & | & 3/2 \end{bmatrix}$, use $-2R_2 + R_1 \to R_1$

and $2R_2 \to R_2$ to get

$\begin{bmatrix} 2 & 0 & | & -4 \\ 0 & 1 & | & 3 \end{bmatrix}$, use $\dfrac{1}{2}R_2 \to R_2$ to get

$\begin{bmatrix} 1 & 0 & | & -2 \\ 0 & 1 & | & 3 \end{bmatrix}$.

Secondly, by matrix inversion note $A^{-1} =$

$\begin{bmatrix} 2 & -1 \\ -3 & 2 \end{bmatrix}$ and $A^{-1}\begin{bmatrix} -1 \\ 0 \end{bmatrix} = \begin{bmatrix} -2 \\ 3 \end{bmatrix}$.

Thirdly, by Cramer's Rule note

$$D = \begin{vmatrix} 2 & 1 \\ 3 & 2 \end{vmatrix} = 1, \ D_x = \begin{vmatrix} -1 & 1 \\ 0 & 2 \end{vmatrix} = -2,$$

and $D_y = \begin{vmatrix} 2 & -1 \\ 3 & 0 \end{vmatrix} = 3.$

So $\dfrac{D_x}{D} = -2, \dfrac{D_y}{D} = 3.$

33. Solution set is $\{(x, y) | x - 5y = 9\}$.

Solution is by Gaussian elimination.

On $\begin{bmatrix} 1 & -5 & | & 9 \\ -2 & 10 & | & -18 \end{bmatrix}$, use $2R_1 + R_2 \to R_2$

to get $\begin{bmatrix} 1 & -5 & | & 9 \\ 0 & 0 & | & 0 \end{bmatrix}$. Dependent system.

This cannot be solved by matrix inversion

since $A^{-1} = \begin{bmatrix} 1 & -5 \\ -2 & 10 \end{bmatrix}^{-1}$ does not exist.

Nor can it be solved by Cramer's rule

since $|D| = \begin{vmatrix} 1 & -5 \\ -2 & 10 \end{vmatrix} = 0.$

35. Solution set is $\emptyset$ as seen by an application of the Gaussian elimination method.

On $\begin{bmatrix} 0.05 & 0.1 & | & 1 \\ 10 & 20 & | & 20 \end{bmatrix}$, use

$-200R_1 + R_2 \to R_2$ to get

$\begin{bmatrix} 0.05 & 0.1 & | & 1 \\ 0 & 0 & | & -180 \end{bmatrix}$, which is inconsistent.

This cannot be solved by matrix inversion since

$A^{-1} = \begin{bmatrix} 0.05 & 0.1 \\ 10 & 20 \end{bmatrix}^{-1}$ does not exist.

Nor can it be solved by Cramer's rule

since $|D| = \begin{vmatrix} 0.05 & 0.1 \\ 10 & 20 \end{vmatrix} = 0.$

37. Solution set is $\{(1, 2, 3)\}$.

First solution is by Gaussian elimination.

On $\begin{bmatrix} 1 & 1 & -2 & | & -3 \\ -1 & 2 & -1 & | & 0 \\ -1 & -1 & 3 & | & 6 \end{bmatrix}$, use

$R_1 + R_2 \to R_2$ to get

$\begin{bmatrix} 1 & 1 & -2 & | & -3 \\ 0 & 3 & -3 & | & -3 \\ 0 & 0 & 1 & | & 3 \end{bmatrix}$, use

$\dfrac{1}{3}R_2 \to R_2$ to get

$\begin{bmatrix} 1 & 1 & -2 & | & -3 \\ 0 & 1 & -1 & | & -1 \\ 0 & 0 & 1 & | & 3 \end{bmatrix}$, use

$R_3 + R_2 \to R_2$ and $2R_3 + R_1 \to R_1$ to get

$\begin{bmatrix} 1 & 1 & 0 & | & 3 \\ 0 & 1 & 0 & | & 2 \\ 0 & 0 & 1 & | & 3 \end{bmatrix}$, use $-1R_2 + R_1 \to R_1$

to get $\begin{bmatrix} 1 & 0 & 0 & | & 1 \\ 0 & 1 & 0 & | & 2 \\ 0 & 0 & 1 & | & 3 \end{bmatrix}$.

Secondly, by matrix inversion note $A^{-1} =$

$\begin{bmatrix} 5/3 & -1/3 & 1 \\ 4/3 & 1/3 & 1 \\ 1 & 0 & 1 \end{bmatrix}$ and $A^{-1}\begin{bmatrix} -3 \\ 0 \\ 6 \end{bmatrix} = \begin{bmatrix} 1 \\ 2 \\ 3 \end{bmatrix}$.

Thirdly, by Cramer's Rule note

$$D = \begin{vmatrix} 1 & 1 & -2 \\ -1 & 2 & -1 \\ -1 & -1 & 3 \end{vmatrix} = 3,$$

$$D_x = \begin{vmatrix} -3 & 1 & -2 \\ 0 & 2 & -1 \\ 6 & -1 & 3 \end{vmatrix} = 3,$$

$$D_y = \begin{vmatrix} 1 & -3 & -2 \\ -1 & 0 & -1 \\ -1 & 6 & 3 \end{vmatrix} = 6,$$

and $D_z = \begin{vmatrix} 1 & 1 & -3 \\ -1 & 2 & 0 \\ -1 & -1 & 6 \end{vmatrix} = 9.$

Then $\dfrac{D_x}{D} = 1$, $\dfrac{D_y}{D} = 2$, and $\dfrac{D_z}{D} = 3$.

39. Solution set is $\{(-3, 4, 1)\}$.

First solution is by Gaussian elimination.

On $\begin{bmatrix} 0 & 1 & -3 & | & 1 \\ 1 & 2 & 0 & | & 5 \\ 1 & 0 & 4 & | & 1 \end{bmatrix}$, use

$-1R_3 + R_2 \to R_2$ and $R_1 \leftrightarrow R_3$ to get

$\begin{bmatrix} 1 & 0 & 4 & | & 1 \\ 0 & 2 & -4 & | & 4 \\ 0 & 1 & -3 & | & 1 \end{bmatrix}$, use $\dfrac{1}{2}R_2 \to R_2$ to get

$\begin{bmatrix} 1 & 0 & 4 & | & 1 \\ 0 & 1 & -2 & | & 2 \\ 0 & 1 & -3 & | & 1 \end{bmatrix}$, use $-1R_2 + R_3 \to R_3$

to get

$\begin{bmatrix} 1 & 0 & 4 & | & 1 \\ 0 & 1 & -2 & | & 2 \\ 0 & 0 & -1 & | & -1 \end{bmatrix}$, use

$-2R_3 + R_2 \to R_2$, $4R_3 + R_1 \to R_1$, and

$-1R_3 \to R_3$ to get

$\begin{bmatrix} 1 & 0 & 0 & | & -3 \\ 0 & 1 & 0 & | & 4 \\ 0 & 0 & 1 & | & 1 \end{bmatrix}.$

Secondly, by matrix inversion

$$A^{-1} = \begin{bmatrix} 4 & -2 & 3 \\ -2 & 3/2 & -3/2 \\ -1 & 1/2 & -1/2 \end{bmatrix}$$

and $A^{-1} \begin{bmatrix} 1 \\ 5 \\ 1 \end{bmatrix} = \begin{bmatrix} -3 \\ 4 \\ 1 \end{bmatrix}.$

Thirdly, by Cramer's Rule note

$$D = \begin{vmatrix} 0 & 1 & -3 \\ 1 & 2 & 0 \\ 1 & 0 & 4 \end{vmatrix} = 2,$$

$$D_x = \begin{vmatrix} 1 & 1 & -3 \\ 5 & 2 & 0 \\ 1 & 0 & 4 \end{vmatrix} = -6,$$

$$D_y = \begin{vmatrix} 0 & 1 & -3 \\ 1 & 5 & 0 \\ 1 & 1 & 4 \end{vmatrix} = 8,$$

and $D_z = \begin{vmatrix} 0 & 1 & 1 \\ 1 & 2 & 5 \\ 1 & 0 & 1 \end{vmatrix} = 2.$

Then $\dfrac{D_x}{D} = -3$, $\dfrac{D_y}{D} = 4$, and $\dfrac{D_z}{D} = 1$.

41. Solution set is

$$\left\{ \left(\frac{3y+3}{2}, y, \frac{1-y}{2} \right) \mid y \text{ is any real number} \right\}.$$

Solution is by Gaussian elimination.

On $\begin{bmatrix} 1 & -1 & 1 & | & 2 \\ 1 & -2 & -1 & | & 1 \\ 2 & -3 & 0 & | & 3 \end{bmatrix}$, use

$-1R_1 + R_2 \to R_2$ and $-2R_1 + R_3 \to R_3$ to get

$\begin{bmatrix} 1 & -1 & 1 & | & 2 \\ 0 & -1 & -2 & | & -1 \\ 0 & -1 & -2 & | & -1 \end{bmatrix}$, use

$-1R_2 + R_3 \to R_3$, $-1R_2 + R_1 \to R_1$,

and $-1R_2 \to R_2$ to get

$\begin{bmatrix} 1 & 0 & 3 & | & 3 \\ 0 & 1 & 2 & | & 1 \\ 0 & 0 & 0 & | & 0 \end{bmatrix}$. Since $z = \dfrac{1-y}{2}$, we get

$$x = 3 - 3z = 3 - 3\left(\frac{1-y}{2} \right) = \frac{3y+3}{2}.$$

This cannot be solved by matrix inversion

since $A^{-1} = \begin{bmatrix} 1 & -1 & 1 \\ 1 & -2 & -1 \\ 2 & -3 & 0 \end{bmatrix}^{-1}$ does not exist.

Nor can it be solved by Cramer's rule

since $|D| = \begin{vmatrix} 1 & -1 & 1 \\ 1 & -2 & -1 \\ 2 & -3 & 0 \end{vmatrix} = 0.$

43. Solution set is $\emptyset$ as seen by an application of the Gaussian elimination method.

On $\begin{bmatrix} 1 & -3 & -1 & | & 2 \\ 1 & -3 & -1 & | & 1 \\ 1 & -3 & -1 & | & 0 \end{bmatrix}$, use

$-1R_1 + R_2 \to R_2$ and $-1R_1 + R_3 \to R_3$ to get

$\begin{bmatrix} 1 & -3 & -1 & | & 2 \\ 0 & 0 & 0 & | & -1 \\ 0 & 0 & 0 & | & -2 \end{bmatrix}$. Inconsistent.

This cannot be solved by matrix inversion

since $A^{-1} = \begin{bmatrix} 1 & -3 & -1 \\ 1 & -3 & -1 \\ 1 & -3 & -1 \end{bmatrix}^{-1}$ does not exist.

Nor can it be solved by Cramer's rule

since $|D| = \begin{vmatrix} 1 & -3 & -1 \\ 1 & -3 & -1 \\ 1 & -3 & -1 \end{vmatrix} = 0.$

45. Using $x = 9$ in $x + y = -3$, we find $9 + y = -3$ and $y = -12$. Solution set is $\{(9, -12)\}$.

47.

Note $\begin{bmatrix} x \\ y \end{bmatrix} = \begin{bmatrix} 1 & 1 \\ 2 & 1 \end{bmatrix}^{-1} \begin{bmatrix} 6 \\ 8 \end{bmatrix} =$

$\begin{bmatrix} -1 & 1 \\ 2 & -1 \end{bmatrix} \begin{bmatrix} 6 \\ 8 \end{bmatrix} = \begin{bmatrix} 2 \\ 4 \end{bmatrix}.$

The solution set is $\{(2, 4)\}$.

49. System of equations can be written as

$$\begin{aligned} x + y &= -3 \\ -x &= 0. \end{aligned}$$

Using $x = 0$ in $x + y = -3$, we get $y = -3$. Solution set is $\{(0, -3)\}$.

51. System of equations can be written as

$\begin{bmatrix} 1 & 1 & 0 \\ 0 & 1 & 1 \\ 1 & 0 & 1 \end{bmatrix} \begin{bmatrix} x \\ y \\ z \end{bmatrix} = \begin{bmatrix} 1 \\ 1 \\ 1 \end{bmatrix}.$

By using matrix inversion, we obtain

$\begin{bmatrix} x \\ y \\ z \end{bmatrix} = \begin{bmatrix} 1/2 & -1/2 & 1/2 \\ 1/2 & 1/2 & -1/2 \\ -1/2 & 1/2 & 1/2 \end{bmatrix} \begin{bmatrix} 1 \\ 1 \\ 1 \end{bmatrix} =$

$\begin{bmatrix} 1/2 \\ 1/2 \\ 1/2 \end{bmatrix}$. The solution set is $\{(1/2, 1/2, 1/2)\}$.

53. By using the inverse of the coefficient matrix, we get

$\begin{bmatrix} x \\ y \\ z \end{bmatrix} = \begin{bmatrix} 3/5 & -3/5 & 2/5 \\ 2/5 & 3/5 & -2/5 \\ -1/5 & 1/5 & 1/5 \end{bmatrix} \begin{bmatrix} -1 \\ 7 \\ 17 \end{bmatrix} =$

$\begin{bmatrix} 2 \\ -3 \\ 5 \end{bmatrix}$. The solution set is $\{(2, -3, 5)\}$.

55. Let x and y be the number of gallons of pollutant A and pollutant B, respectively. Then

$$\begin{aligned} 10x + 6y &= 4060 \\ \frac{3}{4} &= \frac{x}{y}. \end{aligned}$$

Substitute $x = \frac{3}{4}y$ in $10x + 6y = 4060$. Solving for x, one finds the quantities discharged are $x \approx 225.56$ gallons of pollutant A and $y \approx 300.74$ gallons of pollutant B.

57. Let $x, y,$ and z be the expenses including tax for water, gas, and electricity, respectively.

$$\begin{aligned} x + y + z &= 189.83 \\ \frac{x}{1.04} + \frac{y}{1.05} + \frac{z}{1.06} &= 180 \\ z &= 2y \end{aligned}$$

Solving the system, one finds the expenses including taxes are $x = \$22.88$ for water, $y = \$55.65$ for gas, and $z = \$111.30$ for electricity.

Chapter 6 Test

1.

On $\begin{bmatrix} 2 & -3 & | & 1 \\ 1 & 9 & | & 4 \end{bmatrix}$, use $-2R_2 + R_1 \to R_2$

to get

$\begin{bmatrix} 2 & -3 & | & 1 \\ 0 & -21 & | & -7 \end{bmatrix}$, use $\frac{1}{2}R_1 \to R_1$ and

$-\frac{1}{21}R_2 \to R_2$ to get

$\begin{bmatrix} 1 & -3/2 & | & 1/2 \\ 0 & 1 & | & 1/3 \end{bmatrix}$, use $\frac{3}{2}R_2 + R_1 \to R_1$

to get

$$\left[\begin{array}{cc|c} 1 & 0 & 1 \\ 0 & 1 & 1/3 \end{array}\right]. \text{ Solution set is } \{(1, 1/3)\}.$$

2.

On $\left[\begin{array}{ccc|c} 2 & -1 & 1 & 5 \\ 1 & -2 & -1 & -2 \\ 3 & -1 & -1 & 6 \end{array}\right]$, use

$-\dfrac{1}{2}R_1 + R_2 \to R_2$ and

$-\dfrac{3}{2}R_1 + R_3 \to R_3$ to get

$$\left[\begin{array}{ccc|c} 2 & -1 & 1 & 5 \\ 0 & -3/2 & -3/2 & -9/2 \\ 0 & 1/2 & -5/2 & -3/2 \end{array}\right], \text{ use}$$

$-\dfrac{2}{3}R_2 \to R_2$ and $3R_3 + R_2 \to R_3$ to get

$$\left[\begin{array}{ccc|c} 2 & -1 & 1 & 5 \\ 0 & 1 & 1 & 3 \\ 0 & 0 & -9 & -9 \end{array}\right],$$

use $-\dfrac{1}{9}R_3 \to R_3$ and $R_2 + R_1 \to R_1$ to get

$$\left[\begin{array}{ccc|c} 2 & 0 & 2 & 8 \\ 0 & 1 & 1 & 3 \\ 0 & 0 & 1 & 1 \end{array}\right], \text{ use}$$

$-1R_3 + R_2 \to R_2$ and $\dfrac{1}{2}R_1 \to R_1$ to get

$$\left[\begin{array}{ccc|c} 1 & 0 & 1 & 4 \\ 0 & 1 & 0 & 2 \\ 0 & 0 & 1 & 1 \end{array}\right], \text{ use } -1R_3 + R_1 \to R_1$$

to get $\left[\begin{array}{ccc|c} 1 & 0 & 0 & 3 \\ 0 & 1 & 0 & 2 \\ 0 & 0 & 1 & 1 \end{array}\right].$

The solution set is $\{(3, 2, 1)\}$.

3.

On $\left[\begin{array}{ccc|c} 1 & -1 & -1 & 1 \\ 2 & 1 & -1 & 0 \\ 5 & -2 & -4 & 3 \end{array}\right]$, use

$-2R_1 + R_2 \to R_2$ and

$-5R_1 + R_3 \to R_3$ to get

$$\left[\begin{array}{ccc|c} 1 & -1 & -1 & 1 \\ 0 & 3 & 1 & -2 \\ 0 & 3 & 1 & -2 \end{array}\right], \text{ use}$$

$-1R_3 + R_2 \to R_3$ and $\dfrac{1}{3}R_2 \to R_2$ to get

$$\left[\begin{array}{ccc|c} 1 & -1 & -1 & 1 \\ 0 & 1 & 1/3 & -2/3 \\ 0 & 0 & 0 & 0 \end{array}\right], \text{ use}$$

$R_2 + R_1 \to R_1$ to get

$$\left[\begin{array}{ccc|c} 1 & 0 & -2/3 & 1/3 \\ 0 & 1 & 1/3 & -2/3 \\ 0 & 0 & 0 & 0 \end{array}\right].$$

Since $x = \dfrac{2}{3}z + \dfrac{1}{3}$, we find $3x = 2z + 1$ and

$z = \dfrac{3x-1}{2}$. Since $y = -\dfrac{1}{3}z - \dfrac{2}{3}$, we get

$$y = -\frac{1}{3}\left(\frac{3x-1}{2}\right) - \frac{2}{3}$$
$$y = \frac{-3x-3}{6}$$
$$y = \frac{-x-1}{2}$$

The solution set is

$$\left\{\left(x, \frac{-x-1}{2}, \frac{3x-1}{2}\right) \mid x \text{ is any real number}\right\}.$$

4.

$A + B = \left[\begin{array}{cc} 3 & -4 \\ -6 & 10 \end{array}\right]$

5.

$2A - B = \left[\begin{array}{cc} 2 & -2 \\ -4 & 8 \end{array}\right] - \left[\begin{array}{cc} 2 & -3 \\ -4 & 6 \end{array}\right] =$

$\left[\begin{array}{cc} 0 & 1 \\ 0 & 2 \end{array}\right]$

6.

$AB = \left[\begin{array}{cc} 2+4 & -3-6 \\ -4-16 & 6+24 \end{array}\right] = \left[\begin{array}{cc} 6 & -9 \\ -20 & 30 \end{array}\right]$

7.

$AC = \left[\begin{array}{c} -2-1 \\ 4+4 \end{array}\right] = \left[\begin{array}{c} -3 \\ 8 \end{array}\right]$

8. CB is undefined

9. $FG = [-2 \quad 3-2 \quad 1+1] = [-2 \quad 1 \quad 2]$

10.

$EF = \left[\begin{array}{ccc} 2(1) & 2(0) & 2(-1) \\ 3(1) & 3(0) & 3(-1) \\ -1(1) & -1(0) & -1(-1) \end{array}\right]$

$= \left[\begin{array}{ccc} 2 & 0 & -2 \\ 3 & 0 & -3 \\ -1 & 0 & 1 \end{array}\right]$

11.

On $\begin{bmatrix} 1 & -1 & | & 1 & 0 \\ -2 & 4 & | & 0 & 1 \end{bmatrix}$, use

$2R_1 + R_2 \rightarrow R_2$ to get

$\begin{bmatrix} 1 & -1 & | & 1 & 0 \\ 0 & 2 & | & 2 & 1 \end{bmatrix}$, use

$\frac{1}{2}R_2 + R_1 \rightarrow R_1$ and $\frac{1}{2}R_2 \rightarrow R_2$

to get $\begin{bmatrix} 1 & 0 & | & 2 & 1/2 \\ 0 & 1 & | & 1 & 1/2 \end{bmatrix}$.

Then $A^{-1} = \begin{bmatrix} 2 & 1/2 \\ 1 & 1/2 \end{bmatrix}$.

12.

On $\begin{bmatrix} -2 & 3 & 1 & | & 1 & 0 & 0 \\ -3 & 1 & 3 & | & 0 & 1 & 0 \\ 0 & 2 & -1 & | & 0 & 0 & 1 \end{bmatrix}$,

use $-\frac{1}{2}R_1 \rightarrow R_1$ and $-\frac{3}{2}R_1 + R_2 \rightarrow R_2$ to get

$\begin{bmatrix} 1 & -3/2 & -1/2 & | & -1/2 & 0 & 0 \\ 0 & -7/2 & 3/2 & | & -3/2 & 1 & 0 \\ 0 & 2 & -1 & | & 0 & 0 & 1 \end{bmatrix}$, use

$\frac{4}{7}R_2 + R_3 \rightarrow R_3$, $-\frac{3}{7}R_2 + R_1 \rightarrow R_1$,

and $-\frac{2}{7}R_2 \rightarrow R_2$ to get

$\begin{bmatrix} 1 & 0 & -8/7 & | & 1/7 & -3/7 & 0 \\ 0 & 1 & -3/7 & | & 3/7 & -2/7 & 0 \\ 0 & 0 & -1/7 & | & -6/7 & 4/7 & 1 \end{bmatrix}$, use

$-3R_3 + R_2 \rightarrow R_2$, $-8R_3 + R_1 \rightarrow R_1$,

and $-7R_3 \rightarrow R_3$ to get

$\begin{bmatrix} 1 & 0 & 0 & | & 7 & -5 & -8 \\ 0 & 1 & 0 & | & 3 & -2 & -3 \\ 0 & 0 & 1 & | & 6 & -4 & -7 \end{bmatrix}$.

Then $G^{-1} = \begin{bmatrix} 7 & -5 & -8 \\ 3 & -2 & -3 \\ 6 & -4 & -7 \end{bmatrix}$.

13. $|A| = 4 - 2 = 2$

14. $|B| = 12 - 12 = 0$

15. Expanding about the third row, we find

$|G| = -2 \begin{vmatrix} -2 & 1 \\ -3 & 3 \end{vmatrix} + (-1) \begin{vmatrix} -2 & 3 \\ -3 & 1 \end{vmatrix} =$

$-2(-3) + (-1)(7) = -1$

16.

One finds $D = \begin{vmatrix} 1 & -1 \\ -2 & 4 \end{vmatrix} = 2$,

$D_x = \begin{vmatrix} 2 & -1 \\ 2 & 4 \end{vmatrix} = 10, D_y = \begin{vmatrix} 1 & 2 \\ -2 & 2 \end{vmatrix} = 6.$

Then $\frac{D_x}{D} = 5$ and $\frac{D_y}{D} = 3$.

The solution set is $\{(5,3)\}$.

17. Cramer's Rule is not applicable since

$D = \begin{vmatrix} 2 & -3 \\ -4 & 6 \end{vmatrix} = 0$. Rather, multiply first

equation by 2 and add to the second one.

$$\begin{array}{rcl} 4x - 6y & = & 12 \\ -4x + 6y & = & 1 \\ \hline 0 & = & 13 \end{array}$$

Inconsistent and the solution set is $\emptyset$.

18.

One finds $D = \begin{vmatrix} -2 & 3 & 1 \\ -3 & 1 & 3 \\ 0 & 2 & -1 \end{vmatrix} = -1$,

$D_x = \begin{vmatrix} -2 & 3 & 1 \\ -4 & 1 & 3 \\ 0 & 2 & -1 \end{vmatrix} = -6,$

$D_y = \begin{vmatrix} -2 & -2 & 1 \\ -3 & -4 & 3 \\ 0 & 0 & -1 \end{vmatrix} = -2,$ and

$D_z = \begin{vmatrix} -2 & 3 & -2 \\ -3 & 1 & -4 \\ 0 & 2 & 0 \end{vmatrix} = -4.$

Then $\frac{D_x}{D} = 6$, $\frac{D_y}{D} = 2$, and $\frac{D_z}{D} = 4$.

The solution set is $\{(6,2,4)\}$.

19. The inverse of coefficient matrix is given

by $A^{-1} = \begin{bmatrix} 2 & 1/2 \\ 1 & 1/2 \end{bmatrix}$. Since

$A^{-1} \begin{bmatrix} 1 \\ -8 \end{bmatrix} = \begin{bmatrix} -2 \\ -3 \end{bmatrix}$, the solution

set is $\{(-2,-3)\}$.

20. The inverse of coefficient matrix was found in Exercise 12: $A^{-1} = \begin{bmatrix} 7 & -5 & -8 \\ 3 & -2 & -3 \\ 6 & -4 & -7 \end{bmatrix}$.

Since $A^{-1} \begin{bmatrix} 1 \\ 0 \\ -1 \end{bmatrix} = \begin{bmatrix} 15 \\ 6 \\ 13 \end{bmatrix}$, the solution set is $\{(15, 6, 13)\}$.

21. Corresponding system of equations is

$$\begin{aligned} x - y &= 12 \\ 35y - 10x &= 730. \end{aligned}$$

Solving this system, one finds $x = 46$ copies were bought and $y = 34$ copies were sold.

22. Substitute $(0, 3)$, $(1, -1/2)$, and $(4, 3)$ into $y = ax^2 + b\sqrt{x} + c$. Then we obtain

$$\begin{aligned} c &= 3 \\ a + b + c &= -\frac{1}{2} \\ 16a + 2b + c &= 3. \end{aligned}$$

Solving this system, one finds $a = 0.5$, $b = -4$, $c = 3$, and the graph is given by

$$y = 0.5x^2 - 4\sqrt{x} + 3.$$

Tying It All Together

1. Simplify the left-hand side.

$$\begin{aligned} 2x + 6 - 5x &= 7 \\ -3x &= 1 \end{aligned}$$

Solution set is $\{-1/3\}$.

2. Multiply each side of the equation by 30.

$$\begin{aligned} \frac{1}{2}x - \frac{1}{6} &= \frac{4}{5} \\ 15x - 5 &= 24 \\ 15x &= 29 \end{aligned}$$

Solution set is $\{29/15\}$.

3. Multiply $\frac{1}{2}$ to $(2x - 2)$.

$$\begin{aligned} (x - 1)(6x - 8) &= 4 \\ 6x^2 - 14x + 8 &= 4 \\ 6x^2 - 14x + 4 &= 0 \\ 2(3x^2 - 7x + 2) &= 0 \\ 2(3x - 1)(x - 2) &= 0 \end{aligned}$$

Solution set is $\{1/3, 2\}$.

4. Simplify left-hand side.

$$\begin{aligned} 1 - (4x - 2) &= 9 \\ -4x + 3 &= 9 \\ -4x &= 6 \end{aligned}$$

Solution set is $\{-3/2\}$.

5. The solution set is $\{(4, -2)\}$ as can be seen from the point of intersection.

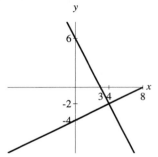

6. Using $y = 1 - 2x$ in $2x + 6y = 2$, we get

$$\begin{aligned} 2x + 6(1 - 2x) &= 2 \\ 2x + 6 - 12x &= 2 \\ -10x &= -4. \end{aligned}$$

Using $x = \frac{2}{5}$ in $y = 1 - 2x$, we have

$y = 1 - \frac{4}{5} = \frac{1}{5}$. Solution set is $\left\{ \left(\frac{2}{5}, \frac{1}{5} \right) \right\}$.

7. Multiply second equation by 6 and add to the first one.

$$\begin{aligned} 2x - 0.06y &= 20 \\ \underline{18x + 0.06y} &= \underline{120} \\ 20x &= 140 \end{aligned}$$

Using $x = 7$ in $2x - 0.06y = 20$, we find

$$\begin{aligned} 14 - 0.06y &= 20 \\ -0.06y &= 6 \\ y &= -100. \end{aligned}$$

The solution set is $\{(7, -100)\}$.

8.

On $\begin{bmatrix} 2 & -1 & | & -1 \\ 1 & 3 & | & -11 \end{bmatrix}$, use $-2R_2 + R_1 \to R_2$

to get

$\begin{bmatrix} 2 & -1 & | & -1 \\ 0 & -7 & | & 21 \end{bmatrix}$, use $-\dfrac{1}{7}R_2 \to R_2$

and $\dfrac{1}{2}R_1 \to R_1$ to get

$\begin{bmatrix} 1 & -1/2 & | & -1/2 \\ 0 & 1 & | & -3 \end{bmatrix}$, use $\dfrac{1}{2}R_2 + R_1 \to R_1$

to get $\begin{bmatrix} 1 & 0 & | & -2 \\ 0 & 1 & | & -3 \end{bmatrix}$.

The solution set is $\{(-2, -3)\}$.

9. Inverse of the coefficient matrix A is

$A^{-1} = \begin{bmatrix} -1/2 & -5/2 \\ -1/2 & -3/2 \end{bmatrix}$.

Since $A^{-1}\begin{bmatrix} -7 \\ 1 \end{bmatrix} = \begin{bmatrix} 1 \\ 2 \end{bmatrix}$,

the solution set is $\{(1, 2)\}$.

10.

One finds $D = \begin{vmatrix} 4 & -3 \\ 3 & -5 \end{vmatrix} = -11$,

$D_x = \begin{vmatrix} 5 & -3 \\ 1 & -5 \end{vmatrix} = -22$, and

$D_y = \begin{vmatrix} 4 & 5 \\ 3 & 1 \end{vmatrix} = -11$.

Then $\dfrac{D_x}{D} = 2$ and $\dfrac{D_y}{D} = 1$.

The solution set is $\{(2, 1)\}$.

11. Note, $6 = 3x - 5y$ and $1 = 3x - 5y$. So, we get $6 = 1$, a false statement. Thus, the solution set is $\emptyset$.

12. Multiply the first equation by 2 and add this to the second equation.

$$\begin{array}{rcr} 2x - 4y & = & 6 \\ -2x + 4y & = & -6 \\ \hline 0 & = & 0 \end{array}$$

A dependent system and the solution set is

$$\{(x, y) \mid x - 2y = 3\}.$$

13. Substitute $x = y - 1$ in $x^2 + y^2 = 25$. Then

$$\begin{array}{rcl} (y^2 - 2y + 1) + y^2 & = & 25 \\ 2y^2 - 2y - 24 & = & 0 \\ 2(y^2 - y - 12) & = & 0 \\ 2(y - 4)(y + 3) & = & 0. \end{array}$$

If $y = 4$, then $x = y - 1 = 4 - 1 = 3$.
If $y = -3$, then $x = (-3) - 1 = -4$.
The solution set is $\{(3, 4), (-4, -3)\}$.

14. Substituting $y = x^2 - 1$ in $x + y = 1$, we get

$$\begin{array}{rcl} x + (x^2 - 1) & = & 1 \\ x^2 + x - 2 & = & 0 \\ (x + 2)(x - 1) & = & 0. \end{array}$$

If $x = -2$, then $y = x^2 - 1 = (-2)^2 - 1 = 3$.
If $x = 1$, then $y = 1^2 - 1 = 0$.
Solution set is $\{(-2, 3), (1, 0)\}$.

15. Adding the given equations, we obtain

$$\begin{array}{rcr} x^2 - y^2 & = & 1 \\ x^2 + y^2 & = & 3 \\ \hline 2x^2 & = & 4 \\ x^2 & = & 2 \\ x & = & \pm\sqrt{2}. \end{array}$$

Substitute $x = \pm\sqrt{2}$ into $x^2 + y^2 = 3$. Then $2 + y^2 = 3$ and so $y = \pm 1$. The solution set is $\{(\sqrt{2}, \pm 1), (-\sqrt{2}, \pm 1)\}$.

16. Multiply the first equation by -2 and add it to the second equation.

$$\begin{array}{rcr} -2x^2 + 2y^2 & = & -2 \\ 2x^2 + 3y^2 & = & 2 \\ \hline 5y^2 & = & 0 \\ y & = & 0 \end{array}$$

Substitute $y = 0$ into $x^2 - y^2 = 1$. Then $x^2 = 1$ and so $x = \pm 1$. The solution set is $\{(\pm 1, 0)\}$.

For Thought

1. False, vertex is $(0, -1/2)$.   **2.**   True

3. True, since $p = 3/2$ and the focus is $(4, 5)$, vertex is $(4 - 3/2, 5) = (5/2, 5)$.

4. False, focus is at $(0, 1/4)$ since parabola opens upward.   **5.**   True, $p = 1/4$.

6. False. Since $p = 4$ and the vertex is $(2, -1)$, equation of parabola is $y = \dfrac{1}{16}(x - 2)^2 - 1$ and x-intercepts are $(6, 0)$, $(-2, 0)$.

7. False, if $x = 0$ then $y = 0$ and y-intercept is $(0, 0)$.   **8.**   True

9. False. Since $p = 1/4$ and the vertex is $(5, 4)$, the focus is $(5, 4 + 1/4) = (-5, 17/4)$.

10. False, it opens to the left.

7.1 Exercises

1. Note $p = 1$.

Vertex $(0, 0)$, focus $(0, 1)$, and directrix $y = -1$

3. Note $p = -1/2$. Vertex $(1, 2)$, focus $(1, 3/2)$, and directrix $y = 5/2$

5. Note $p = 3/4$. Vertex $(3, 1)$, focus $(15/4, 1)$, and directrix $x = 9/4$

7. Since the vertex is equidistant from $y = -2$ and $(0, 2)$, the vertex is $(0, 0)$ and $p = 2$. Then $a = \dfrac{1}{4p} = \dfrac{1}{8}$ and an equation is $y = \dfrac{1}{8}x^2$.

9. Since the vertex is equidistant from $y = 3$ and $(0, -3)$, the vertex is $(0, 0)$ and $p = -3$. Then $a = \dfrac{1}{4p} = -\dfrac{1}{12}$ and an equation is $y = -\dfrac{1}{12}x^2$.

11. One finds $p = \dfrac{3}{2}$. So $a = \dfrac{1}{4p} = \dfrac{1}{6}$ and vertex is $\left(3, 5 - \dfrac{3}{2}\right) = \left(3, \dfrac{7}{2}\right)$. Parabola is given by $y = \dfrac{1}{6}(x - 3)^2 + \dfrac{7}{2}$.

13. One finds $p = -\dfrac{5}{2}$. So $a = \dfrac{1}{4p} = -\dfrac{1}{10}$ and vertex is $\left(1, -3 + \dfrac{5}{2}\right) = \left(1, -\dfrac{1}{2}\right)$. Parabola is given by $y = -\dfrac{1}{10}(x - 1)^2 - \dfrac{1}{2}$.

15. One finds $p = 0.2$. So $a = \dfrac{1}{4p} = 1.25$ and vertex is $(-2, 1.2 - 0.2) = (-2, 1)$. Parabola is given by $y = 1.25(x + 2)^2 + 1$.

17. Since $p = 1$, we get $a = \dfrac{1}{4p} = \dfrac{1}{4}$.

An equation is $y = \dfrac{1}{4}x^2$.

19. Since $p = -\dfrac{1}{4}$, we get $a = \dfrac{1}{4p} = -1$.

An equation is $y = -x^2$

21. Note, vertex is $(1, 0)$ and since $a = \dfrac{1}{4p} = 1$, we find $p = \dfrac{1}{4}$. The focus is $(1, 0 + p) = \left(1, \dfrac{1}{4}\right)$ and the directrix is $y = 0 - p$ or $y = -\dfrac{1}{4}$.

23. Note, vertex is $(3, 0)$ and since $a = \dfrac{1}{4p} = \dfrac{1}{4}$, we find $p = 1$. The focus is $(3, 0 + p) = (3, 1)$ and the directrix is $y = 0 - p$ or $y = -1$.

25. Note, vertex is $(3, 4)$ and since $a = \dfrac{1}{4p} = -2$, we find $p = -\dfrac{1}{8}$. The focus is $(3, 4 + p) = \left(3, \dfrac{31}{8}\right)$ and the directrix is $y = 4 - p$ or $y = \dfrac{33}{8}$.

27. Completing the square, we obtain

$$y = (x^2 - 8x + 16) - 16 + 3$$
$$y = (x - 4)^2 - 13.$$

Since $\dfrac{1}{4p} = 1$, $p = 0.25$.

Since vertex is $(4, -13)$, focus is $(4, -13 + 0.25) = (4, -51/4)$, and directrix is $y = -13 - p = -53/4$.

29. Completing the square, we obtain

$$\begin{aligned} y &= 2(x^2 + 6x + 9) + 5 - 18 \\ y &= 2(x + 3)^2 - 13. \end{aligned}$$

Since $\dfrac{1}{4p} = 2$, $p = 1/8$.

Since vertex is $(-3, -13)$, the focus is $(-3, -13 + 1/8) = (-3, -103/8)$, and directrix is $y = -13 - 1/8 = -105/8$.

31. Completing the square, we get

$$\begin{aligned} y &= -2(x^2 - 3x + 9/4) + 1 + 9/2 \\ y &= -2(x - 3/2)^2 + 11/2. \end{aligned}$$

Since $\dfrac{1}{4p} = -2$, $p = -1/8$.

Since vertex is $(3/2, 11/2)$, the focus is $(3/2, 11/2 - 1/8) = (3/2, 43/8)$, and directrix is $y = 11/2 + 1/8 = 45/8$.

33. Completing the square,

$$\begin{aligned} y &= 5(x^2 + 6x + 9) - 45 \\ y &= 5(x + 3)^2 - 45. \end{aligned}$$

Since $\dfrac{1}{4p} = 5$, we have $p = 0.05$.

Since vertex is $(-3, -45)$, the focus is $(-3, -45 + 0.05) = (-3, -44.95)$, and directrix is $y = -45 - .05 = -45.05$.

35. Completing the square, we get

$$\begin{aligned} y &= \frac{1}{8}(x^2 - 4x + 4) + \frac{9}{2} - \frac{1}{2} \\ y &= \frac{1}{8}(x - 2)^2 + 4. \end{aligned}$$

Since $\dfrac{1}{4p} = 1/8$, we find $p = 2$.

Since vertex is $(2, 4)$, the focus is $(2, 4 + 2) = (2, 6)$, and directrix is $y = 4 - 2$ or $y = 2$.

37. Since $a = 1$ and $b = -4$, we find $x = \dfrac{-b}{2a} = \dfrac{4}{2} = 2$. Since $\dfrac{1}{4p} = a = 1$, we obtain $p = 1/4$.

Substituting $x = 2$, we get $y = 2^2 - 4(2) + 3 = -1$. Thus, the vertex is $(2, -1)$, the focus is $(2, -1 + p) = (2, -3/4)$, the directrix is $y = -1 - p = -5/4$, and the parabola opens upward since $a > 0$.

39. Note, $a = -1$, $b = 2$. So $x = \dfrac{-b}{2a} = \dfrac{-2}{-2} = 1$ and since $\dfrac{1}{4p} = a = -1$, $p = -1/4$.

Substituting $x = 1$, $y = -(1)^2 + 2(1) - 5 = -4$. The vertex is $(1, -4)$ and

focus is $(1, -4 + p) = (1, -17/4)$,

directrix is $y = -4 - p = -15/4$, and parabola opens down since $a < 0$.

41. Note, $a = 3$, $b = -6$. So $x = \dfrac{-b}{2a} = \dfrac{6}{6} = 1$ and since $\dfrac{1}{4p} = a = 3$, $p = 1/12$. Substituting $x = 1$, $y = 3(1)^2 - 6(1) + 1 = -2$.

The vertex is $(1, -2)$ and

focus is $(1, -2 + p) = (1, -23/12)$,

directrix is $y = -2 - p = -25/12$, and parabola opens up since $a > 0$.

43. Note, $a = -1/2$, $b = -3$. So $x = \dfrac{-b}{2a} = \dfrac{3}{-1} = -3$ and since $\dfrac{1}{4p} = a = -1/2$, $p = -1/2$. Substituting $x = -3$, we have $y = -\dfrac{1}{2}(-3)^2 - 3(-3) + 2 = \dfrac{13}{2}$. The vertex is $\left(-3, \dfrac{13}{2}\right)$ and focus is $\left(-3, \dfrac{13}{2} + p\right) = (-3, 6)$, directrix is $y = \dfrac{13}{2} - p = 7$, and parabola opens down since $a < 0$.

45. Note, $y = \dfrac{1}{4}x^2 + 5$ is of the form $y = a(x - h)^2 + k$. So $h = 0$, $k = 5$, and $\dfrac{1}{4p} = a = \dfrac{1}{4}$ from which we have $p = 1$. The

vertex is $(h,k) = (0,5)$, focus is $(0,5+p) = (0,6)$, directrix is $y = 5 - p = 4$ and parabola opens up since $a > 0$.

47. From the given focus and directrix, one

finds $p = 1/4$. So $a = \dfrac{1}{4p} = 1$, vertex is

$(h,k) = (1/2, -2 - p) = (1/2, -9/4)$,

axis of symmetry is $x = 1/2$, and parabola is given by $y = a(x-h)^2 + k = \left(x - \dfrac{1}{2}\right)^2 - \dfrac{9}{4}$.

If $y = 0$ then $x - \dfrac{1}{2} = \pm\dfrac{3}{2}$ or $x = 2, -1$.

The x-intercepts are $(2,0), (-1,0)$.

If $x = 0$ then $y = \left(0 - \dfrac{1}{2}\right)^2 - \dfrac{9}{4} = -2$ and

y-intercept is $(0,-2)$.

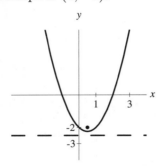

49. From the given focus and directrix one

finds $p = -1/4$. So $a = \dfrac{1}{4p} = -1$, vertex is

$(h,k) = (-1/2, 6 - p) = (-1/2, 25/4)$,

axis of symmetry is $x = -1/2$, and parabola is given by

$$y = a(x-h)^2 + k = -\left(x + \dfrac{1}{2}\right)^2 + \dfrac{25}{4}.$$

If $y = 0$ then $x + \dfrac{1}{2} = \pm\dfrac{5}{2}$ or $x = -3, 2$.

The x-intercepts are $(-3,0), (2,0)$.

If $x = 0$ then $y = -\left(0 + \dfrac{1}{2}\right)^2 + \dfrac{25}{4} = 6$

and y-intercept is $(0,6)$.

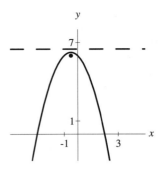

51. Since $\dfrac{1}{2}(x+2)^2 + 2$ is of the form $a(x-h)^2 + k$,

vertex is $(h,k) = (-2,2)$ and axis of symmetry is $x = -2$. If $y = 0$ then $0 = \dfrac{1}{2}(x+2)^2 + 2$; this has no solution since left-hand side is always positive. No x-intercept. If $x = 0$ then

$$y = \dfrac{1}{2}(0+2)^2 + 2 = 4.$$ y-intercept is $(0,4)$.

Since $\dfrac{1}{4p} = a = \dfrac{1}{2}$, $p = \dfrac{1}{2}$, focus is $(h, k+p) = (-2, 5/2)$, and directrix is $y = k - p = 3/2$.

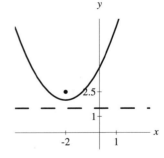

53. Since $-\dfrac{1}{4}(x+4)^2 + 2$ is of the form $a(x-h)^2 + k$,

vertex is $(h,k) = (-4,2)$ and axis of symmetry is $x = -4$. If $y = 0$ then

$$\begin{aligned} \dfrac{1}{4}(x+4)^2 &= 2 \\ x + 4 &= \pm\sqrt{8}. \end{aligned}$$

x-intercepts are $(-4 \pm 2\sqrt{2}, 0)$. If $x = 0$, then $y = -\dfrac{1}{4}(0+4)^2 + 2 = -2$. The y-intercept is $(0,-2)$. Since $\dfrac{1}{4p} = a = -\dfrac{1}{4}$, $p = -1$,

focus is $(h, k+p) = (-4,1)$, and directrix is $y = k - p = 3$.

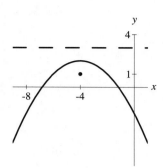

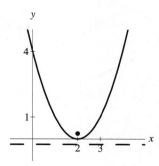

55. Since $\frac{1}{2}x^2 - 2$ is of the form $a(x-h)^2 + k$,

vertex is $(h, k) = (0, -2)$ and axis of symmetry is $x = 0$. If $y = 0$, then

$$\begin{aligned} \frac{1}{2}x^2 &= 2 \\ x^2 &= 4. \end{aligned}$$

x-intercepts are $(\pm 2, 0)$. If $x = 0$ then

$y = \frac{1}{2}(0)^2 - 2 = -2$. The y-intercept is $(0, -2)$.

Since $\frac{1}{4p} = a = \frac{1}{2}$, $p = 1/2$, focus is

$(h, k+p) = (0, -3/2)$, and directrix is $y = k - p = -5/2$.

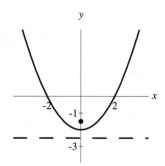

57. Since $y = (x-2)^2$ is of the form $a(x-h)^2 + k$, vertex is $(h, k) = (2, 0)$, and axis of symmetry is $x = 2$. If $y = 0$ then $(x-2)^2 = 0$ and x-intercept is $(2, 0)$. If $x = 0$ then $y = (0-2)^2 = 4$ and y-intercept is $(0, 4)$. Since $\frac{1}{4p} = a = 1$,

$p = 1/4$, focus is $(h, k+p) = (2, 1/4)$, and directrix is $y = k - p = -1/4$.

59. By completing the square, we obtain

$$y = \frac{1}{3}(x - 3/2)^2 - 3/4.$$

Vertex is $(h, k) = (3/2, -3/4)$ and axis of symmetry is $x = 3/2$. If $y = 0$, then

$$\begin{aligned} \frac{1}{3}\left(x - \frac{3}{2}\right)^2 &= \frac{3}{4} \\ \left(x - \frac{3}{2}\right)^2 &= \frac{9}{4} \\ x &= \frac{3}{2} \pm \frac{3}{2}. \end{aligned}$$

x-intercepts are $(3, 0), (0, 0)$. If $x = 0$, then

$y = \frac{1}{3}\left(0 - \frac{3}{2}\right)^2 - \frac{3}{4} = 0$ and y-intercept is

$(0, 0)$. Since $\frac{1}{4p} = a = 1/3$, $p = 3/4$, focus

is $(h, k+p) = (3/2, 0)$, and directrix is $y = k - p = -3/2$ or $y = -3/2$.

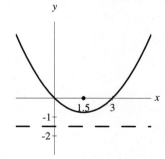

61. Since $x = -y^2$ is of the form $x = a(y-h)^2 + k$, vertex is $(k, h) = (0, 0)$ and axis of symmetry is $y = 0$. If $y = 0$ then $x = -0^2 = 0$ and x-intercept is $(0, 0)$. If $x = 0$ then $0 = -y^2$ and y-intercept is $(0, 0)$. Since $\frac{1}{4p} = a = -1$,

$p = -1/4$, focus is $(k+p, h) = (-1/4, 0)$, and directrix is $x = k - p = 1/4$.

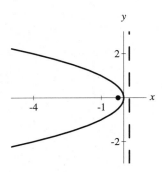

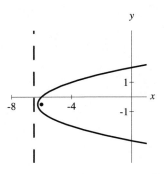

63. Since $x = -\dfrac{1}{4}y^2 + 1$ is of the form

$$x = a(y - h)^2 + k,$$

vertex is $(k, h) = (1, 0)$ and axis of symmetry is $y = 0$. If $y = 0$ then $x = 1$ and x-intercept is $(1, 0)$. If $x = 0$ then $\dfrac{1}{4}y^2 = 1$, $y^2 = 4$, and y-intercepts are $(0, \pm 2)$. Since $\dfrac{1}{4p} = a = -\dfrac{1}{4}$, $p = -1$, focus is $(k + p, h) = (0, 0)$, and directrix is $x = k - p = 2$ or $x = 2$.

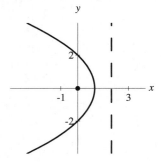

65. By completing the square, we obtain

$$x = (y + 1/2)^2 - 25/4.$$

Vertex is $(k, h) = (-25/4, -1/2)$ and axis of symmetry is $y = -1/2$. If $y = 0$, then $x = 1/4 - 25/4 = -6$. The x-intercept is $(-6, 0)$. If $x = 0$, then

$$\left(y + \frac{1}{2}\right)^2 = \frac{25}{4}$$
$$y = -\frac{1}{2} \pm \frac{5}{2}$$
$$y = 2, -3.$$

y-intercepts are $(0, 2), (0, -3)$. Since $\dfrac{1}{4p} = a = 1$, $p = 1/4$, focus is $(k + p, h) = (-6, -1/2)$, and directrix is $x = k - p = -13/2$.

67. By completing the square, we get

$$x = -\frac{1}{2}(y + 1)^2 - 7/2.$$

Vertex is $(k, h) = (-7/2, -1)$ and axis of symmetry is $y = -1$. If $y = 0$, then $x = -1/2 - 7/2 = -4$. The x-intercept is $(-4, 0)$. If $x = 0$, then

$$0 = -\frac{1}{2}(y + 1)^2 - 7/2 < 0$$

which is inconsistent and so there is no y-intercept. Since $\dfrac{1}{4p} = a = -1/2$, $p = -1/2$, focus is $(k + p, h) = (-4, -1)$, and directrix is $x = k - p = -3$.

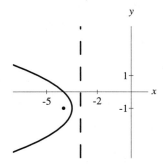

69. Since $x = 2(y - 1)^2 + 3$ is of the form $a(y - h)^2 + k$, we find that the vertex is $(k, h) = (3, 1)$ and axis of symmetry is $y = 1$. If $y = 0$ then $x = 2(-1)^2 + 3 = 5$ and x-intercept is $(5, 0)$. If $x = 0$, we obtain $2(y - 1)^2 + 3 = 0$ which is inconsistent since the left-hand side is always positive. No y-intercept.

Since $\dfrac{1}{4p} = a = 2$, $p = 1/8$, focus is $(k + p, h) = (25/8, 1)$, and directrix is $x = k - p = 23/8$.

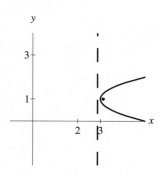

71. Since $x = -\dfrac{1}{2}(y+2)^2 + 1$ is of the form $a(y-h)^2 + k$, vertex is $(k,h) = (1,-2)$ and axis of symmetry is $y = -2$. If $y = 0$, then $x = -\dfrac{1}{2}(2)^2 + 1 = -1$ and x-intercept is $(-1,0)$. If $x = 0$, then

$$\begin{aligned} \frac{1}{2}(y+2)^2 &= 1 \\ (y+2)^2 &= 2 \\ y &= -2 \pm \sqrt{2}. \end{aligned}$$

The y-intercepts are $(0, -2 \pm \sqrt{2})$. Since $\dfrac{1}{4p} = a = -\dfrac{1}{2}$, we find $p = -\dfrac{1}{2}$, focus is $(k+p, h) = (1/2, -2)$, and directrix is $x = k - p = 3/2$.

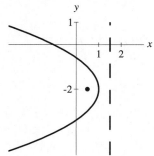

73. Since focus is 1 unit above the vertex $(1,4)$, we obtain $p = 1$. Then $a = \dfrac{1}{4p} = \dfrac{1}{4}$ and parabola is given by $y = \dfrac{1}{4}(x-1)^2 + 4$.

75. Since vertex $(0,0)$ is 2 units to the right of the directrix, we find $p = 2$ and parabola opens to the right. Then $a = \dfrac{1}{4p} = \dfrac{1}{8}$ and the parabola is given by $x = \dfrac{1}{8}y^2$.

77. Since the parabola opens up, $p = 55(12)$ inches, and the vertex is $(0,0)$. Note, $\dfrac{1}{4p} = \dfrac{1}{2640}$. Thus, the parabola is given by $y = \dfrac{1}{2640}x^2$. Thickness at the outside edge is $23 + \dfrac{1}{2640}(100)^2 \approx 26.8$ in.

79. $y = x^2$ has vertex $(0,0)$ and opens up. The second parabola can be written as

$$y = 2(x-1)^2 + 3$$

and its vertex is $(1,3)$. In the given viewing window these graphs look alike.

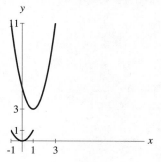

81. Two functions are $f_1(x) = \sqrt{-x}$ and $f_2(x) = -\sqrt{-x}$ where $x \le 0$.

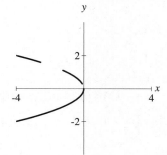

For Thought

1. False, y-intercepts are $(\pm 3, 0)$.

2. True, since it can be written as $\dfrac{x^2}{1/2} + y^2 = 1$.

3. True, length of the major axis is $2a = 2(5) = 10$.

4. True, if $y = 0$ then $x^2 = \dfrac{1}{0.5} = 2$ and $x = \pm\sqrt{2}$.

5. True, if $x = 0$ then $y^2 = 3$ and $y = \pm\sqrt{3}$.

6. False, the center is not a point on the circle.

7. True   **8.** False, $(3, -1)$ satisfies equation.

9. False. No point satisfies the equation since the left-hand side is always positive.

10. False, since the circle can be written as $(x - 2)^2 + (y + 1/2)^2 = 53/4$ and so the radius is $\sqrt{53}/2$.

7.2 Exercises

1. Foci $(\pm\sqrt{5}, 0)$, vertices $(\pm 3, 0)$, center $(0, 0)$

3. Foci $(2, 1 \pm \sqrt{5})$, vertices $(2, 4)$ and $(2, -2)$, center $(2, 1)$

5. Since $c = 2$ and $b = 3$, we get
$$a^2 = b^2 + c^2 = 9 + 4 = 13 \text{ and } a = \sqrt{13}.$$

Ellipse is given by $\dfrac{x^2}{13} + \dfrac{y^2}{9} = 1.$

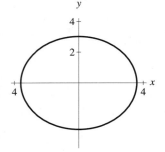

7. Since $c = 4$ and $a = 5$, we find
$$b^2 = a^2 - c^2 = 25 - 16 = 9 \text{ and } b = 3.$$

The ellipse is given by
$$\frac{x^2}{25} + \frac{y^2}{9} = 1.$$

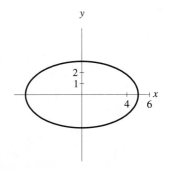

9. Since $c = 2$ and $b = 2$, we obtain
$$a^2 = b^2 + c^2 = 4 + 4 = 8 \text{ and } a = \sqrt{8}.$$

The ellipse is given by
$$\frac{x^2}{4} + \frac{y^2}{8} = 1.$$

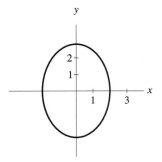

11. Since $c = 4$ and $a = 7$, we obtain
$$b^2 = a^2 - c^2 = 49 - 16 = 33 \text{ and } b = \sqrt{33}.$$

The ellipse is given by
$$\frac{x^2}{33} + \frac{y^2}{49} = 1.$$

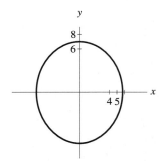

13. Since $c = \sqrt{a^2 - b^2} = \sqrt{16 - 4} = 2\sqrt{3}$, the foci are $(\pm 2\sqrt{3}, 0)$

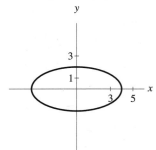

15. Since $c = \sqrt{a^2 - b^2} = \sqrt{36 - 9} = \sqrt{27}$, the foci are $(0, \pm 3\sqrt{3})$

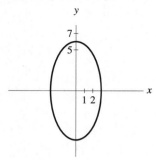

17. Since $c = \sqrt{a^2 - b^2} = \sqrt{25 - 1} = \sqrt{24}$, the foci are $(\pm 2\sqrt{6}, 0)$

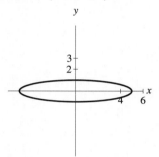

19. Since $c = \sqrt{a^2 - b^2} = \sqrt{25 - 9} = \sqrt{16}$, the foci are $(0, \pm 4)$

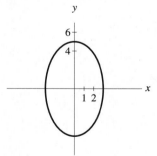

21. From $x^2 + \dfrac{y^2}{9} = 1$, one finds $c = \sqrt{a^2 - b^2}$

$= \sqrt{9 - 1} = \sqrt{8}$ and foci are $(0, \pm 2\sqrt{2})$.

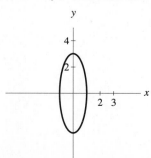

23. From $\dfrac{x^2}{9} + \dfrac{y^2}{4} = 1$, one finds $c = \sqrt{a^2 - b^2}$

$= \sqrt{9 - 4} = \sqrt{5}$ and foci are $(\pm\sqrt{5}, 0)$.

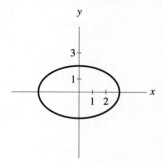

25. Since $c = \sqrt{a^2 - b^2} = \sqrt{16 - 9} = \sqrt{7}$, the foci are $(1 \pm c, -3) = (1 \pm \sqrt{7}, -3)$.

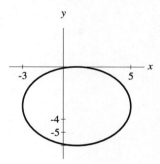

27. Since $c = \sqrt{a^2 - b^2} = \sqrt{25 - 9} = \sqrt{16} = 4$, the foci are $(3, -2 \pm c)$, or $(3, 2)$ and $(3, -6)$.

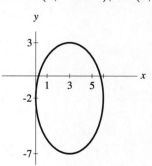

29. Since $\dfrac{(x + 4)^2}{36} + (y + 3)^2 = 1$, we get

$$c = \sqrt{a^2 - b^2} = \sqrt{36 - 1} = \sqrt{35}$$

and the foci are

$$(-4 \pm \sqrt{35}, -3).$$

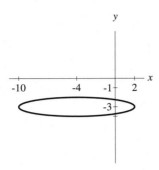

31. If one applies the method of completing the square, one obtains

$$9(x^2 - 2x + 1) + 4(y^2 + 4y + 4) = 11 + 9 + 16$$
$$9(x - 1)^2 + 4(y + 2)^2 = 36$$
$$\frac{(x-1)^2}{4} + \frac{(y+2)^2}{9} = 1.$$

From $a^2 = b^2 + c^2$ with $a = 3$ and $b = 2$, one finds $c = \sqrt{5}$. The foci are $(1, -2 \pm \sqrt{5})$ and a sketch of the ellipse is given.

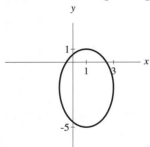

33. Applying the method of completing the square, we find

$$9(x^2 - 6x + 9) + 4(y^2 + 4y + 4) = -61 + 81 + 16$$
$$9(x - 3)^2 + 4(y + 2)^2 = 36$$
$$\frac{(x-3)^2}{4} + \frac{(y+2)^2}{9} = 1.$$

From $a^2 = b^2 + c^2$ with $a = 3$ and $b = 2$, we find $c = \sqrt{5}$. The foci are $(3, -2 \pm \sqrt{5})$ and a sketch of the ellipse is given.

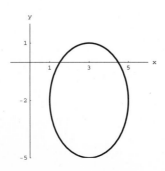

35. Since $\dfrac{x^2}{16} + \dfrac{y^2}{4} = 1$, we get $c = \sqrt{a^2 - b^2}$ $= \sqrt{16 - 4} = \sqrt{12}$, and the foci are $(\pm 2\sqrt{3}, 0)$.

37. Since $\dfrac{(x+1)^2}{4} + \dfrac{(y+2)^2}{16} = 1$, we get $c = \sqrt{a^2 - b^2} = \sqrt{16 - 4} = \sqrt{12}$, and the foci are $(-1, -2 \pm 2\sqrt{3})$.

39. $x^2 + y^2 = 4$

41. Since $r = \sqrt{(4-0)^2 + (5-0)^2} = \sqrt{41}$, the circle is given by

$$x^2 + y^2 = 41.$$

43. Since $r = \sqrt{(4-2)^2 + (1+3)^2} = \sqrt{20}$, the circle is given by

$$(x - 2)^2 + (y + 3)^2 = 20.$$

45. Since center is $\left(\dfrac{3-1}{2}, \dfrac{4+2}{2}\right) = (1, 3)$ and $r = \sqrt{(3-1)^2 + (4-3)^2} = \sqrt{5}$, the circle is given by

$$(x - 1)^2 + (y - 3)^2 = 5.$$

47. center $(0, 0)$, radius 10

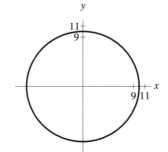

49. center $(1, 2)$, radius 2

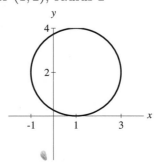

51. center $(-2, -2)$, radius $\sqrt{8}$ or $2\sqrt{2}$

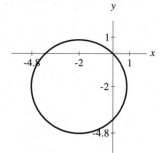

53. Completing the square, we obtain

$$\begin{aligned} x^2 + (y^2 + 2y + 1) &= 8 + 1 \\ x^2 + (y + 1)^2 &= 9. \end{aligned}$$

The center is $(0, -1)$ with radius 3.

55. Completing the square, we find

$$\begin{aligned} x^2 + 8x + 16 + y^2 - 10y + 25 &= 16 + 25 \\ (x + 4)^2 + (y - 5)^2 &= 41. \end{aligned}$$

The center is $(-4, 5)$ with radius $\sqrt{41}$.

57. Completing the square, we find

$$\begin{aligned} (x^2 + 4x + 4) + y^2 &= 5 + 4 \\ (x + 2)^2 + y^2 &= 9. \end{aligned}$$

The center is $(-2, 0)$ with radius 3.

59. Completing the square, we have

$$\begin{aligned} x^2 - x + \frac{1}{4} + y^2 + y + \frac{1}{4} &= \frac{1}{2} + \frac{1}{4} + \frac{1}{4} \\ (x - 1/2)^2 + (y + 1/2)^2 &= 1. \end{aligned}$$

The center is $(0.5, -0.5)$ with radius 1.

61. Completing the square, we get

$$\begin{aligned} x^2 + \frac{2}{3}x + \frac{1}{9} + y^2 + \frac{1}{3}y + \frac{1}{36} &= \frac{1}{9} + \frac{1}{9} + \frac{1}{36} \\ (x + 1/3)^2 + (y + 1/6)^2 &= 1/4. \end{aligned}$$

The center is $(-1/3, -1/6)$ with radius $1/2$.

63. Divide equation by 2 and complete the square.

$$\begin{aligned} x^2 + 2x + y^2 &= 1/2 \\ (x^2 + 2x + 1) + y^2 &= 1/2 + 1 \\ (x + 1)^2 + y^2 &= 3/2 \end{aligned}$$

The center is $(-1, 0)$ with radius $\sqrt{\dfrac{3}{2}}$ or $\dfrac{\sqrt{6}}{2}$.

65. Completing the square, we get

$$\begin{aligned} y^2 - y + x^2 &= 0 \\ y^2 - y + \frac{1}{4} + x^2 &= \frac{1}{4} \\ (y - 1/2)^2 + x^2 &= \frac{1}{4}, \end{aligned}$$

which is a circle.

67. Divide equation by 4.

$$\begin{aligned} x^2 + 3y^2 &= 1 \\ x^2 + \frac{y^2}{1/3} &= 1 \end{aligned}$$

This is an ellipse.

69. Solve for y and complete the square.

$$\begin{aligned} y &= -2x^2 - 4x + 4 \\ y &= -2(x^2 + 2x) + 4 \\ y &= -2(x^2 + 2x + 1) + 4 + 2 \\ y &= -2(x + 1)^2 + 6 \end{aligned}$$

We find a parabola.

71. Note, $(y - 2)^2 = (2 - y)^2$. Then we solve for x.

$$\begin{aligned} 2 - x &= (y - 2)^2 \\ x &= -(y - 2)^2 + 2 \end{aligned}$$

This is a parabola.

73. Simplify and note $(x - 4)^2 = (4 - x)^2$.

$$\begin{aligned} 2(x - 4)^2 &= 4 - y^2 \\ 2(x - 4)^2 + y^2 &= 4 \\ \frac{(x - 4)^2}{2} + \frac{y^2}{4} &= 1 \end{aligned}$$

We find an ellipse.

75. Divide given equation by 9 to get the circle given by $x^2 + y^2 = \dfrac{1}{9}$.

77. From the foci, we obtain $c = 2$ and $a^2 = b^2 + c^2 = b^2 + 4$. Equation of the ellipse is of the form $\dfrac{x^2}{b^2 + 4} + \dfrac{y^2}{b^2} = 1$.

Substitute $x = 2$ and $y = 3$.

$$\frac{4}{b^2 + 4} + \frac{9}{b^2} = 1$$
$$4b^2 + 9(b^2 + 4) = b^2(b^2 + 4)$$
$$0 = b^4 - 9b^2 - 36$$
$$0 = (b^2 - 12)(b^2 + 3)$$

So $b^2 = 12$ and $a^2 = 12 + 4 = 16$.

The ellipse is given by $\dfrac{x^2}{16} + \dfrac{y^2}{12} = 1$.

79. If c is the distance between the center and focus $(0,0)$ then the other focus is $(2c, 0)$. Since the distance between the x-intercepts is $6 + 2c$, which is also the length of the major axis, then $6 + 2c = 2a$. Since $2a$ is the sum of the distances of $(0, 5)$ from the foci,

$$5 + \sqrt{25 + 4c^2} = 2a$$
$$5 + \sqrt{25 + 4c^2} = 6 + 2c$$
$$\sqrt{25 + 4c^2} = 1 + 2c$$
$$25 + 4c^2 = 1 + 4c + 4c^2$$
$$24 = 4c$$
$$6 = c.$$

The other focus is $(2c, 0) = (12, 0)$.

81. Since the sun is a focus of the elliptical orbit, the length of the major axis is $2a = 521$ (the sum of the shortest distance, $P = 1$ AU, and longest distance, $A = 520$ AU, between the orbit and the sun, respectively). In addition, $c = 259.5$ AU (which is the distance from the center to a focus). The eccentricity is given by $e = \dfrac{c}{a} = \dfrac{259.5}{260.5} \approx .996$. Orbit's equation is

$$\frac{x^2}{260.5^2} + \frac{y^2}{520} = 1.$$

83. If $2a$ is the sum of the distances from Haley's comet to the two foci and c is the distance from the sun to the center of the ellipse then, $c = a - 8 \times 10^7$.

Since $0.97 = c/a$, we get

$$0.97 = \frac{a - 8 \times 10^7}{a}$$

and the solution of this equation is $a \approx 2.667 \times 10^9$. So $c = a(0.97) \approx 2.587 \times 10^9$. The maximum distance from the sun is $c + a \approx 5.25 \times 10^9$ km.

85. Solving for y, one finds

$$y^2 = 6360^2 - x^2$$
$$y = \pm\sqrt{6360^2 - x^2}.$$

A sketch of the circle is given.

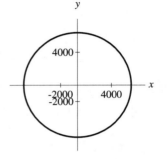

87.

a) As derived below, an equation of the tangent line is given by

$$\frac{x_1 x}{a^2} + \frac{y_1 y}{b^2} = 1$$
$$\frac{-4x}{25} + \frac{\frac{9}{5}y}{9} = 1$$
$$\frac{1}{5}y = 1 + \frac{4x}{25}$$
$$y = \frac{4x}{5} + 5.$$

b) The tangent line and the ellipse intersect at the point $\left(-4, \dfrac{9}{5}\right)$ as shown.

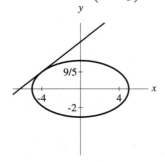

89. A parabolic reflector is preferable since otherwise one would have to place the moving quarterback on one focus of the ellipse.

91. If (x, y) is any point on the ellipse and the sum of the distances from the two foci $(0, \pm c)$ is $2a$, then

$$\sqrt{x^2 + (y - c)^2} + \sqrt{x^2 + (y + c)^2} = 2a.$$

Simplify the inside of the radical and transpose a term to the right-hand side.

$$\sqrt{x^2 + y^2 - 2yc + c^2} = 2a - \sqrt{x^2 + y^2 + 2cy + c^2}$$

Squaring both sides, we find $x^2 + y^2 - 2yc + c^2 = 4a^2 - 4a\sqrt{x^2 + y^2 + 2cy + c^2} + x^2 + y^2 + 2cy + c^2$. Cancel like terms and simplify.

$$
\begin{aligned}
-4yc - 4a^2 &= -4a\sqrt{x^2 + y^2 + 2cy + c^2} \\
yc + a^2 &= a\sqrt{x^2 + y^2 + 2cy + c^2} \\
y^2c^2 + 2yca^2 + a^4 &= a^2(x^2 + y^2 + 2cy + c^2) \\
y^2c^2 + a^4 &= a^2(x^2 + y^2 + c^2) \\
a^4 - a^2c^2 &= a^2x^2 + y^2(a^2 - c^2)
\end{aligned}
$$

Let $b^2 = a^2 - c^2$. So $a^2b^2 = a^2x^2 + b^2y^2$ and

$\dfrac{x^2}{b^2} + \dfrac{y^2}{a^2} = 1$ is an ellipse with foci $(0, \pm c)$ and

one finds that the x-intercepts are $(\pm b, 0)$.

For Thought

1. False, it is a parabola.

2. False, there is no y-intercept.

3. True **4.** True

5. False, $y = \dfrac{b}{a}x$ is an asymptote.

6. True **7.** True, $c = \sqrt{16 + 9} = 5$.

8. True, $c = \sqrt{3 + 5} = \sqrt{8}$.

9. False, $y = \dfrac{2}{3}x$ is an asymptote.

10. False, it is a circle centered at $(0, 0)$.

7.3 Exercises

1. Vertices $(\pm 1, 0)$, foci $(\pm\sqrt{2}, 0)$, asymptotes $y = \pm x$

3. Vertices $(1, \pm 3)$, foci $(1, \pm\sqrt{10})$

Since the slope of the asymptotes are ± 3, the equations of the asymptotes are of the form $y = \pm 3x + b$. If we substitute $(1, 0)$ into $y = \pm 3x + b$, then $0 = \pm 3(1) + b$ or $b = \mp 3$. Thus, the asymptotes are $y = 3x - 3$ and $y = -3x + 3$.

5. Note, $c = \sqrt{a^2 + b^2} = \sqrt{2^2 + 3^2} = \sqrt{13}$.

Foci $(\pm\sqrt{13}, 0)$, asymptotes $y = \pm\dfrac{b}{a}x = \pm\dfrac{3}{2}x$

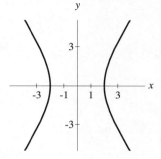

7. Note, $c = \sqrt{a^2 + b^2} = \sqrt{2^2 + 5^2} = \sqrt{29}$.

Foci $(0, \pm\sqrt{29})$, asymptotes $y = \pm\dfrac{a}{b}x = \pm\dfrac{2}{5}x$

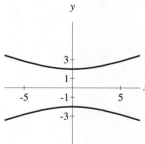

9. Note, $c = \sqrt{a^2 + b^2} = \sqrt{2^2 + 1^2} = \sqrt{5}$.

Foci $(\pm\sqrt{5}, 0)$, asymptotes $y = \pm\dfrac{b}{a}x = \pm\dfrac{1}{2}x$

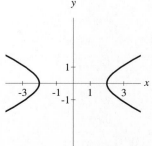

11. Note, $c = \sqrt{a^2 + b^2} = \sqrt{1^2 + 3^2} = \sqrt{10}$.

Foci $(\pm\sqrt{10}, 0)$, asymptotes $y = \pm\dfrac{b}{a}x = \pm 3x$

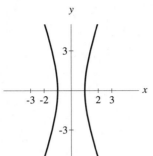

13. Dividing by 144, we get $\dfrac{x^2}{9} - \dfrac{y^2}{16} = 1$.

Note, $c = \sqrt{a^2 + b^2} = \sqrt{3^2 + 4^2} = 5$.

Foci $(\pm 5, 0)$, asymptotes $y = \pm\dfrac{b}{a}x = \pm\dfrac{4}{3}x$

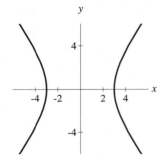

15. Note, $c = \sqrt{a^2 + b^2} = \sqrt{1^2 + 1^2} = \sqrt{2}$.

Foci $(\pm\sqrt{2}, 0)$, asymptotes $y = \pm\dfrac{b}{a}x = \pm x$

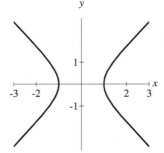

17. Note, $c = \sqrt{a^2 + b^2} = \sqrt{2^2 + 3^2} = \sqrt{13}$.
Since the center is $(-1, 2)$, we find that the foci are $(-1 \pm \sqrt{13}, 2)$. Solving for y in $y - 2 = \pm\dfrac{3}{2}(x + 1)$, we obtain that the asymptotes are $y = \dfrac{3}{2}x + \dfrac{7}{2}$ and $y = -\dfrac{3}{2}x + \dfrac{1}{2}$.

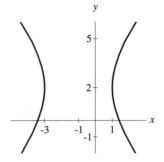

19. Note, $c = \sqrt{a^2 + b^2} = \sqrt{2^2 + 1^2} = \sqrt{5}$.
Since the center is $(-2, 1)$, the foci are $(-2, 1 \pm \sqrt{5})$. Solving for y in $y - 1 = \pm 2(x + 2)$, we obtain that the asymptotes are $y = 2x + 5$ and $y = -2x - 3$.

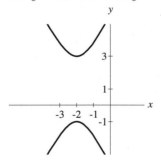

21. Note, $c = \sqrt{a^2 + b^2} = \sqrt{4^2 + 3^2} = 5$.
Since the center is $(-2, 3)$, the foci are $(3, 3)$ and $(-7, 3)$. Solving for y in $y - 3 = \pm\dfrac{3}{4}(x + 2)$, we find that the asymptotes are $y = \dfrac{3}{4}x + \dfrac{9}{2}$ and $y = -\dfrac{3}{4}x + \dfrac{3}{2}$.

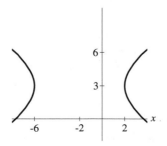

23. Note, $c = \sqrt{a^2 + b^2} = \sqrt{1^2 + 1^2} = \sqrt{2}$.
Since the center is $(3, 3)$, the foci are $(3, 3 \pm \sqrt{2})$. Solving for y in $y - 3 = \pm(x - 3)$, we obtain that the asymptotes are $y = x$ and $y = -x + 6$.

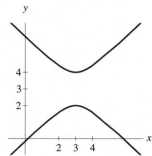

25. Since the x-intercepts are $(\pm 6, 0)$, the hyperbola is given by $\dfrac{x^2}{6^2} - \dfrac{y^2}{b^2} = 1$.
From the asymptotes one gets $\dfrac{1}{2} = \dfrac{b}{6}$ and $b = 3$. An equation is $\dfrac{x^2}{36} - \dfrac{y^2}{9} = 1$.

27. Since the x-intercepts are $(\pm 3, 0)$, the hyperbola is given by $\dfrac{x^2}{3^2} - \dfrac{y^2}{b^2} = 1$.
From the foci, $c = 5$ and $b^2 = c^2 - a^2 = 5^2 - 3^2 = 16$. An equation is $\dfrac{x^2}{9} - \dfrac{y^2}{16} = 1$.

29. By using the vertices of the fundamental rectangle and since it opens sideways, one gets $a = 3$, $b = 5$, and the center is at the origin. An equation is $\dfrac{x^2}{9} - \dfrac{y^2}{25} = 1$.

31. $\dfrac{x^2}{9} - \dfrac{y^2}{16} = 1$

33. $\dfrac{y^2}{9} - \dfrac{(x-1)^2}{9} = 1$

35. Completing the square,

$$\begin{aligned}
y^2 - (x^2 - 2x) &= 2 \\
y^2 - (x^2 - 2x + 1) &= 2 - 1 \\
y^2 - (x - 1)^2 &= 1
\end{aligned}$$

we obtain a hyperbola.

37. $y = x^2 + 2x$ is a parabola.

39. Simplifying,

$$\begin{aligned}
25x^2 + 25y^2 &= 2500 \\
x^2 + y^2 &= 100
\end{aligned}$$

we obtain a circle.

41. Simplifying,

$$\begin{aligned}
25x &= -100y^2 + 2500 \\
x &= -4y^2 + 100
\end{aligned}$$

we find a parabola.

43. Completing the square,

$$\begin{aligned}
2(x^2 - 2x) + 2(y^2 - 4y) &= -9 \\
2(x^2 - 2x + 1) + 2(y^2 - 4y + 4) &= -9 + 2 + 8 \\
2(x - 1)^2 + 2(y - 2)^2 &= 1 \\
(x - 1)^2 + (y - 2)^2 &= \frac{1}{2}
\end{aligned}$$

we find a circle.

45. Completing the square,

$$\begin{aligned}
2(x^2 + 2x) + y^2 + 6y &= -7 \\
2(x^2 + 2x + 1) + y^2 + 6y + 9 &= -7 + 2 + 9 \\
2(x + 1)^2 + (y + 3)^2 &= 4 \\
\frac{(x + 1)^2}{2} + \frac{(y + 3)^2}{4} &= 1
\end{aligned}$$

we get an ellipse.

47. Completing the square, we find

$$\begin{aligned}
25(x^2 - 6x + 9) - 4(y^2 + 2y + 1) &= -121 + 225 - 4 \\
25(x - 3)^2 - 4(y + 1)^2 &= 100 \\
\frac{(x - 3)^2}{4} - \frac{(y + 1)^2}{25} &= 1.
\end{aligned}$$

We have a hyperbola.

49. From the center $(0, 0)$ and vertex $(0, 8)$ one gets $a = 8$. From the foci $(0, \pm 10)$, $c = 10$. So $b^2 = c^2 - a^2 = 10^2 - 8^2 = 36$.

Hyperbola is given by $\dfrac{y^2}{64} - \dfrac{x^2}{36} = 1$.

51. Multiply $16y^2 - x^2 = 16$ by 9 and add to $9x^2 - 4y^2 = 36$.

$$
\begin{array}{rcl}
-9x^2 + 144y^2 &=& 144 \\
9x^2 - 4y^2 &=& 36 \\
\hline
140y^2 &=& 180 \\
y^2 &=& \dfrac{9}{7} \\
y &=& \dfrac{3\sqrt{7}}{7}
\end{array}
$$

Using $y^2 = \dfrac{9}{7}$ in $x^2 = 16(y^2 - 1)$, we get

$$x^2 = 16\left(\frac{2}{7}\right) \text{ or } x = \frac{4\sqrt{14}}{7}.$$

The exact location is $\left(\dfrac{4\sqrt{14}}{7}, \dfrac{3\sqrt{7}}{7}\right)$.

53. Since $c^2 = a^2 + b^2 = 1^2 + 1^2 = 2$, the foci of $x^2 - y^2 = 1$ are $A(\sqrt{2}, 0)$ and $B(-\sqrt{2}, 0)$. Note, $y^2 = x^2 - 1$. Suppose (x, y) is a point on the hyperbola whose distance from B is twice the distance between (x, y) and A. Then

$$
\begin{array}{rcl}
2\sqrt{(x - \sqrt{2})^2 + y^2} &=& \sqrt{(x + \sqrt{2})^2 + y^2} \\
4((x - \sqrt{2})^2 + y^2) &=& (x + \sqrt{2})^2 + y^2
\end{array}
$$

$$
\begin{array}{rcl}
4(x - \sqrt{2})^2 - (x + \sqrt{2})^2 + 3y^2 &=& 0 \\
3x^2 - 10\sqrt{2}x + 6 + 3y^2 &=& 0 \\
3x^2 - 10\sqrt{2}x + 6 + 3(x^2 - 1) &=& 0 \\
6x^2 - 10\sqrt{2}x + 3 &=& 0.
\end{array}
$$

Solving for x, one finds $x = \dfrac{3\sqrt{2}}{2}$ and $x = \dfrac{\sqrt{2}}{6}$; the second value must be excluded since it is out of the domain. Substituting $x = \dfrac{3\sqrt{2}}{2}$ into $y^2 = x^2 - 1$, one obtains $y = \pm\dfrac{\sqrt{14}}{2}$.

By the symmetry of the hyperbola, there are four points that are twice as far from one focus as they are from the other focus. Namely, these points are

$$\left(\pm\frac{3\sqrt{2}}{2}, \pm\frac{\sqrt{14}}{2}\right).$$

55. Note, the asymptotes are $y = \pm x$. The difference is

$$50 - \sqrt{50^2 - 1} \approx 0.01$$

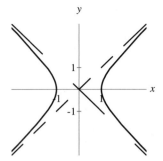

59. Since the asymptotes are $y = \pm x$, the lines parallel to the asymptotes and passing through (x_1, y_1) are $f(x) = x + (y_1 - x_1)$ and $g(x) = -x + (x_1 + y_1)$. Similarly, the lines through (x_2, y_2) and parallel to $y = \pm x$ are $s(x) = x + (y_2 - x_2)$ and $t(x) = -x + (x_2 + y_2)$. To find the point of intersection between lines f and t, we solve a system of equations. Then

$$
\begin{array}{rcl}
x + (y_1 - x_1) &=& -x + (x_2 + y_2) \\
x &=& \dfrac{(x_2 + y_2) - (y_1 - x_1)}{2}.
\end{array}
$$

Substituting into f, one finds

$$
\begin{array}{rcl}
y &=& \dfrac{(x_2 + y_2) - (y_1 - x_1)}{2} + (y_1 - x_1) \\
y &=& \dfrac{(x_2 + y_2) + (y_1 - x_1)}{2}.
\end{array}
$$

The point of intersection between f and t is

$$A = \left(\frac{(x_2 + y_2) - (y_1 - x_1)}{2}, \frac{(x_2 + y_2) + (y_1 - x_1)}{2}\right).$$

Similarly, one can show that the point of intersection between g and s is

$$B = \left(\frac{(x_1 + y_1) - (y_2 - x_2)}{2}, \frac{(y_2 - x_2) + (x_1 + y_1)}{2}\right).$$

Let $A = (A_1, A_2)$ and $B = (B_1, B_2)$. To show that the line joining A and B passes through the origin it will be sufficient to show $\dfrac{A_2}{A_1} = \dfrac{B_2}{B_1}$. Rewriting $\dfrac{A_2}{A_1}$ and noting that $x_2^2 - y_2^2 = 1$, one obtains

$$\frac{A_2}{A_1} = \frac{(x_2 + y_2) + (y_1 - x_1)}{(x_2 + y_2) - (y_1 - x_1)}$$

$$= \frac{(x_2 + y_2) + (y_1 - x_1)}{(x_2 + y_2) - (y_1 - x_1)} \cdot \frac{x_2 - y_2}{x_2 - y_2}$$

$$= \frac{(x_2 + y_2)(x_2 - y_2) + (y_1 - x_1)(x_2 - y_2)}{(x_2 + y_2)(x_2 - y_2) - (y_1 - x_1)(x_2 - y_2)}$$

$$= \frac{1 + (y_1 - x_1)(x_2 - y_2)}{1 - (y_1 - x_1)(x_2 - y_2)}$$

$$= \frac{1 + (x_1 - y_1)(y_2 - x_2)}{1 - (x_1 - y_1)(y_2 - x_2)}$$

$$= \frac{(x_1 - y_1)(y_2 - x_2) + 1}{1 - (x_1 - y_1)(y_2 - x_2)}$$

$$= \frac{(y_2 - x_2) + (x_1 + y_1)}{(x_1 + y_1) - (y_2 - x_2)} \cdot \frac{x_1 - y_1}{x_1 - y_1}$$

$$\frac{A_2}{A_1} = \frac{B_2}{B_1}.$$

Hence, the line through A and B passes through the origin.

Review Exercises

1. If $y = 0$, then by factoring we get
$0 = (x + 6)(x - 2)$ and the x-intercepts are
$(2, 0)$ and $(-6, 0)$.

If $x = 0$, then $y = -12$ and the y-intercept is
$(0, -12)$.

By completing the square, we obtain
$y = (x + 2)^2 - 16$, vertex $(h, k) = (-2, -16)$,
and the axis of symmetry is $x = -2$.

Since $p = \dfrac{1}{4a} = \dfrac{1}{4}$, the focus is

$(h, k + p) = (-2, -63/4)$ and
directrix is $y = k - p$ or $y = -65/4$.

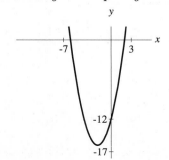

3. If $y = 0$, then by factoring we find $0 = x(6 - 2x)$
and the x-intercepts are $(0, 0)$ and $(3, 0)$.

If $x = 0$, then $y = 0$ and the y-intercept is
$(0, 0)$.

By completing the square, one gets
$y = -2(x - 3/2)^2 + 9/2$, with
vertex $(h, k) = (3/2, 9/2)$,
and axis of symmetry is $x = 3/2$.

Since $p = \dfrac{1}{4a} = -\dfrac{1}{8}$, the focus is
$(h, k + p) = (3/2, 35/8)$ and
the directrix is $y = k - p$ or $y = 37/8$.

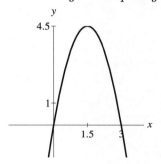

5. By completing the square, we have
$x = (y + 2)^2 - 10$. If $x = 0$, then $y + 2 = \pm\sqrt{10}$
and the y-intercepts are $(0, -2 \pm \sqrt{10})$.

If $y = 0$, then $x = -6$ and the x-intercept is
$(-6, 0)$. Since $x = (y + 2)^2 - 10$ is of the form
$x = a(y - h)^2 + k$, the vertex is
$(k, h) = (-10, -2)$, and the axis
of symmetry is $y = -2$.

Since $p = \dfrac{1}{4a} = \dfrac{1}{4}$, the focus is

$$(k + p, h) = (-39/4, -2)$$

and directrix is $x = k - p$ or
$x = -41/4$.

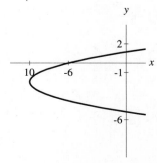

7. Since $c = \sqrt{a^2 - b^2} = \sqrt{36 - 16} = 2\sqrt{5}$, the foci are $(0, \pm 2\sqrt{5})$

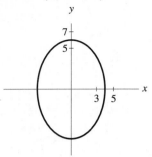

9. Since $c = \sqrt{a^2 - b^2} = \sqrt{24 - 8} = 4$, the foci are $(1, 1 \pm 4)$, or $(1, 5)$ and $(1, -3)$.

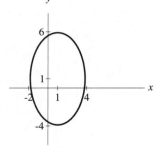

11. Since $c = \sqrt{a^2 - b^2} = \sqrt{10 - 8} = \sqrt{2}$, the foci are $(1, -3 \pm \sqrt{2})$.

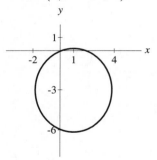

13. center $(0, 0)$, radius 9

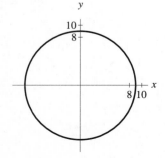

15. center $(-1, 0)$, radius 2

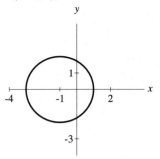

17. Completing the square, we have

$$x^2 + 5x + \frac{25}{4} + y^2 = -\frac{1}{4} + \frac{25}{4}$$
$$\left(x + \frac{5}{2}\right)^2 + y^2 = 6,$$

and so center is $(-5/2, 0)$ and radius is $\sqrt{6}$.

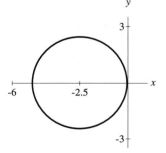

19. $x^2 + (y + 4)^2 = 9$

21. $(x + 2)^2 + (y + 7)^2 = 6$

23. Since $c = \sqrt{a^2 + b^2} = \sqrt{8^2 + 6^2} = 10$, the foci are $(\pm 10, 0)$ and the asymptotes are $y = \pm \dfrac{b}{a} = \pm \dfrac{3}{4} x$

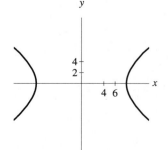

25. Since $c = \sqrt{a^2 + b^2} = \sqrt{8^2 + 4^2} = 4\sqrt{5}$,

the foci are $(4, 2 \pm 4\sqrt{5})$. Solving

for y in $y - 2 = \pm\dfrac{8}{4}(x - 4)$,

asymptotes are $y = 2x - 6$ and $y = -2x + 10$.

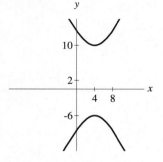

27. Completing the square, we have

$$
\begin{aligned}
x^2 - 4x - 4(y^2 - 8y) &= 64 \\
x^2 - 4x + 4 - 4(y^2 - 8y + 16) &= 64 + 4 - 64 \\
(x - 2)^2 - 4(y - 4)^2 &= 4 \\
\frac{(x - 2)^2}{4} - (y - 4)^2 &= 1,
\end{aligned}
$$

and so $c = \sqrt{a^2 + b^2} = \sqrt{2^2 + 1^2} = \sqrt{5}$, and

the foci are $(2 \pm \sqrt{5}, 4)$. Solving for y

in $y - 4 = \pm\dfrac{1}{2}(x - 2)$, we get that the

asymptotes are $y = \dfrac{1}{2}x + 3$ and $y = -\dfrac{1}{2}x + 5$.

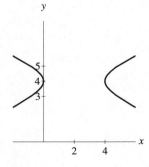

29. Hyperbola

31. Ellipse

33. Parabola

35. Hyperbola

37. $x^2 + y^2 = 4$ is a circle

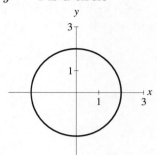

39. Since $4y = x^2 - 4$, $y = \dfrac{1}{4}x^2 - 1$.

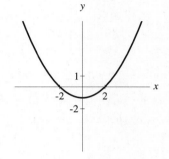

41. Since $x^2 + 4y^2 = 4$, $\dfrac{x^2}{4} + y^2 = 1$.

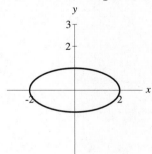

43. Since $x^2 - 4x + 4 + 4y^2 = 4$, we find

$$(x - 2)^2 + 4y^2 = 4 \text{ and } \frac{(x - 2)^2}{4} + y^2 = 1.$$

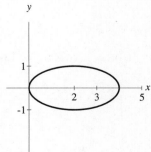

45. Since the vertex is midway between the focus $(1,3)$ and directrix $x = \dfrac{1}{2}$, the vertex is $\left(\dfrac{3}{4}, 3\right)$ and $p = \dfrac{1}{4}$. Since $a = \dfrac{1}{4p} = 1$, parabola is given by

$$x = (y-3)^2 + \dfrac{3}{4}.$$

47. From the foci and vertices one gets $c = 4$ and $a = 6$, respectively. Since $b^2 = a^2 - c^2 = 36 - 16 = 20$, the ellipse is given by $\dfrac{x^2}{36} + \dfrac{y^2}{20} = 1$.

49. Radius is $\sqrt{(-1-1)^2 + (-1-3)^2} = \sqrt{20}$. Equation is $(x-1)^2 + (y-3)^2 = 20$.

51. From the foci and x-intercepts one gets $c = 3$ and $a = 2$, respectively. Since $b^2 = c^2 - a^2 = 9 - 4 = 5$, the hyperbola is given by $\dfrac{x^2}{4} - \dfrac{y^2}{5} = 1$.

53. Since the center of the circle is $(-2, 3)$ and the raidus is 3, we have $(x+2)^2 + (y-3)^2 = 9$.

55. Note, the center of the ellipse is $(-2, 1)$. Using the lengths of the major axis, we get $a = 3$ and $b = 1$. Thus, an equation is

$$\dfrac{(x+2)^2}{9} + (y-1)^2 = 1.$$

57. Note, the center of the hyperbola is $(2, 1)$. By using the fundamental rectangle, we find $a = 3$ and $b = 2$. Thus, an equation is

$$\dfrac{(y-1)^2}{9} - \dfrac{(x-2)^2}{4} = 1.$$

59. The equation is of the form $\dfrac{x^2}{100^2} - \dfrac{y^2}{b^2} = 1$.

Since the graph passes through $(120, 24\sqrt{11})$, we get

$$
\begin{aligned}
\dfrac{120^2}{100^2} - \dfrac{(24\sqrt{11})^2}{b^2} &= 1 \\
1.44 - \dfrac{6336}{b^2} &= 1 \\
b^2 &= \dfrac{6336}{0.44} \\
b^2 &= 120^2.
\end{aligned}
$$

Equation is $\dfrac{x^2}{100^2} - \dfrac{y^2}{120^2} = 1$.

61. Note $c = 30$ and $a = 34$. Then an equation we can use is of the form $\dfrac{x^2}{34^2} + \dfrac{y^2}{b^2} = 1$. Since $b^2 = a^2 - c^2 = 34^2 - 30^2 = 16^2$, the equation is $\dfrac{x^2}{34^2} + \dfrac{y^2}{16^2} = 1$.
To find h, let $x = 32$.

$$
\begin{aligned}
\dfrac{32^2}{34^2} + \dfrac{y^2}{16^2} &= 1 \\
y^2 &= \left(1 - \dfrac{32^2}{34^2}\right) 16^2 \\
y &\approx 5.407.
\end{aligned}
$$

Thus, $h = 2y \approx 10.81$ feet.

Chapter 7 Test

1. Circle $x^2 + y^2 = 8$

2. Ellipse $\dfrac{x^2}{9} + \dfrac{y^2}{100} = 1$

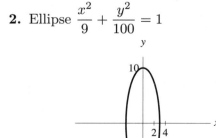

3. Parabola $y = x^2 + 6x + 8$

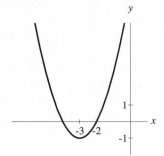

4. Hyperbola $\dfrac{y^2}{25} - \dfrac{x^2}{9} = 1$

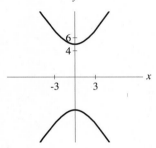

5. Circle $(x+3)^2 + (y-1)^2 = 10$

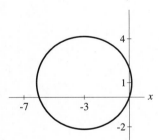

6. Hyperbola $\dfrac{(x-2)^2}{9} - \dfrac{(y+3)^2}{4} = 1$

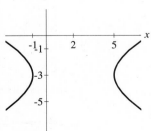

7. By completing the square, the equation can be written as $(x-4)^2 - 16 = y^2$. This is a hyperbola.

8. By completing the square, we can rewrite the equation as $(x-4)^2 - 16 = y$. This is a parabola.

9. By completing the square, we can rewrite the equation as $-(x-4)^2 + 16 = y^2$. This is a circle.

10. By completing the square, the equation can be written as $-(x-4)^2 + 16 = 8y^2$. This is an ellipse.

11. Since $(2\sqrt{3})^2 = 12$, the circle is given by $(x+3)^2 + (y-4)^2 = 12$.

12. Midway between the focus $(2,0)$ and directrix $x = -2$ is the vertex $(0,0)$. Since $p = 2$,
$$a = \frac{1}{4p} = \frac{1}{8}. \text{ Equation is } x = \frac{1}{8}y^2.$$

13. Equation is of the form $\dfrac{x^2}{2^2} + \dfrac{y^2}{a^2} = 1$.

Since $a^2 = b^2 + c^2 = 2^2 + \sqrt{6}^2 = 10$,

equation is $\dfrac{x^2}{4} + \dfrac{y^2}{10} = 1$.

14. From the foci and vertices one gets $c = 8$ and $a = 6$. Since $b^2 = c^2 - a^2 = 8^2 - 6^2 = 28$,

equation is $\dfrac{x^2}{36} - \dfrac{y^2}{28} = 1$.

15. Complete the square to get $y = (x-2)^2 - 4$.
So vertex is $(h,k) = (2,-4)$, axis of

symmetry $x = 2$, $a = 1$, and $p = \dfrac{1}{4a} = \dfrac{1}{4}$.

Focus is $(h, k+p) = (2, -15/4)$ and directrix is $y = k - p = -17/4$.

16. Since $\dfrac{x^2}{16} + \dfrac{y^2}{4} = 1$, we get
$c = \sqrt{a^2 - b^2} = \sqrt{4^2 - 2^2} = 2\sqrt{3}$.
Foci are $(\pm 2\sqrt{3}, 0)$, length of major axis is $2a = 8$, length of minor axis is $2b = 4$.

17. Since $y^2 - \dfrac{x^2}{16} = 1$, we obtain

$c = \sqrt{a^2 + b^2} = \sqrt{1^2 + 4^2} = \sqrt{17}$.
Foci are $(0, \pm\sqrt{17})$, vertices $(0, \pm 1)$,

asymptotes $y = \pm\dfrac{1}{4}x$, length of transverse

axis is $2a = 2$, length of conjugate axis is
$2b = 8$.

18. Completing the square, we find

$$x^2 + x + \frac{1}{4} + y^2 - 3y + \frac{9}{4} \;=\; -\frac{1}{4} + \frac{1}{4} + \frac{9}{4}$$
$$\left(x + \frac{1}{2}\right)^2 + \left(y - \frac{3}{2}\right)^2 \;=\; \frac{9}{4}.$$

The center is $\left(-\dfrac{1}{2}, \dfrac{3}{2}\right)$ and the radius is $\dfrac{3}{2}$.

19. Since $\dfrac{x^2}{225} + \dfrac{y^2}{81} = 1$, we find $a = 15$ and $b = 9$.

Note, $c = \sqrt{a^2 - c^2} = \sqrt{15^2 - 9^2} = 12$.
Since the foci are $(\pm 12, 0)$, the distance from
the point of generation of the waves to the
kidney stones is $2c = 24$ cm.

Tying It All Together

1. Parabola $y = 6x - x^2$

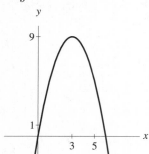

2. Line $y = 6x$

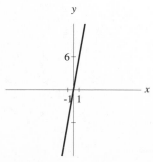

3. Parabola $y = 6 - x^2$

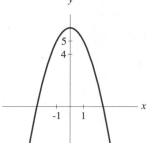

4. Circle $x^2 + y^2 = 6$

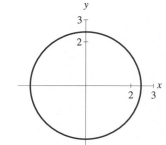

5. Line $y = x + 6$

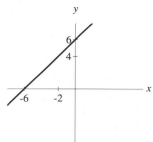

6. Parabola $x = 6 - y^2$

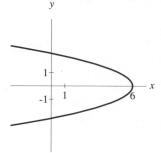

7. Parabola $y = (6 - x)^2$

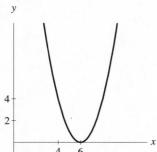

8. $y = |x + 6|$ goes through $(-6, 0)$, $(-5, 1)$

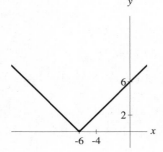

9. $y = 6^x$ goes through $(0, 1)$, $(1, 6)$

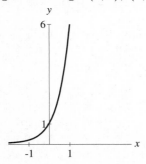

10. $y = \log_6(x)$ goes through $(6, 1)$, $(1/6, -1)$

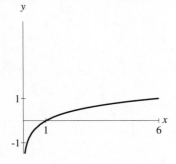

11. $y = \dfrac{1}{x^2 - 6}$ has asymptotes $x = \pm\sqrt{6}$

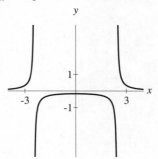

12. Ellipse $\dfrac{x^2}{9} + \dfrac{y^2}{4} = 1$

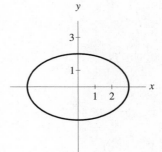

13. Hyperbola $\dfrac{x^2}{9} - \dfrac{y^2}{4} = 1$

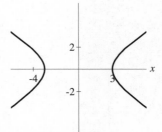

14. $y = x(6 - x^2)$ goes through $(0, 0)$, $(\pm\sqrt{6}, 0)$

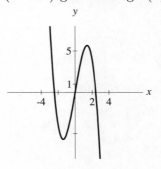

15. Line $2x + 3y = 6$

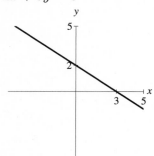

16. By completing the square, the equation can be written as

$$\frac{(x-3)^2}{3} - \frac{y^2}{3} = 1.$$

This is a hyperbola.

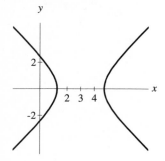

17. Solving for x, we get

$$\begin{aligned} 3x - 9 + 5 &= -9 \\ 3x &= -5. \end{aligned}$$

The solution set is $\{-5/3\}$.

18. Solving for x, we find

$$\begin{aligned} 5x - 8x + 12 &= 17 \\ -3x &= 5. \end{aligned}$$

The solution set is $\{-5/3\}$.

19. Solving for x, we obtain

$$\begin{aligned} 2x - \frac{2}{3} - \frac{3}{2} + 3x &= \frac{3}{2} \\ 5x - \frac{13}{6} &= \frac{3}{2} \\ 5x &= \frac{11}{3}. \end{aligned}$$

The solution set is $\{11/15\}$.

20. Solving for x, we have

$$\begin{aligned} -4x + 2 &= 3x + 2 \\ -7x &= 0. \end{aligned}$$

The solution set is $\{0\}$.

21. Solving for x, we find

$$\begin{aligned} \frac{1}{2}x - \frac{1}{4}x &= \frac{3}{2} + \frac{1}{3} \\ \frac{1}{4}x &= \frac{11}{6} \\ x &= \frac{44}{6}. \end{aligned}$$

Solution set is $\{22/3\}$.

22. Multiplying both sides by 100, we obtain

$$\begin{aligned} 5(x - 20) + 2(x + 10) &= 270 \\ 5x - 100 + 2x + 20 &= 270 \\ 7x &= 350. \end{aligned}$$

The solution set is $\{50\}$.

23. By using the quadratic formula to solve

$$2x^2 + 31x - 51 = 0$$

one obtains

$$\begin{aligned} x &= \frac{-31 \pm \sqrt{31^2 - 4(2)(-51)}}{4} \\ &= \frac{-31 \pm \sqrt{1369}}{4} \\ &= \frac{-31 \pm 37}{4} \\ &= \frac{3}{2}, -17. \end{aligned}$$

The solution set is $\left\{\frac{3}{2}, -17\right\}$.

24. Since $x(2x + 31) = 0$, the solution set is

$$\left\{ 0, -\frac{31}{2} \right\}.$$

25. By using the quadratic formula to solve $x^2 - 34x + 286 = 0$, one obtains

$$x = \frac{34 \pm \sqrt{34^2 - 4(1)(286)}}{2}$$
$$x = \frac{34 \pm \sqrt{12}}{2}$$
$$x = \frac{34 \pm 2\sqrt{3}}{2}$$
$$x = 17 \pm \sqrt{3}.$$

The solution set is $\left\{ 17 \pm \sqrt{3} \right\}$.

26. In solving $x^2 - 34x + 290 = 0$, one can use the quadratic formula. Then

$$x = \frac{34 \pm \sqrt{34^2 - 4(1)(290)}}{2}$$
$$x = \frac{34 \pm \sqrt{-4}}{2}$$
$$x = \frac{34 \pm 2i}{2}$$
$$x = 17 \pm i.$$

The solution set is $\{17 \pm i\}$.

For Thought

1. True

2. False, the domain of a finite sequence is $\{1, 2, ..., n\}$ for some positive integer n.

3. True

4. False, n is the independent variable.

5. False, the first four terms are $1, -8, 27, -64$.

6. False, $a_5 = -3 + 4 \cdot 6 = 21$.

7. False, the common difference is $d = 4 - 7 = -3$.

8. False, there is no common difference.

9. False, since if $14 = 4 + 2d$ then $d = 5$ and $a_4 = 4 + 3 \cdot 5 = 19$.   **10.**   True

8.1 Exercises

1. $a_1 = 1^2 = 1$, $a_2 = 2^2 = 4$,
$a_3 = 3^2 = 9$, $a_4 = 4^2 = 16$, $a_5 = 5^2 = 25$,
$a_6 = 6^2 = 36$, $a_7 = 7^2 = 49$.

First seven terms are $1, 4, 9, 16, 25, 36, 49$.

3. $b_1 = \dfrac{(-1)^2}{2} = \dfrac{1}{2}$, $b_2 = \dfrac{(-1)^3}{3} = -\dfrac{1}{3}$,

$b_3 = \dfrac{(-1)^4}{4} = \dfrac{1}{4}$, $b_4 = \dfrac{(-1)^5}{5} = -\dfrac{1}{5}$,

$b_5 = \dfrac{(-1)^6}{6} = \dfrac{1}{6}$, $b_6 = \dfrac{(-1)^7}{7} = -\dfrac{1}{7}$,

$b_7 = \dfrac{(-1)^8}{8} = \dfrac{1}{8}$, $b_8 = \dfrac{(-1)^9}{9} = -\dfrac{1}{9}$.

First eight terms are

$\dfrac{1}{2}, -\dfrac{1}{3}, \dfrac{1}{4}, -\dfrac{1}{5}, \dfrac{1}{6}, -\dfrac{1}{7}, \dfrac{1}{8}, -\dfrac{1}{9}$.

5. $c_1 = (-2)^0 = 1$, $c_2 = (-2)^1 = -2$,
$c_3 = (-2)^2 = 4$, $c_4 = (-2)^3 = -8$,
$c_5 = (-2)^4 = 16$, $c_6 = (-2)^5 = -32$.
First six terms are $1, -2, 4, -8, 16, -32$

7. $a_1 = 2^1 = 2$, $a_2 = 2^0 = 1$,

$a_3 = 2^{-1} = \dfrac{1}{2}$, $a_4 = 2^{-2} = \dfrac{1}{4}$, $a_5 = 2^{-3} = \dfrac{1}{8}$.

First five terms are $2, 1, \dfrac{1}{2}, \dfrac{1}{4}, \dfrac{1}{8}$.

9. $a_1 = -6 + 0 = -6$, $a_2 = -6 - 4 = -10$,
$a_3 = -6 - 8 = -14$, $a_4 = -6 - 12 = -18$,
$a_5 = -6 - 16 = -22$. First five
terms are $-6, -10, -14, -18, -22$.

11. $b_1 = 5 + 0 = 5$, $b_2 = 5 + 0.5 = 5.5$,
$b_3 = 5 + 1 = 6$, $b_4 = 5 + 1.5 = 6.5$,
$b_5 = 5 + 2 = 7$, $b_6 = 5 + 2.5 = 7.5$,
$b_7 = 5 + 3 = 8$. First seven
terms are $5, 5.5, 6, 6.5, 7, 7.5, 8$.

13. $a_1 = -0.1 + 9 = 8.9$,
$a_2 = -0.2 + 9 = 8.8$, $a_3 = -0.3 + 9 = 8.7$,
$a_4 = -0.4 + 9 = 8.6$, $a_{10} = -1 + 9 = 8$.
First four terms are $8.9, 8.8, 8.7, 8.6$ and
and 10th term is 8.

15. $a_1 = 8 + 0 = 8$, $a_2 = 8 - 3 = 5$,
$a_3 = 5 - 3 = 2$, $a_4 = 2 - 3 = -1$, and
$a_{10} = 8 - 27 = -19$. First four terms
are $8, 5, 2, -1$ and 10th term is -19.

17. One gets $a_1 = \dfrac{4}{2 + 1} = \dfrac{4}{3}$,

$a_2 = \dfrac{4}{4 + 1} = \dfrac{4}{5}$, $a_3 = \dfrac{4}{6 + 1} = \dfrac{4}{7}$,

$a_4 = \dfrac{4}{8 + 1} = \dfrac{4}{9}$, $a_{10} = \dfrac{4}{20 + 1} = \dfrac{4}{21}$

The first four terms are

$\dfrac{4}{3}, \dfrac{4}{5}, \dfrac{4}{7}, \dfrac{4}{9}$ and 10th term is $\dfrac{4}{21}$.

19. One gets $a_1 = \dfrac{(-1)^1}{2 \cdot 3} = -\dfrac{1}{6}$,

$a_2 = \dfrac{(-1)^2}{3 \cdot 4} = \dfrac{1}{12}$, $a_3 = \dfrac{(-1)^3}{4 \cdot 5} = -\dfrac{1}{20}$,

$a_4 = \dfrac{(-1)^4}{5 \cdot 6} = \dfrac{1}{30}$, $a_{10} = \dfrac{(-1)^{10}}{11 \cdot 12} = \dfrac{1}{132}$.

The first four terms are

$-\dfrac{1}{6}, \dfrac{1}{12}, -\dfrac{1}{20}, \dfrac{1}{30}$ and 10th term is $\dfrac{1}{132}$.

21. $a_1 = 2! = 2$, $a_2 = 4! = 24$, $a_3 = 6! = 720$,
$a_4 = 8! = 40,320$, and $a_5 = 10! = 3,628,800$

23. $a_1 = \dfrac{2^1}{1!} = 2$, $a_2 = \dfrac{2^2}{2!} = 2$, $a_3 = \dfrac{2^3}{3!} = \dfrac{4}{3}$,

$a_4 = \dfrac{2^4}{4!} = \dfrac{2}{3}$, and $a_5 = \dfrac{2^5}{5!} = \dfrac{4}{15}$

25. $b_1 = \dfrac{1!}{0!} = 1$, $b_2 = \dfrac{2!}{1!} = 2$, $b_3 = \dfrac{3!}{2!} = 3$,

$b_4 = \dfrac{4!}{3!} = 4$, and $b_5 = \dfrac{5!}{4!} = \dfrac{5 \cdot 4!}{4!} = 5$

27. $c_1 = \dfrac{(-1)^1}{2!} = -\dfrac{1}{2}$, $c_2 = \dfrac{(-1)^2}{3!} = \dfrac{1}{6}$,

$c_3 = \dfrac{(-1)^3}{4!} = -\dfrac{1}{24}$, $c_4 = \dfrac{(-1)^4}{5!} = \dfrac{1}{120}$,

and $c_5 = \dfrac{(-1)^5}{6!} = -\dfrac{1}{720}$.

29. Note $t_n = \dfrac{1}{(n+2)!e^n}$. Then we obtain

$t_1 = \dfrac{1}{3!e} = \dfrac{1}{6e}$, $t_2 = \dfrac{1}{4!e^2} = \dfrac{1}{24e^2}$,

$t_3 = \dfrac{1}{5!e^3} = \dfrac{1}{120e^3}$, $t_4 = \dfrac{1}{6!e^4} = \dfrac{1}{720e^4}$,

and $t_5 = \dfrac{1}{7!e^5} = \dfrac{1}{5040e^5}$.

31. $a_n = 2n$

33. $a_n = 2n + 7$

35. $a_n = (-1)^{n+1}$

37. $a_n = n^3$

39. $a_n = e^n$

41. $a_n = \dfrac{1}{2^{n-1}}$

43. $a_2 = 3a_1 + 2 = 3(-4) + 2 = -10$,
$a_3 = 3a_2 + 2 = 3(-10) + 2 = -28$,
$a_4 = 3a_3 + 2 = 3(-28) + 2 = -82$,
$a_5 = 3a_4 + 2 = 3(-82) + 2 = -244$,
$a_6 = 3a_5 + 2 = 3(-244) + 2 = -730$,
$a_7 = 3a_6 + 2 = 3(-730) + 2 = -2188$,
$a_8 = 3a_7 + 2 = 3(-2188) + 2 = -6562$.
First four terms $-4, -10, -28, -82$ and 8th term is -6562.

45. One finds $a_2 = a_1^2 - 3 = 2^2 - 3 = 1$,
$a_3 = a_2^2 - 3 = 1^2 - 3 = -2$,
$a_4 = a_3^2 - 3 = (-2)^2 - 3 = 1$.

By a repeating pattern, $a_8 = 1$. First four terms are $2, 1, -2, 1$ and 8th term is 1.

47. $a_2 = a_1 + 7 = (-15) + 7 = -8$,
$a_3 = a_2 + 7 = (-8) + 7 = -1$,
$a_4 = a_3 + 7 = (-1) + 7 = 6$.
There is a common difference of 7.
So $a_8 = 34$. First four terms are $-15, -8, -1, 6$ and 8th term is 34.

49. First four terms are $6, 3, 0, -3$ and 10th term is $a_{10} = 6 + 9(-3) = -21$.

51. First four terms are $1, 0.9, 0.8, 0.7$ and 10th term is $c_{10} = 1 + 9(-0.1) = 0.1$.

53. One finds $w_1 = -\dfrac{1}{3}(1) + 5 = \dfrac{14}{3}$,

$w_2 = -\dfrac{1}{3}(2) + 5 = \dfrac{13}{3}$,

$w_3 = -\dfrac{1}{3}(3) + 5 = \dfrac{12}{3}$,

$w_4 = -\dfrac{1}{3}(4) + 5 = \dfrac{11}{3}$, and

$w_{10} = -\dfrac{1}{3}(10) + 5 = \dfrac{5}{3}$.

First four terms are $\dfrac{14}{3}, \dfrac{13}{3}, \dfrac{12}{3}, \dfrac{11}{3}$

and 10th term is $\dfrac{5}{3}$.

55. Yes, $d = 1$ is the common difference.

57. No, there is no common difference.

59. No, there is no common difference.

61. Yes, $d = \dfrac{\pi}{4}$ is the common difference.

63. Since $d = 5$ and $a_1 = 1$, $a_n = a_1 + (n-1)d$
$= 1 + (n-1)5$. Then $a_n = 5n - 4$.

65. Since $d = 2$ and $a_1 = 0$, $a_n = a_1 + (n-1)d$
$= 0 + (n-1)2$. So $a_n = 2n - 2$.

67. Since $d = -4$ and $a_1 = 5$, $a_n = a_1 + (n-1)d$
$= 5 + (n-1)(-4)$. So $a_n = -4n + 9$.

69. Since $d = 0.1$ and $a_1 = 1$, $a_n = a_1 + (n-1)d$
$= 1 + (n-1)(0.1)$. So $a_n = 0.1n + 0.9$.

71. Since $d = \dfrac{\pi}{6}$ and $a_1 = \dfrac{\pi}{6}$, $a_n = a_1 + (n-1)d$

$= \dfrac{\pi}{6} + (n-1)\dfrac{\pi}{6}$. So $a_n = \dfrac{\pi}{6}n$.

73. Since $d = 15$ and $a_1 = 20$, we have

$a_n = a_1 + (n-1)d = 20 + (n-1)15$.

Then $a_n = 15n + 5$.

75. Since $a_n = a_1 + (n-1)d$, we get

$a_8 = (-3) + (7)5 = 32$.

77. Since $a_n = a_1 + (n-1)d$, we find

$$a_1 + 2d = 6$$
$$a_1 + 6d = 18$$

Subtracting the first equation from the second, we get $4d = 12$ and $d = 3$. From the first equation, $a_1 + 6 = 6$ and $a_1 = 0$. Then $a_{10} = a_1 + 9d = 0 + 9 \cdot 3 = 27$. So $a_{10} = 27$.

79. Since $a_n = a_1 + (n-1)d$, we get $a_{21} = 12 + 20d = 96$. So $20d = 84$ and the common difference is $d = 4.2$.

81. Since $a_n = a_1 + (n-1)d$, we find

$$a_1 + 2d = 10$$
$$a_1 + 6d = 20.$$

Subtracting the first equation from the second, we obtain $4d = 10$ and $d = 2.5$. From the first equation, we get $a_1 + 5 = 10$ and $a_1 = 5$. Then $a_n = 5 + (n-1)(2.5) = 2.5n + 2.5$. So $a_n = 2.5n + 2.5$.

83. $a_n = a_{n-1} + 9$, $a_1 = 3$

85. $a_n = 3a_{n-1}$, $a_1 = \dfrac{1}{3}$

87. $a_n = \sqrt{a_{n-1}}$, $a_1 = 16$

89. A formula for the MSRP is

$$a_n = 35,265(1.06)^n$$

where n is the number of years since 2002. For the years 2003-2007, the MSRP are $a_1 = \$37,381$, $a_2 = \$39,624$, $a_3 = \$42,001$, $a_4 = \$44,521$, and $a_5 = \$47,193$.

91. The number of pages read on the nth day of November is $a_n = 5 + 3(n-1)$ or $a_n = 3n + 2$. On November 30, they will read $a_{30} = 92$ pages.

93. In year n, the annual cost is

$$a_n = 51,000 + 1800(n - 2005).$$

In 2014, the cost is projected to be $a_{2014} = 51,000 + 1800(2014 - 2005) = \$67,200$.

95. Since there are four corners, $C_n = 4$. If the corner tiles are in place, the number of edge tiles needed for each side is $2(n-1)$. Since there are four sides, $E_n = 8(n-1)$.

If the corner tiles and edge tiles are in place, the interior tiles will occupy an $(n-1)$-by-$(n-1)$ square.

Note, the area of a tile is $\dfrac{1}{4}$ ft^2. Then the number of interior tiles is $I_n = \dfrac{(n-1)^2}{\frac{1}{4}}$ or equivalently $I_n = 4n^2 - 8n + 4$ for $n = 1, 2, 3, ...$

For Thought

1. True, $\displaystyle\sum_{i=1}^{3} (-2)^i = (-2)^1 + (-2)^2 + (-2)^3 = -6$

2. True, $\displaystyle\sum_{i=1}^{6} (0 \cdot i + 5) = (0 \cdot 1 + 5) + (0 \cdot 2 + 5) +$

$(0 \cdot 3 + 5) + (0 \cdot 4 + 5) + (0 \cdot 5 + 5) + (0 \cdot 6 + 5) = 30.$

3. True, since $\displaystyle\sum_{i=1}^{k} (5i) = (5 \cdot 1) + ... + (5 \cdot k) =$

$5(1 + 2 + ... + k) = 5\displaystyle\sum_{i=1}^{k} (i).$

4. True, $\displaystyle\sum_{i=1}^{k} (i^2 + 1) = (1^2 + 1) + ... + (k^2 + 1) =$

$(1^2 + 2^2 + ... + k^2) + \underbrace{(1 + ... + 1)}_{k \ times} = \displaystyle\sum_{i=1}^{k} i^2 + k.$

5. False, there are ten terms.

6. True, if $j = i - 1$ and $i = 2$ then $j = 1$.

7. True

8. True, $1 + 2 + \ldots + n = \dfrac{n}{2}(1 + n)$ represents an arithmetic series.

9. False, for if $a_1 = 8$, $a_n = 68$, $d = 2$ then $68 = 8 + (n-1)2$ and $n = 31$. The sum of the arithmetic series is $S_{31} = \dfrac{31}{2}(8 + 68)$.

10. False, $\sum\limits_{i=1}^{10} i^2$ is not an arithmetic series.

8.2 Exercises

1. $1^2 + 2^2 + 3^2 + 4^2 + 5^2 = 55$

3. $2^0 + 2^1 + 2^2 + 2^3 = 15$

5. $1 + 1 + 2 + 6 + 24 = 34$

7. $\underbrace{3 + \ldots + 3}_{10 \ times} = 30$

9. $(2 + 5) + (4 + 5) + (6 + 5) + (8 + 5) + (10 + 5)$ $+ (12 + 5) + (14 + 5) = 91$

11. $[(-1)^7 + (-1)^8] + \ldots + [(-1)^{43} + (-1)^{44}] =$ $0 + \ldots + 0 = 0$

13. $0 + 0 + 0 + 3(2)(1) + 4(3)(2) + 5(4)(3) = 90$

15. $\sum\limits_{i=1}^{6} i$

17. $\sum\limits_{i=1}^{5}(-1)^i(2i-1)$

19. $\sum\limits_{i=1}^{5} i^2$

21. $\sum\limits_{i=1}^{\infty}\left(-\dfrac{1}{2}\right)^{i-1}$

23. $\sum\limits_{i=1}^{\infty} \ln(x_i)$

25. $\sum\limits_{i=1}^{11} ar^{i-1}$

27. Let $i = j - 1$. Since $1 \le i = j - 1 \le 3$, we find $2 \le j \le 4$. Then we find

$$\sum_{i=1}^{3} i^2 = \sum_{i=2}^{4}(j-1).$$

The equation is true.

29. We find $\sum\limits_{x=1}^{5}(2x-1) = 25$ and $\sum\limits_{y=3}^{7}(2y-1) = 45$. The equation is false.

31. Let $x + 5 = j + 6$. Since $1 \le x = j + 1 \le 5$, we find $0 \le j \le 4$. Then we obtain

$$\sum_{i=1}^{5}(x+5) = \sum_{j=0}^{4}(j+6).$$

The equation is true.

33. Let $j = i - 1$. New series is $\sum\limits_{j=0}^{31}(-1)^{j+1}$

35. Let $j = i - 3$. New series is $\sum\limits_{j=1}^{10}(2j+7)$

37. Let $j = x - 2$. New series is $\sum\limits_{j=0}^{8} \dfrac{10!}{(j+2)!(8-j)!}$

39. Let $j = n + 3$. New series is $\sum\limits_{j=5}^{\infty} \dfrac{5^{j-3} \cdot e^{-5}}{(j-3)!}$

41. $0.5 + 0.5r + 0.5r^2 + 0.5r^3 + 0.5r^4 + 0.5r^5$

43. $a^4 + a^3b + a^2b^2 + ab^3 + b^4$

45. $a^2 + 2ab + b^2$

47. $\dfrac{6 + 23 + 45}{3} = \dfrac{74}{3}$

49. $\dfrac{-6 + 0 + 3 + 4 + 3 + 92}{6} = 16$

51. $\dfrac{\sqrt{2} + \pi + 33.6 - 19.4 + 52}{5} \approx 14.151$

53. Since $a_1 = -3$, $a_{12} = 6(12) - 9 = 63$, and $n = 12$, the sum is $S_{12} = \dfrac{12}{2}(-3 + 63) = 360$.

55. Since $a_3 = 0.7$, $a_{15} = -0.5$, and there are $n = 13$ terms in the series, the sum is $S_{13} = \dfrac{13}{2}(0.7 - 0.5) = 1.3$.

57. Note $a_1 = 1$, $a_n = 47$, $n = 47$. Then the sum is $S_{47} = \dfrac{47}{2}(1 + 47) = 1128$.

59. Since $95 = 5 + (n-1)5$, we get $n = 19$. Then the sum is $S_{19} = \dfrac{19}{2}(5 + 95) = 950$.

61. Since $-238 = 10 + (n-1)(-1)$, we get $n = 249$.

The sum is $S_{249} = \dfrac{249}{2}(10 - 238) = -28{,}386$.

63. Since $-16 = 8 + (n-1)(-3)$, we get $n = 9$.

Thus, the sum is $S_9 = \dfrac{9}{2}(8 + (-16)) = -36$.

65. Since $55 = 3 + (n-1)4$, we obtain $n = 14$.

The sum is $S_{14} = \dfrac{14}{2}(3 + 55) = 406$.

67. Since $5 = \dfrac{1}{2} + (n-1)\dfrac{1}{4}$, we get $n = 19$.

Thus, the sum is $S_{19} = \dfrac{19}{2}(1/2 + 5) = 52.25$.

69. Note, the mean of an arithmetic sequence with n terms is $\overline{a} = \dfrac{a_1 + a_n}{2}$. Thus, $\overline{a} = \dfrac{2 + 16}{2} = 9$.

71. Note, the mean of an arithmetic sequence with n terms is $\overline{a} = \dfrac{a_1 + a_n}{2}$. Then $\overline{a} = \dfrac{2 + 142}{2} = 72$.

73. Since $a_1 = 30{,}000$, $d = 1{,}000$, and $a_{30} = 30{,}000 + 29(1{,}000) = 59{,}000$, we find his total salary for 30 years of work is

$$S_{30} = \dfrac{30}{2}(30{,}000 + 59{,}000) = \$1{,}335{,}000.$$

The mean annual salary for 30 years is

$$\dfrac{S_{30}}{30} = \$44{,}500.$$

75. In the nth level, the number of cans is $a_n = 12(10-n)$. In the mountain of cans there are $\sum\limits_{i=1}^{9}(120 - 12i) = \sum\limits_{j=1}^{9}(120 - 12(10 - j)) =$

$$\sum_{j=1}^{9} 12j = \dfrac{9}{2}(12 + 108) = 540 \text{ cans.}$$

77. At 5% compounded annually, the future value of $1000 after i years is $\$1000(1.05)^i$. Then the amount in the account on January 1, 2000 is $\sum\limits_{i=1}^{10} 1000(1.05)^i$.

79. With a calculator, one finds $200 + 200(.63) + 200(.63)^2 + 200(.63)^3 \approx 455$ mg.

81. Since $a_9 = 101$, $a_{60} = 356$, and there are $n = 52$ terms from a_9 to a_{60}, we get that the mean of the 9th through the 60th terms

is $\dfrac{1}{52}\sum\limits_{i=9}^{60} a_i = \dfrac{1}{52}\dfrac{52}{2}(101 + 356) = 228.5$.

83. One finds $a_2 = a_1^2 - 3 = (-2)^2 - 3 = 1$ and $a_3 = a_2^2 - 3 = 1^2 - 3 = -2$. By a repeating pattern, one derives $a_7 = a_9 = -2$ and $a_8 = a_{10} = 1$. The mean from the 7th term through the 10th terms is

$$\dfrac{-2 + 1 - 2 + 1}{4} = -\dfrac{1}{2}.$$

For Thought

1. False, the ratios $\dfrac{6}{2}$ and $\dfrac{24}{6}$ are not equal

2. True, $a_n = 3 \cdot 2^3 \cdot 2^{-n} = 24\left(\dfrac{1}{2}\right)^n$.

3. True, $a_1 = 5(0.3)^1 = 1.5$.

4. False, since $a_n = \left(\dfrac{1}{5}\right)^n$, the common ratio is $\dfrac{1}{5}$.

5. True

6. False, and note that the ratio 2 does not satisfy $|r| < 1$ and one cannot use the formula $S = \dfrac{a}{1 - r}$.

7. False, $\sum\limits_{i=1}^{9} 3(0.6)^i = \dfrac{1.8(1 - 0.6^9)}{1 - 0.6}$.

8. True, $\sum\limits_{i=0}^{4} 2(10)^i = \dfrac{2(1 - 10^5)}{1 - 10} = 22{,}222$.

9. True, $\sum\limits_{i=1}^{\infty} 3(0.1)^i = \dfrac{0.3}{1 - 0.1} = \dfrac{1}{3}$.

10. True, $\sum\limits_{i=1}^{\infty} \left(\dfrac{1}{2}\right)^i = \dfrac{1/2}{1 - 1/2} = 1$.

8.3 Exercises

1. $a_1 = 3 \cdot 2^0 = 3$, $a_2 = 3 \cdot 2^1 = 6$, $a_3 = 3 \cdot 2^2 = 12$, $a_4 = 3 \cdot 2^3 = 24$

First four terms are $3, 6, 12, 24$, and common ratio is 2.

3. $a_1 = 800 \cdot \left(\frac{1}{2}\right)^1 = 400$, $a_2 = 800 \cdot \left(\frac{1}{2}\right)^2 = 200$,

$a_3 = 800 \cdot \left(\frac{1}{2}\right)^3 = 100$, $a_4 = 800 \cdot \left(\frac{1}{2}\right)^4 = 50$

First four terms are $400, 200, 100, 50$, and

common ratio is $\frac{1}{2}$.

5. $a_1 = \left(-\frac{2}{3}\right)^0 = 1$, $a_2 = \left(-\frac{2}{3}\right)^1 = -\frac{2}{3}$,

$a_3 = \left(-\frac{2}{3}\right)^2 = \frac{4}{9}$, $a_4 = \left(-\frac{2}{3}\right)^3 = -\frac{8}{27}$

First four terms are $1, -\frac{2}{3}, \frac{4}{9}, -\frac{8}{27}$, and

common ratio is $-\frac{2}{3}$.

7. $\frac{1}{2}$ **9.** 10

11. -2 **13.** -1

15. Since $r = \frac{1/3}{1/6} = 2$, we get $a_n = \frac{1}{6}2^{n-1}$.

17. Since $r = \frac{0.09}{0.9} = 0.1$, we find

$a_n = (0.9)(0.1)^{n-1}$.

19. Since $r = \frac{-12}{4} = -3$, we have $a_n = 4 \cdot (-3)^{n-1}$.

21. Arithmetic, since $4 - 2 = 6 - 4 = \ldots = 2$.

23. Neither, since there is no common difference or ratio.

25. Geometric, since $\frac{-4}{2} = \frac{8}{-4} = \ldots = -2$.

27. Arithmetic, since $\frac{1}{3} - \frac{1}{6} = \frac{1}{2} - \frac{1}{3} = \ldots = \frac{1}{6}$.

29. Geometric, since $\frac{1/3}{1/6} = \frac{2/3}{1/3} = \ldots = 2$.

31. Neither, since there is no common difference or ratio.

33. $a_1 = 2 \cdot 1 = 2$, $a_2 = 2 \cdot 2 = 4$,
$a_3 = 2 \cdot 3 = 6$, $a_4 = 2 \cdot 4 = 8$. First four terms are $2, 4, 6, 8$. Arithmetic sequence.

35. One finds $a_1 = 1^2 = 1$, $a_2 = 2^2 = 4$,
$a_3 = 3^2 = 9$, $a_4 = 4^2 = 16$. First four terms are $1, 4, 9, 16$. Neither geometric nor arithmetic.

37. $a_1 = \left(\frac{1}{2}\right)^1 = \frac{1}{2}$, $a_2 = \left(\frac{1}{2}\right)^2 = \frac{1}{4}$,

$a_3 = \left(\frac{1}{2}\right)^3 = \frac{1}{8}$, $a_4 = \left(\frac{1}{2}\right)^4 = \frac{1}{16}$. First four

terms are $\frac{1}{2}, \frac{1}{4}, \frac{1}{8}, \frac{1}{16}$. Geometric sequence.

39. $b_1 = 2^3 = 8$, $b_2 = 2^5 = 32$,
$b_3 = 2^7 = 128$, $b_4 = 2^9 = 512$. First four terms are $8, 32, 128, 512$. Geometric sequence.

41. $c_2 = -3c_1 = -3 \cdot 3 = -9$,
$c_3 = -3c_2 = -3 \cdot (-9) = 27$,
$c_4 = -3c_3 = -3 \cdot 27 = -81$. First four terms are $3, -9, 27, -81$. Geometric sequence.

43. Since $a_n = a_1 r^{n-1}$, we obtain

$$\frac{3}{1024} = 3\left(\frac{1}{2}\right)^{n-1}$$
$$\frac{1}{1024} = \left(\frac{1}{2}\right)^{n-1}$$
$$\left(\frac{1}{2}\right)^{10} = \left(\frac{1}{2}\right)^{n-1}$$

So $n - 1 = 10$ and the number of terms is $n = 11$.

45. Since $a_n = a_1 r^{n-1}$, we have $\frac{1}{81} = a_1 \left(\frac{1}{3}\right)^5$.

Solving, one finds the first term is $a_1 = 3$.

47. Since $a_n = a_1 r^{n-1}$, we obtain $6 = \frac{2}{3}r^2$.

So $9 = r^2$ and the common ratio is $r = \pm 3$.

49. Since $a_6 = a_3 r^3$, we obtain $96 = -12r^3$ and $r = -2$. Since $a_3 = a_1 r^2$, we get $-12 = a_1(-2)^2$. Then $a_1 = -3$ and $a_n = (-3)(-2)^{n-1}$.

51. Since $a = 1$, $r = 2$, and $n = 5$, we find that

the sum is $S = \frac{1(1 - 2^5)}{1 - 2} = 31$.

53. Since $a = 2$, $r = 2$, and $n = 6$, we obtain that the sum is $S = \dfrac{2(1 - 2^6)}{1 - 2} = 126$.

55. Since $a = 9$, $r = 1/3$, and $n = 6$, the sum is $S = \dfrac{9(1 - (1/3)^6)}{1 - 1/3} = 364/27$.

57. Since $a = 6$, $r = \dfrac{1}{3}$, and $n = 5$, we find that the sum is $S = \dfrac{6\left(1 - \left(\dfrac{1}{3}\right)^5\right)}{1 - \dfrac{1}{3}} = \dfrac{242}{27}$.

59. $\displaystyle\sum_{i=1}^{8} 1.5\,(-2)^{i-1} = \dfrac{1.5(1 - (-2)^8)}{1 - (-2)} = -127.5$

61. $\dfrac{2(1 - 1.05^{12})}{1 - 1.05} \approx 31.8343$

63. $\dfrac{200(1 - 1.01^8)}{1 - 1.01} \approx 1657.13$

65. $\displaystyle\sum_{n=1}^{5} 3\left(-\dfrac{1}{3}\right)^{n-1}$

67. $\displaystyle\sum_{n=1}^{\infty} 0.6(0.1)^{n-1}$

69. $\displaystyle\sum_{n=1}^{\infty} (-4.5)\left(-\dfrac{1}{3}\right)^{n-1}$

71. $S = \dfrac{a_1}{1 - r} = \dfrac{3}{1 - (-1/3)} = \dfrac{3}{4/3} = \dfrac{9}{4}$

73. $S = \dfrac{a_1}{1 - r} = \dfrac{0.9}{1 - 0.1} = \dfrac{0.9}{0.9} = 1$

75. $S = \dfrac{a_1}{1 - r} = \dfrac{-9.9}{1 - (-1/3)} = \dfrac{-9.9}{4/3} =$
$-\dfrac{99}{10} \cdot \dfrac{3}{4} = -\dfrac{297}{40}$

77. $S = \dfrac{a_1}{1 - r} = \dfrac{0.34}{1 - 0.01} = \dfrac{0.34}{0.99} = \dfrac{34}{99}$

79. No sum, since $|r| = |-1.06| > 1$.

81. $S = \dfrac{a_1}{1 - r} = \dfrac{0.6}{1 - 0.1} = \dfrac{0.6}{0.9} = \dfrac{2}{3}$

83. $S = \dfrac{a_1}{1 - r} = \dfrac{34}{1 - (-0.7)} = \dfrac{34}{1.7} = 20$

85. $\displaystyle\sum_{i=2}^{\infty} 4(0.1)^i = \dfrac{0.04}{1 - 0.1} = \dfrac{0.04}{0.9} = \dfrac{4}{90} = \dfrac{2}{45}$

87. $8.2 + 0.05454... = 8.2 + 0.1(0.5454...) =$

$8.2 + 0.1 \displaystyle\sum_{i=0}^{\infty} 0.54(0.01)^i =$

$8.2 + (0.1)\dfrac{0.54}{1 - 0.01} = 8.2 + \dfrac{0.054}{0.99} =$

$8.2 + \dfrac{54}{990} = \dfrac{8118 + 54}{990} = \dfrac{8172}{990} = \dfrac{454}{55}$

89. Using a formula for S_n, the sum is

$$\sum_{n=1}^{25} 100(.69)^{n-1} \;=\; \dfrac{100\left(1 - (.69)^{25}\right)}{1 - .69}$$
$$\approx\; 322.55 \;\; \text{mg}.$$

91. The total number of subscriptions in June is

$$\sum_{i=1}^{30} 2^{i-1} = \dfrac{1 - 2^{30}}{1 - 2} = 1,073,741,823.$$

93. At the end of the nth quarter the amount is $a_n = 4000(1.02)^n$. At the end of the 37th quarter, it is $a_{37} = 4000(1.02)^{37} = \$8,322.74$.

95. At the end of the 12th month, the amount in the account is $\displaystyle\sum_{i=1}^{12} 200(1.01)^i =$

$\dfrac{200(1.01)(1 - 1.01^{12})}{1 - 1.01} = \$2561.87$.

97. The value of the annuity immediately after the last payment is $\displaystyle\sum_{i=0}^{359} 100\left(1 + \dfrac{0.09}{12}\right)^i =$

$\dfrac{100(1 - 1.0075^{360})}{1 - 1.0075} = \$183,074.35$.

99. Assume the ball has a small radius. Approximately, the distance it travels before it comes to a rest is

$$9 + \dfrac{2}{3} \cdot 9 + \dfrac{2}{3} \cdot 9 + \left(\dfrac{2}{3}\right)^2 \cdot 9 + \left(\dfrac{2}{3}\right)^2 \cdot 9 + ... =$$

$$9 + 18\left[\dfrac{2}{3} + \left(\dfrac{2}{3}\right)^2 + \left(\dfrac{2}{3}\right)^3 + ...\right] =$$

$$9 + 18\dfrac{2/3}{1 - 2/3} = 45 \text{ feet}.$$

101. The first amount spent in Hammond is $2 million, then 75% of $2 million is the next amount spent, and so on. So the total economic impact is

$$\sum_{i=0}^{\infty} 2(10^6)(0.75)^i = \frac{2,000,000}{1-0.75} =$$

$8,000,000$.

103. Yes, since if $a = d = 0$ and r is any positive number, then the arithmetic sequence is also a geometric sequence. This sequence is $0, 0, \ldots$

105. If $0 < r < 1$, then $r^x \to 0$ as $x \to \infty$. If $r > 1$, then $|r^x| \to \infty$ as $x \to \infty$.

The graphs of $y = 2^x$ and $y = \left(\frac{1}{2}\right)^x$ are shown below.

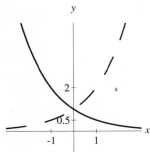

For Thought

1. False, there are $90 \cdot 26$ different codes.

2. False, there are 24^3 different possible fraternity names.

3. False, there are $3 \cdot 5 \cdot 3 \cdot 1 = 45$ different outfits.

4. True, $5! = 120$.

5. True, since

$$P(10, 2) = \frac{10!}{8!} = 10 \cdot 9 = 90.$$

6. False, the number of ways is 4^{20}

7. True

8. True, since $\dfrac{1000(999)998!}{998!} = 999,000.$

9. False, since

$$P(10, 1) = \frac{10!}{9!} = 10$$

and

$$P(10, 9) = \frac{10!}{1!} = 10!.$$

10. False, $P(29, 1) = \dfrac{29!}{28!} = \dfrac{29 \cdot 28!}{28!} = 29.$

8.4 Exercises

1. The number of possible outcomes is $2 \cdot 2$ or 4.

3. The number of possible outcomes is $2 \cdot 2 \cdot 2$ or 8.

5. Number of routes is $2 \cdot 4 = 8$.

7. There are $3! = 6$ different schedules. These schedules are given by:

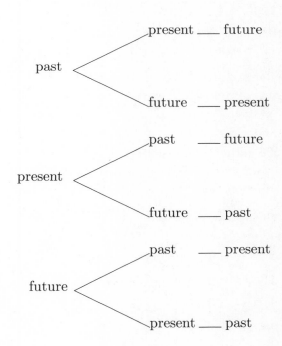

9. There are $4^5 = 1024$ different possible hands.

11. There are 8 optional items since $2^8 = 256$.

13. $\dfrac{7(6)(5)4!}{4!} = 7(6)(5) = 210$

15. $\dfrac{5!}{0!} = \dfrac{120}{1} = 120$

17. $\dfrac{78(77!)}{77!} = 78$

19. $\dfrac{4!}{2!} = 12$

21. $\dfrac{5!}{2!} = 60$

23. $\dfrac{7!}{4!} = 210$

25. $\dfrac{9!}{4!} = 15,120$

27. $\dfrac{5!}{0!} = 120$

29. $\dfrac{11!}{8!} = 990$

31. $\dfrac{99!}{99!} = 1$

33. $\dfrac{105!}{103!} = \dfrac{105(104)(103!)}{103!} = 105(104) = 10,920$

35. Hercules can perform 12 tasks in $12! = 479,001,600$ ways.

37. The number of ways a first, second, and a third restaurant can be chosen is
$$P(15,3) = \dfrac{15!}{12!} = 2730 \text{ ways.}$$

39. Since there are 8 half-hour shows from 6:00 p.m. to 10:00 p.m., the number of possible different schedules is $P(26,8) \approx 6.3 \times 10^{10}$.

41. Since each question has 4 possible answers, the number of ways to answer 6 questions is 4^6 or 4096.

43. There are $4 \cdot 3 \cdot 6 = 72$ possible prizes.

45. There are $P(26,3) = 15,600$ possible passwords.

47. There are $3 \cdot 10^4 = 30,000$ possible phone numbers.

49. Superman has $4! = 24$ ways to arrange these rescues.

51. Since the true-false questions can be answered in 2^5 ways and the multiple choice questions in 4^6 ways, the test can be answered in $2^5 \cdot 4^6 = 131,072$ ways.

53. On her list will be $P(7,3) = 210$ possible words.

For Thought

1. True, the number of ways is $C(5,3) = C(5,2) = 10$.

2. False, the number of ways is 3^5.

3. True, $P(5,2) = 20$.

4. False, it contains $n+1$ terms.   **5.**   True

6. True, $C(n,r)$ is the number of subsets of size r, and a set with n elements has 2^n subsets.

7. False, $P(8,3) = 336$ and $C(8,3) = 56$.

8. True   **9.**   False, since $P(8,3) = 336$ and $P(8,5) = 6720$.   **10.**   True

8.5 Exercises

1. 10   **3.**   7

5. $\dfrac{5!}{4!1!} = 5$

7. $\dfrac{8!}{4!4!} = 70$

9. $\dfrac{7!}{3!4!} = 35$

11. $\dfrac{10!}{10!0!} = 1$

13. $\dfrac{12!}{0!12!} = 1$

15. There are $C(4,2) = 6$ possible selections. They are Alice and Brenda, Alice and Carol, Alice and Dolores, Brenda and Carol, Brenda and Dolores, Carol and Dolores.

17. There are $C(5,2) = 10$ possible selections.

19. $C(5,3) = 10$ ways for an in-depth interview

21. $C(49,6) = 13,983,816$ ways to choose six numbers

23. $C(52,5) = 2,598,960$ possible poker hands

25. These assignments can be done in
$$\frac{10!}{5!3!2!} = 2520 \text{ ways.}$$

27. Since A occurs 4 times, the number of permutations is $\dfrac{7!}{4!} = 210$.

29. There are 3, 6, 10, and $C(n, 2)$ distinct chords, respectively,.

31. There are $\dfrac{10!}{3!2!5!} = 2520$ possible assignments.

33. There are $3^3 = 27$ ways to watch.

35. $C(8, 3) = 56$ ways to make the selection

37. Since 3 men can be selected in $C(9, 3)$ ways and 2 women in $C(6, 2)$ ways, the team can be selected in $C(9, 3) \cdot C(6, 2) = 1260$ ways.

39. $12! = 479,001,600$ ways to return the papers

41. $6 \cdot 6 = 36$ possible outcomes

43. Since there are 4! ways to arrange the bands in a line and 3! ways to line up the floats, the parade can be lead in $4!3! = 144$ ways.

45. Use the numbers 1,2,1 from Pascal's triangle. Then $(x+y)^2 = 1x^2 + 2xy + 1y^2 = x^2 + 2xy + y^2$.

47. Use the numbers 1,2,1 from Pascal's triangle. Then $(2a + (-3))^2 =$
$$1(2a)^2 + 2(2a)(-3) + 1(-3)^2 = 4a^2 - 12a + 9.$$

49. Use the numbers 1,3,3,1 from Pascal's triangle. Then $(a - 2)^3 =$
$$1(a^3) + 3(a^2)(-2) + 3(a)(-2)^2 + 1(-2)^3 =$$
$$a^3 - 6a^2 + 12a - 8.$$

51. Use the numbers 1,3,3,1 from Pascal's triangle. Thus, $(2a + b^2)^3 =$
$$1(2a)^3 + 3(2a)^2(b^2) + 3(2a)(b^2)^2 + 1(b^2)^3 =$$
$$8a^3 + 12a^2b^2 + 6ab^4 + b^6.$$

53. Use the numbers 1,4,6,4,1 from Pascal's triangle. Then we get $(x - 2y)^4 =$
$$1(x)^4 + 4(x)^3(-2y) + 6(x)^2(-2y)^2 + 4(x)(-2y)^3$$
$$+1(-2y)^4 =$$
$$x^4 - 8x^3y + 24x^2y^2 - 32xy^3 + 16y^4.$$

55. Use the numbers 1,4,6,4,1 from Pascal's triangle. Then we obtain $(x^2 + 1)^4 =$
$$1(x^2)^4 + 4(x^2)^3(1) + 6(x^2)^2(1)^2 + 4(x^2)(1)^3$$
$$+1(1)^4 = x^8 + 4x^6 + 6x^4 + 4x^2 + 1.$$

57. Use the numbers 1,5,10,10,5,1 from Pascal's triangle. Then we obtain $(a - 3)^5 =$
$$1(a)^5 - 5a^4(3) + 10a^3(3)^2 - 10a^2(3)^3 +$$
$$5a(3)^4 - 1(3)^4 =$$
$$a^5 - 15a^4 + 90a^3 - 270a^2 + 405a - 243.$$

59. Use the numbers 1,6,15,20,15,6,1 from Pascal's triangle. Then we obtain $(x + 2a)^6 =$
$$1(x)^6 + 6x^5(2a) + 15x^4(2a)^2 + 20x^3(2a)^3 +$$
$$15x^2(2a)^4 + 6x(2a)^5 + 1(2a)^6 =$$
$$x^6 + 12ax^5 + 60a^2x^4 + 160a^3x^3 +$$
$$240a^4x^2 + 192a^5x + 64a^6.$$

61. Use the Binomial Theorem. So $(x + y)^9 =$
$$\binom{9}{0}x^9 + \binom{9}{1}x^8y + \binom{9}{2}x^7y^2 + \ldots =$$
$$x^9 + 9x^8y + 36x^7y^2 + \ldots$$

63. Use the Binomial Theorem. Then $(2x - y)^{12} =$
$$\binom{12}{0}(2x)^{12} + \binom{12}{1}(2x)^{11}(-y) +$$
$$\binom{12}{2}(2x)^{10}(-y)^2 + \ldots =$$
$$4096x^{12} - 24,576x^{11}y + 67,584x^{10}y^2 + \ldots$$

65. Use the Binomial Theorem. So $(2s - 0.5t)^8 =$
$$\binom{8}{0}(2s)^8 + \binom{8}{1}(2s)^7(-0.5t) +$$
$$\binom{8}{2}(2s)^6(-0.5t)^2 + \ldots =$$
$$256s^8 - 512s^7t + 448s^6t^2 + \ldots$$

67. Use the Binomial Theorem. Thus, $(m^2 - 2w^3)^9 =$
$$\binom{9}{0}(m^2)^9 + \binom{9}{1}(m^2)^8(-2w^3) +$$
$$\binom{9}{2}(m^2)^7(-2w^3)^2 + \ldots =$$
$$m^{18} - 18m^{16}w^3 + 144m^{14}w^6 + \ldots$$

69. $\begin{pmatrix} 8 \\ 5 \end{pmatrix} = 56$

71. $\begin{pmatrix} 13 \\ 5 \end{pmatrix}(-2)^5 = -41,184$

73. By looking at a particular term, one finds

$$(a + [b + c])^{12} = \ldots + \begin{pmatrix} 12 \\ 10 \end{pmatrix}a^2[b + c]^{10} + \ldots =$$

$$\ldots + \begin{pmatrix} 12 \\ 10 \end{pmatrix}a^2\left[\ldots + \begin{pmatrix} 10 \\ 6 \end{pmatrix}b^4c^6 + \ldots\right] + \ldots$$

So the coefficient of $a^2b^4c^6$ is

$$\begin{pmatrix} 12 \\ 10 \end{pmatrix}\begin{pmatrix} 10 \\ 6 \end{pmatrix} = \frac{12!}{2!4!6!} = 13,860.$$

75. The coefficient of a^3b^7 in $(a + b + 2c)^{10}$ is the number of rearrangements of aaabbbbbbb, which is $\dfrac{10!}{3!7!} = 120.$

For Thought

1. False, $P(E) = \dfrac{n(E)}{n(S)}.$

2. False, there are $2^4 = 16$ outcomes.

3. True, since possible outcomes are (H, H), (H, T), (T, H), and (T, T) then

$$P(\text{at least one tail}) = \frac{3}{4} = 0.75 .$$

4. True, since $\dfrac{5}{6}$ is the probability of not getting a four when a die is tossed then $P(\text{at least one four}) = 1 - P(\text{no four}) =$

$$1 - \frac{5}{6} \cdot \frac{5}{6} = \frac{11}{36}.$$

5. False, since $\dfrac{1}{2}$ is the probability of getting a tail when a coin is tossed, $P(\text{at least one head}) = 1 - P(\text{four tails})$

$$= 1 - \left(\frac{1}{2}\right)^4 = \frac{15}{16}.$$

6. False, the complement is getting either no head or exactly one head.

7. True, the complement of getting exactly 3 tails in a toss of three coins is getting at least one head.

8. False, the odds in favor of snow is

$$\frac{P(\text{snow})}{P(\text{no snow})} = \frac{7/10}{3/10} = \frac{7}{3} \text{ i.e. 7 to 3.}$$

9. False, $P(E) = 3/7.$

10. False, it is equivalent to 1 to 4.

8.6 Exercises

1. $\{(H, H), (H, T), (T, H), (T, T)\}$

3. $\{(H, 1), (H, 2), (H, 3), (H, 4), (H, 5), (H, 6),$ $(T, 1), (T, 2), (T, 3), (T, 4), (T, 5), (T, 6)\}$

5. Note, (H, T) and (T, H) are the only outcomes with exactly one head. Since the sample space has 4 elements, the probability of getting exactly one head is $\dfrac{n\{(H, T), (T, H)\}}{4} = \dfrac{1}{2}.$

7. Note, $(6, 6)$ is the only outcome with exactly two sixes. Since the sample space has 36 elements, the probability of getting exactly two sixes is $\dfrac{n\{(6, 6)\}}{36} = \dfrac{1}{36}.$

9. Note, $(H, 5)$ is the only outcome with heads and a five. Since the sample space has 12 elements, the probability of getting heads and a five is $\dfrac{n\{(H, 5)\}}{12} = \dfrac{1}{12}.$

11. Note, (T, T, T) is the only outcome with all tails. Since the sample space has 8 elements, the probability of getting all tails is

$$\frac{n\{(T, T, T)\}}{8} = \frac{1}{8}.$$

13. (a) $\dfrac{n(\{3, 4, 5, 6\})}{6} = \dfrac{4}{6} = \dfrac{2}{3},$

(b) $\dfrac{n(\{1, 2, 3, 4, 5, 6\})}{6} = \dfrac{6}{6} = 1,$

(c) $\dfrac{n(\{1, 2, 3, 5, 6\})}{6} = \dfrac{5}{6},$

(d) 0, (e) $\dfrac{n(\{1\})}{6} = \dfrac{1}{6}$

15. (a) $\dfrac{n\{(T,T)\}}{4} = 1/4,$

(b) $1 - P(2 \; tails) = 1 - 1/4 = 3/4,$

(c) $\dfrac{n\{(H,H)\}}{4} = 1/4,$

(d) $\dfrac{n\{(T,T),(H,T),(T,H)\}}{4} = 3/4$

17. Note there are 36 possible outcomes.

(a) $\dfrac{n\{(3,3)\}}{36} = 1/36,$

(b) Since there are 11 possible outcomes with at least one 3, $P(at \; least \; one \; 3) = 11/36,$

(c) Since there are 5 possible outcomes with a sum of 6, $P(sum \; is \; 6) = 5/36,$

(d) $1 - P(sum \; is \; 2) = 1 - 1/36 = 35/36,$

(e) $P(sum \; is \; 2) = 1/36$

19. (a) 1/3, (b) 2/3, (c) 0

21. (a) 3/13, (b) 9/13, (c) 9/13,

(d) $P(marble \; is \; yellow) = 4/13$

23. There are 72 possible outcomes.

(a) $\dfrac{n\{(1,9)\}}{72} = 1/72,$

(b) $\dfrac{n\{(1,3),(3,1)\}}{72} = 2/72 = 1/36,$

(c) $\dfrac{n\{(1,4),(4,1),(2,3),(3,2)\}}{72} = 4/72 = 1/18$

25. (a) $\dfrac{1}{C(52,5)} = \dfrac{1}{2,598,960}$

(b) By using the counting principle, the probability of one 3, one 4, one 5, one 6, and one 7 is $\dfrac{4^5}{C(52,5)} = \dfrac{1024}{2,598,960}.$

27. $P(A \cup B) = P(A) + P(B) - P(A \cap B) = 0.8 + 0.6 - 0.5 = 0.9$

29. Since $P(E \cup F) = P(E) + P(F) - P(E \cap F),$ we get $0.8 = 0.2 + 0.7 - P(E \cap F).$ Then $P(E \cap F) = 0.2 + 0.7 - 0.8 = 0.1$

31. $P(A \cup B) = P(A) + P(B) - P(A \cap B) = 0.2 + 0.3 - 0 = 0.5$

33. Yes, since an outcome cannot have a sum that is both 4 and 5.

35. No, since the sum in the outcome $(1,1)$ is 2 (which is less than 5) and is even.

37. Yes, since there is no outcome with a two and with a sum greater than nine. Note, in $(2,6)$, the highest possible sum is 8.

39. $P(high-risk \; or \; a \; woman) = P(high-risk) + P(woman) - P(high - risk \; and \; a \; woman) = 38\% + 64\% - 24\% = 78\%.$

41. One finds the probabilities $P(3 \; boys) = P(3 \; girls) = 1/8.$ By the addition rule for mutually exclusive events, we get $P(3 \; boys \; or \; 3 \; girls) = 1/8 + 1/8 = 1/4.$

43. One obtains the probabilities $P(heart) = 13/52,$ $P(king) = 4/52,$ $P(heart \; and \; king) = 1/52.$ By the addition rule, we get $P(heart \; or \; king) = 13/52 + 4/52 - 1/52 = 4/13.$

45. (a) 34%, (b) 22% + 18% = 40%, (c) 34% + 22% + 18% = 74%

47. $1 - P(surviving) = 1 - 0.001 = 0.999$

49. Note there are 36 possible outcomes.

(a) $\dfrac{n\{(4,4)\}}{36} = 1/36,$

(b) $1 - P(pair \; of \; 4's) = 1 - 1/36 = 35/36,$

(c) 35/36, since this is the same event as (b)

51. Since $\dfrac{4/5}{1/5} = 4,$ the odds in favor of rain is 4 to 1.

53. Odds in favor of the eye of the hurricane coming ashore is $\dfrac{0.8}{0.2} = 4,$ i.e., 4 to 1

55. (a) since $\dfrac{1/4}{3/4} = \dfrac{1}{3},$ the odds in favor of stock market going up is 1 to 3

(b) odds against market going up is 3 to 1

57. If p is the probability of rain today, then

$$\frac{p}{1-p} = \frac{4}{1}$$
$$p = 4 - 4p$$
$$p = \frac{4}{5}.$$

The probability of rain today is $\frac{4}{5}$ or 80%.

59. (a) 1 to 9 (b) 9/10

61. Since $P(2\ heads) = \frac{C(4,2)}{2^4} = 3/8$ and

$\frac{3/8}{5/8} = 3/5$, the odds in favor of getting

2 heads in four tosses is 3 to 5

63. Since $\frac{P(sum\ of\ 7)}{P(sum\ that\ is\ not\ 7)} = \frac{1/6}{5/6} = 1/5$,

odds in favor of getting a sum of 7 is 1 to 5.

65. 1 to 1,999,999

67. $\frac{1}{1+31} = 1/32$

69. A and B are mutually exclusive if $P(A \cap B) = 0$. They are complementary events if $P(A) + P(B) = 1$

For Thought

1. True, for when $n = 1$ we get $4 \cdot 1 - 2 = 2 \cdot 1^2$ which is a true staetment.

2. False, when $n = 3$ one gets $27 < 12 + 15$, which is inconsistent.

3. False, $\frac{100-1}{100+1} \approx 0.98$.

4. False, it can prove it for the positive integers.

5. True **6.** False, rather $\sum_{i=1}^{k+1} \frac{1}{i(i+1)} = \frac{k+1}{k+2}$.

7. False **8.** False, since $0 > 0$ is inconsistent.

9. True

To see this let S_n be the inequality $n^2 - n > 0$. Note S_2 is true since $2^2 - 2 > 0$ holds. Suppose

S_k is true. One needs to show S_{k+1} is true. Then we have

$$(k+1)^2 - (k+1) = (k^2 + 2k + 1) - k - 1$$
$$= (k^2 - k) + 2k$$
$$> 0 + 2k \quad \text{since } S_k \text{ is true}$$
$$= 2k$$
$$> 0.$$

Thus, S_{k+1} holds and S_n is true for integers $n > 1$.

10. False, when $n = 1$ one gets $1^2 - 1 > 0$, which is inconsistent.

8.7 Exercises

1. If $n = 1$, then $3 - 1 = 2$ and $\frac{3 \cdot 1^2 + 1}{2} = \frac{4}{2}$.

If $n = 2$, then $(3-1) + (6-1) = 7$ and

$$\frac{3 \cdot 2^2 + 2}{2} = \frac{14}{2} = 7.$$

If $n = 3$, then $(3-1) + (6-1) + (9-1) = 15$

and $\frac{3 \cdot 3^2 + 3}{2} = \frac{30}{2} = 15.$

It is true for $n = 1, 2$, and 3.

3. If $n = 1$, then $\frac{1}{1(1+1)} = \frac{1}{1+1}$.

If $n = 2$, then $\frac{1}{2} + \frac{1}{6} = \frac{4}{6}$ and

$$\frac{2}{2+1} = \frac{2}{3}.$$

If $n = 3$, then $\frac{1}{2} + \frac{1}{6} + \frac{1}{12} = \frac{9}{12}$

and $\frac{3}{3+1} = \frac{3}{4}.$

It is true for $n = 1, 2$, and 3.

5. If $n = 1$, one has $1 = 4 - 3$.

If $n = 2$, then $1 + 4 = 5$ and $4(2) - 3 = 5$.

If $n = 3$, then $1 + 4 + 9 = 14$ and $4(3) - 3 = 9$.

It is true for $n = 1, 2$ and false for $n = 3$.

7. If $n = 1$, one has $1 < 1$.

If $n = 2$, one has $4 < 8$. If $n = 3$, $9 < 27$.

It is true for $n = 2, 3$ and false for $n = 1$.

9. $S_1 : 1 = \dfrac{1(1+1)}{2}$ or $1 = \dfrac{2}{2}$

is a true statement.

$S_2 : 1 + 2 = \dfrac{2(2+1)}{2}$ or $3 = \dfrac{6}{2}$

is a true statement.

$S_3 : 1 + 2 + 3 = \dfrac{3(3+1)}{2}$ or $6 = \dfrac{12}{2}$

is a true statement.

$S_4 : 1 + 2 + 3 + 4 = \dfrac{4(4+1)}{2}$ or $10 = \dfrac{20}{2}$

is a true statement.

11. $S_1 : 1^3 = \dfrac{1^2(1+1)^2}{4}$ or $1 = \dfrac{4}{4}$

is a true statement.

$S_2 : 1^3 + 2^3 = \dfrac{2^2(2+1)^2}{4}$ or $9 = \dfrac{36}{4}$

is a true statement.

$S_3 : 1^3 + 2^3 + 3^3 = \dfrac{3^2(3+1)^2}{4}$ or

$36 = \dfrac{144}{4}$ is a true statement.

$S_4 : 1^3 + 2^3 + 3^3 + 4^3 = \dfrac{4^2(4+1)^2}{4}$ or

$100 = \dfrac{400}{4}$ is a true statement.

13. $S_1 : 7^1 - 1$ divisible by 6 is true since the

quotient $\dfrac{7^1 - 1}{6} = \dfrac{6}{6} = 1$

is a whole number.

$S_2 : 7^2 - 1$ divisible by 6 is true since the

quotient $\dfrac{7^2 - 1}{6} = \dfrac{48}{6} = 8$

is a whole number.

$S_3 : 7^3 - 1$ divisible by 6 is true since the

quotient $\dfrac{7^3 - 1}{6} = \dfrac{342}{6} = 57$

is a whole number.

$S_4 : 7^4 - 1$ divisible by 6 is true since the

quotient $\dfrac{7^4 - 1}{6} = \dfrac{2400}{6} = 400$

is a whole number.

15. $S_1 : 2(1) = 1(1+1)$

$S_k : \displaystyle\sum_{i=1}^{k} 2i = k(k+1)$

$S_{k+1} : \displaystyle\sum_{i=1}^{k+1} 2i = (k+1)(k+2)$

17. $S_1 : 2 = 2 \cdot 1^2$
$S_k : 2 + 6 + \ldots + (4k - 2) = 2k^2$

$S_{k+1} :$

$$\begin{aligned} 2 + 6 + \ldots + (4(k+1) - 2) &= \\ 2 + 6 + \ldots + (4k+2) &= 2(k+1)^2 \end{aligned}$$

19. $S_1 : 2 = 2^2 - 2$
$S_k : \displaystyle\sum_{i=1}^{k} 2^i = 2^{k+1} - 2$
$S_{k+1} : \displaystyle\sum_{i=1}^{k+1} 2^i = 2^{k+2} - 2$

21. $S_1 : (ab)^1 = a^1 b^1$
$S_k : (ab)^k = a^k b^k$
$S_{k+1} : (ab)^{k+1} = a^{k+1} b^{k+1}$

23. $S_1 :$ If $0 < a < 1$ then $0 < a^1 < 1$
$S_k :$ If $0 < a < 1$ then $0 < a^k < 1$
$S_{k+1} :$ If $0 < a < 1$ then $0 < a^{k+1} < 1$

25. Let $T_n : 1 + 2 + \ldots + n = \dfrac{n(n+1)}{2}$.

Step 1: If $n = 1$ then $T_1 : 1 = \dfrac{1(2)}{2}$.
So T_1 is true.

Step 2: Assume $T_k : 1 + 2 + \ldots + k = \dfrac{k(k+1)}{2}$
is true. Add $(k+1)$ to both sides. Then we obtain

$$\begin{aligned} 1 + 2 + \ldots + k + (k+1) &= \frac{k(k+1)}{2} + (k+1) \\ &= \frac{k(k+1) + 2(k+1)}{2} \\ &= \frac{(k+1)(k+2)}{2}. \end{aligned}$$

Then the truth of T_k implies the truth of T_{k+1}. T_n is true for every positive integer n.

27. Let $T_n : 3 + 7 + ... + (4n - 1) = n(2n + 1)$.

Step 1: If $n = 1$ then $T_1 : 3 = 1(2 + 1)$.

Thus, T_1 is true.

Step 2: Assume T_k is true i.e.
$3 + 7 + ... + (4k - 1) = k(2k + 1)$.
Note $4(k + 1) - 1 = 4k + 3$. Adding $4k + 3$ to both sides, we get

$3 + ... + (4k - 1) + (4k + 3) =$

$$\begin{aligned} &= k(2k + 1) + (4k + 3) \\ &= 2k^2 + k + 4k + 3 \\ &= 2k^2 + 5k + 3 \\ &= (k + 1)(2k + 3) \\ &= (k + 1)(2(k + 1) + 1). \end{aligned}$$

The truth of T_k implies the truth of T_{k+1}. T_n is true for every positive integer n.

29. Let $T_n : \sum_{i=1}^{n} 2^i = 2^{n+1} - 2$.

Step 1: If $n = 1$ then $T_1 : 2 = 2^2 - 2$.

So T_1 is true.

Step 2: Assume $T_k : \sum_{i=1}^{k} 2^i = 2^{k+1} - 2$ is true.
Then add 2^{k+1} to both sides.

$$\begin{aligned} \sum_{i=1}^{k} 2^i + 2^{k+1} &= 2^{k+1} - 2 + 2^{k+1} \\ &= 2 \cdot 2^{k+1} - 2 \\ &= 2^{k+2} - 2 \\ &= 2^{(k+1)+1} - 2 \end{aligned}$$

Thus, the truth of T_k implies the truth of T_{k+1}. T_n is true for every positive integer n.

31. Let $T_n : \sum_{i=1}^{n} (3i - 1) = \dfrac{3n^2 + n}{2}$.

Step 1: If $n = 1$ then $T_1 : 3 - 1 = \dfrac{3 + 1}{2}$.

So T_1 is true.

Step 2: Assume $T_k : \sum_{i=1}^{k} (3i - 1) = \dfrac{3k^2 + k}{2}$

is true. Note $3(k + 1) - 1 = 3k + 2$. Then add $3k + 2$ to both sides.

$$\sum_{i=1}^{k} (3i - 1) + 3k + 2 = \dfrac{3k^2 + k}{2} + (3k + 2)$$

$$\begin{aligned} &= \dfrac{3k^2 + k + 2(3k + 2)}{2} \\ &= \dfrac{3k^2 + 7k + 4}{2} \\ &= \dfrac{3(k^2 + 2k + 1) + (k + 1)}{2} \\ &= \dfrac{3(k + 1)^2 + (k + 1)}{2} \end{aligned}$$

Thus, the truth of T_k implies the truth of T_{k+1}. T_n is true for every positive integer n.

33. Let $T_n : 1^2 + 2^2 + ... + n^2 = \dfrac{n(n + 1)(2n + 1)}{6}$.

Step 1: If $n = 1$ then $T_1 : 1 = \dfrac{1(2)(3)}{6}$.

So T_1 is true.

Step 2: Assume T_k is true i.e. we

assume $1^2 + 2^2 + ... + k^2 = \dfrac{k(k + 1)(2k + 1)}{6}$

Add $(k + 1)^2$ to both sides.

$1^2 + 2^2 + ... + k^2 + (k + 1)^2 =$

$$\begin{aligned} &= \dfrac{k(k + 1)(2k + 1)}{6} + (k + 1)^2 \\ &= \dfrac{k(k + 1)(2k + 1) + 6(k + 1)^2}{6} \\ &= \dfrac{(k + 1)[k(2k + 1) + 6(k + 1)]}{6} \\ &= \dfrac{(k + 1)[2k^2 + 7k + 6]}{6} \\ &= \dfrac{(k + 1)[k + 2][2k + 3]}{6} \\ &= \dfrac{(k + 1)[(k + 1) + 1][2(k + 1) + 1]}{6} \end{aligned}$$

Thus, the truth of T_k implies the truth of T_{k+1}. T_n is true for every positive integer n.

35. Let $T_n : 1 \cdot 3 + 2 \cdot 4 + ... + n \cdot (n + 2) = \dfrac{n}{6}(n + 1)(2n + 7)$.

Step 1: If $n = 1$ then $T_1 : 1 \cdot 3 = \dfrac{1}{6}(2)(2 + 7)$.

So T_1 is true.

Step 2: Assume T_k is true i.e. we assume

$1 \cdot 3 + 2 \cdot 4 + ... + k \cdot (k + 2) = \dfrac{k}{6}(k + 1)(2k + 7)$.

Add $(k + 1)(k + 3)$ to both sides. Then

$$1 \cdot 3 + 2 \cdot 4 + \ldots + k \cdot (k+2) + (k+1)(k+3) =$$

$$= \frac{k}{6}(k+1)(2k+7) + (k+1)(k+3)$$

$$= \frac{k(k+1)(2k+7) + 6(k+1)(k+3)}{6}$$

$$= \frac{(k+1)(2k^2 + 13k + 18)}{6}$$

$$= \frac{(k+1)(2k+9)(k+2)}{6}$$

$$= \frac{(k+1)}{6}[k+2][2k+9]$$

$$= \frac{(k+1)}{6}[(k+1)+1][2(k+1)+7]$$

So the truth of T_k implies the truth of T_{k+1}. Then T_n is true for every positive integer n.

37. Let $T_n : \dfrac{1}{1 \cdot 3} + \dfrac{1}{3 \cdot 5} + \ldots + \dfrac{1}{(2n-1)(2n+1)} = \dfrac{n}{2n+1}$.

Step 1: If $n = 1$ then $T_1 : \dfrac{1}{1 \cdot 3} = \dfrac{1}{2+1}$.

So T_1 is true.

Step 2: Assume T_k is true i.e. we assume

$$\frac{1}{1 \cdot 3} + \ldots + \frac{1}{(2k-1)(2k+1)} = \frac{k}{2k+1}.$$

Note $\dfrac{1}{[2(k+1)-1][2(k+1)+1]} =$

$\dfrac{1}{(2k+1)(2k+3)}$. Then add

$\dfrac{1}{(2k+1)(2k+3)}$ to both sides.

So $\dfrac{1}{1 \cdot 3} + \ldots$

$$\ldots + \frac{1}{(2k-1)(2k+1)} + \frac{1}{(2k+1)(2k+3)} =$$

$$= \frac{k}{2k+1} + \frac{1}{(2k+1)(2k+3)}$$

$$= \frac{k(2k+3)+1}{(2k+1)(2k+3)}$$

$$= \frac{2k^2 + 3k + 1}{(2k+1)(2k+3)}$$

$$= \frac{(2k+1)(k+1)}{(2k+1)(2k+3)}$$

$$= \frac{k+1}{2(k+1)+1}.$$

Thus, the truth of T_k implies the truth of T_{k+1}. T_n is true for every positive integer n.

39. Let $T_n :$ If $0 < a < 1$ then $0 < a^n < 1$.

Step 1: Clearly, T_1 is true for $n = 1$.

Step 2: Assume $0 < a < 1$ implies $0 < a^k < 1$, i.e., assume T_k is true. Multiply $0 < a^k < 1$ by a to get $0 < a^{k+1} < a$. Since $a < 1$ and by transitivity, we get $0 < a^{k+1} < 1$.

Then the truth of T_k implies the truth of T_{k+1}. T_n is true for every positive integer n.

41. Let $T_n : n < 2^n$.

Step 1: If $n = 1$, then $1 < 2^1$. So T_1 is true.

Step 2: Assume $k < 2^k$, i.e., assume T_k is true. Add 1 to both sides.

$$\begin{aligned} k+1 \;&<\; 2^k + 1 \\ &<\; 2^k + 2^k \\ &=\; 2 \cdot 2^k \\ &=\; 2^{k+1} \end{aligned}$$

Then $k + 1 < 2^{k+1}$.

So, the truth of T_k implies the truth of T_{k+1}. T_n is true for every positive integer n.

43. Let $T_n : 5^n - 1$ is divisible by 4.

Step 1: If $n = 1$, then $5^1 - 1$ is divisible by 4. So T_1 is true.

Step 2: Assume $5^k - 1$ is divisible by 4, i.e., assume T_k is true.
Observe that $5^{k+1} - 1 = 5(5^k - 1) + 4$. Since sums of multiples of 4 are again multiples of 4, we get that $5^{k+1} - 1$ is a multiple of 4.

Then the truth of T_k implies the truth of T_{k+1}. T_n is true for every positive integer n.

45. Let $T_n : (ab)^n = a^n b^n$.

Step 1: If $n = 1$ then $(ab)^1 = a^1 b^1$. So T_1 is true.

Step 2: Assume $(ab)^k = a^k b^k$, i.e., assume T_k is true. Then

$$\begin{aligned} (ab)^{k+1} \;&=\; (ab)^k (ab) \\ &=\; a^k b^k (ab) \quad \text{since } T_k \text{ is true} \\ &=\; a^{k+1} b^{k+1}. \end{aligned}$$

Thus, the truth of T_k implies the truth of T_{k+1}. T_n is true for every positive integer n.

47. Let $T_n : \sum\limits_{i=0}^{n} x^i = \dfrac{x^{n+1} - 1}{x - 1}$, $x \neq 1$.

Step 1: If $n = 1$, then $T_1 : 1 + x = \dfrac{x^2 - 1}{x - 1}$.

T_1 is true since $(x^2 - 1) = (x - 1)(x + 1)$.

Step 2: Assume $T_k : \sum\limits_{i=0}^{k} x^i = \dfrac{x^{k+1} - 1}{x - 1}$ is true. Add x^{k+1} to both sides.

$$\sum_{i=0}^{k} x^i + x^{k+1} = \dfrac{x^{k+1} - 1}{x - 1} + x^{k+1}$$

$$= \dfrac{x^{k+1} - 1 + (x - 1)x^{k+1}}{x - 1}$$

$$= \dfrac{x^{k+1} - 1 + x^{k+2} - x^{k+1}}{x - 1}$$

$$= \dfrac{x^{k+2} - 1}{x - 1}$$

Then the truth of T_k implies the truth of T_{k+1}. T_n is true for every positive integer n.

Review Exercises

1. One gets $a_1 = 2^0 = 1$, $a_2 = 2^1 = 2$, $a_3 = 2^2 = 4$, $a_4 = 2^3 = 8$, $a_5 = 2^4 = 16$

First five terms are $1, 2, 4, 8, 16$

3. $a_1 = \dfrac{(-1)^1}{1!} = -1$, $a_2 = \dfrac{(-1)^2}{2!} = \dfrac{1}{2}$, $a_3 = \dfrac{(-1)^3}{3!} = \dfrac{-1}{6}$, $a_4 = \dfrac{(-1)^4}{4!} = \dfrac{1}{24}$.

First four terms are $-1, \dfrac{1}{2}, -\dfrac{1}{6}, \dfrac{1}{24}$.

5. $a_1 = 3(0.5)^0 = 3$, $a_2 = 3(0.5)^1 = 1.5$, and $a_3 = 3(0.5)^2 = 0.75$.

First three terms are $3, 1.5, 0.75$.

7. $a_1 = -3 + 6 = 3$, $a_2 = -6 + 6 = 0$, and $a_3 = -9 + 6 = -3$.

First three terms are $3, 0, -3$.

9. Since $S = \dfrac{a_1(1 - r^n)}{1 - r}$ is the sum of a geometric

series, $S = \dfrac{0.5(1 - 0.5^4)}{1 - 0.5} = 0.9375$.

11. Since $S = \dfrac{n}{2}(a_1 + a_n)$ is the sum of an

arithmetic series, $S = \dfrac{50}{2}(11 + 207) = 5450$.

13. Since $S = \dfrac{a_1}{1 - r}$ is the sum of a geometric

series, $S = \dfrac{0.3}{1 - 0.1} = \dfrac{0.3}{0.9} = \dfrac{1}{3}$.

15. Since $S = \dfrac{a_1(1 - r^n)}{1 - r}$ is the sum of a geometric

series, $S = \dfrac{1000(1 - 1.05^{20})}{1 - 1.05} \approx 33,065.9541$

17. $a_n = \dfrac{(-1)^n}{n + 2}$

19. $a_n = 6\left(\dfrac{1}{6}\right)^{n-1}$

21. $\sum\limits_{i=1}^{\infty} \dfrac{(-1)^{i+1}}{i + 1}$

23. $\sum\limits_{i=1}^{14} 2i$

25. Note, $a_n = a_1 r^{n-1}$. Since $256 = 4r^6$, we get $64 = r^6$ and the common ratio is $r = \pm 2$.

27. $100\left(1 + \dfrac{0.09}{4}\right)^{40} = \$243.52$

29. $\sum\limits_{i=1}^{10} 1000(1.06)^i = \dfrac{1000(1.06)(1 - 1.06^{10})}{1 - 1.06}$

$= \$13,971.64$

31. If the pattern continues, the total number of Tummy Masters that could be sold is

$\sum\limits_{i=0}^{\infty} 100,000(0.9)^i = \dfrac{100,000}{1 - 0.9} = 1$ million.

33. $a^4 + \begin{pmatrix} 4 \\ 1 \end{pmatrix} a^3(2b) +$

$\begin{pmatrix} 4 \\ 2 \end{pmatrix} a^2(2b)^2 + \begin{pmatrix} 4 \\ 3 \end{pmatrix} a(2b)^3 + (2b)^4 =$

$a^4 + 8a^3b + 24a^2b^2 + 32ab^3 + 16b^4$

35. $(2a)^5 + \begin{pmatrix} 5 \\ 1 \end{pmatrix} (2a)^4(-b) + \begin{pmatrix} 5 \\ 2 \end{pmatrix} (2a)^3(-b)^2 +$

$\begin{pmatrix} 5 \\ 3 \end{pmatrix} (2a)^2(-b)^3 + \begin{pmatrix} 5 \\ 4 \end{pmatrix} (2a)(-b)^4 + (-b)^5 =$

$$32a^5 - 80a^4b + 80a^3b^2 - 40a^2b^3 + 10ab^4 - b^5$$

37. $a^{10} + \begin{pmatrix} 10 \\ 1 \end{pmatrix} a^9 b + \begin{pmatrix} 10 \\ 2 \end{pmatrix} a^8 b^2 + ... =$

$a^{10} + 10a^9b + 45a^8b^2 + ...$

39. $(2x)^8 + \begin{pmatrix} 8 \\ 1 \end{pmatrix} (2x)^7 \left(\dfrac{y}{2} \right) + \begin{pmatrix} 8 \\ 2 \end{pmatrix} (2x)^6 \left(\dfrac{y}{2} \right)^2 +$
...

$= 256x^8 + 512x^7y + 448x^6y^2 + ...$

41. $\begin{pmatrix} 13 \\ 9 \end{pmatrix} = 715$

43. $\dfrac{11!}{2!3!6!} 2^3 = 36,960$

45. 24 terms

47. $5^9 = 1,953,125$ ways to mark the answers

49. $P(7,3) = 210$ possible three-letter 'words'

51. $3 \cdot 5 \cdot 4 = 60$ ways to place advertisements

53. $C(8,5) = 56$ possible vacations where order is not taken into account

55. (a) $C(8,4) = 70$ possible councils,
(b) $C(5,4) = 5$ possible councils from Democrats,
(c) $C(5,2) \cdot C(3,2) = 30$ possible councils with two Democrats and two Republicans

57. For KANSAS, $\dfrac{6!}{2!2!} = 180$ arrangements;

for TEXAS, $5! = 120$ arrangements.

59. $2^7 = 128$ different families

61. Since there are 2^{10} ways to answer the test, the probabilities are

$P(all\ 10\ correct) = \dfrac{1}{1024}$ and

$P(all\ 10\ wrong) = \dfrac{1}{1024}.$

63. (a) 5/13, (b) 8/13, (c) 0, (d) 1

65. Assuming the probabilities of a boy or girl being born are equal, $P(3\ boys) = \dfrac{1}{8}$. Then the odds in favor of 3 boys is $\dfrac{1/8}{7/8}$, i.e., 1 to 7.

67. Odds in favor of catching a perch is $\dfrac{0.9}{0.1}$, i.e., 9 to 1.

69. By the Addition Rule, $P(Math\ or\ English) = 0.7 + 0.6 - 0.4 = 0.9$, i.e., 90%.

71. $40,320$

73. 20

75. 1680

77. $\dfrac{8!}{2!6!} = 28$

79. $\dfrac{8!}{4!} = 1680$

81. $\dfrac{8!}{7!1!} = 8$

83. Let $T_n : 3 + 6 + 9 + ... + 3n = \dfrac{3}{2}(n^2 + n)$.

Step 1: If $n = 1$, then $T_1 : 3 = \dfrac{3}{2}(1 + 1)$.
So T_1 is true.

Step 2: Assume $T_k : 3 + 6 + ... + 3k = \dfrac{3}{2}(k^2 + k)$
is true. Add $3(k+1)$ to both sides.
$3 + 6 + ... + 3k + 3(k+1) =$

$$= \dfrac{3}{2}(k^2 + k) + 3(k + 1)$$
$$= \dfrac{3(k^2 + k) + 6(k + 1)}{2}$$
$$= \dfrac{3k^2 + 9k + 6}{2}$$
$$= \dfrac{3}{2}(k^2 + 3k + 2)$$
$$= \dfrac{3}{2}((k + 1)^2 + (k + 1))$$

Then the truth of T_k implies the truth of T_{k+1}. T_n is true for every positive integer n.

Chapter 8 Test

1. $a_1 = 2.3$, $a_2 = 2.3 + 0.5$,
$a_3 = 2.3 + 1$, and $a_4 = 2.3 + 1.5$.
First four terms are $2.3, 2.8, 3.3, 3.8$.

2. One finds $c_2 = \dfrac{1}{2} \cdot 20 = 10$,

$c_3 = \dfrac{1}{2} \cdot 10 = 5$, and $c_4 = \dfrac{1}{2} \cdot 5 = 2.5$.

First four terms are $20, 10, 5, 2.5$.

3. $a_n = (-1)^{n-1}(n-1)^2$

4. $a_n = 7 + 3(n-1) = 3n + 4$

5. $a_n = \dfrac{1}{3}\left(-\dfrac{1}{2}\right)^{n-1}$

6. Since $S = \dfrac{n}{2}(a_1 + a_n)$, the

sum is $\dfrac{54}{2}(-5 + 154) = 4023$.

7. Since $S = \dfrac{a_1(1-r^n)}{1-r}$, the sum is

$\dfrac{300(1.05)(1 - 1.05^{23})}{1 - 1.05} \approx 13{,}050.5997$.

8. Since $S = \dfrac{a_1}{1-r}$, the sum is $\dfrac{0.98}{1 - 0.98} = 49$.

9. Since $a_n = a_1 + d(n-1)$, $9 = -3 + 8d$.
Solving for d, one gets $d = 1.5$.
So $a_n = -3 + 1.5(n-1)$ or

$$a_n = 1.5n - 4.5.$$

10. $\dfrac{9!}{4!2!2!} = 3780$ possible nine-letter 'words'

11. Since $S = \dfrac{n}{2}(a_1 + a_n)$ is the sum of an

arithmetic series, the mean daily sales is

$\dfrac{1}{30}\displaystyle\sum_{i=0}^{29} 300 + 10i = \dfrac{1}{30}\dfrac{30}{2}(300 + 590) = \$445$

12. There are $10^4 = 10{,}000$ possible secret numbers. The probability of guessing the correct number in 3 tries is the probability of getting it in the 1st try plus the probability of getting it in the 2nd try plus the probability of getting it in the 3rd try. Assuming the person remembers what he tries and tries a different number after a failure, this probability is

$$\dfrac{1}{10{,}000} + \dfrac{9{,}999}{10{,}000} \cdot \dfrac{1}{9{,}999} + \dfrac{9{,}999}{10{,}000} \cdot \dfrac{9{,}998}{9{,}999}.$$

$$\dfrac{1}{9{,}998} = \dfrac{3}{10{,}000}.$$

13. At the end of the 300th month the value of the annuity is $\displaystyle\sum_{i=1}^{300} 700\left(1 + \dfrac{0.06}{12}\right)^i =$

$\displaystyle\sum_{i=1}^{300} 700(1.005)^i = 700(1.005)\left(\dfrac{1 - 1.005^{300}}{1 - 1.005}\right)$

$= \$487{,}521.25.$

The cost of the house at the end of 25 years is

$$120{,}000(1.06)^{25} = \$515{,}024.49.$$

No, they will not have enough money to buy the house.

14. $a^5 + \begin{pmatrix} 5 \\ 1 \end{pmatrix}a^4(-2x) + \begin{pmatrix} 5 \\ 2 \end{pmatrix}a^3(-2x)^2 +$

$\begin{pmatrix} 5 \\ 3 \end{pmatrix}a^2(-2x)^3 + \begin{pmatrix} 5 \\ 4 \end{pmatrix}a(-2x)^4 + (-2x)^5 =$

$a^5 - 10a^4x + 40a^3x^2 - 80a^2x^3 + 80ax^4 - 32x^5$

15. $x^{24} + \begin{pmatrix} 24 \\ 1 \end{pmatrix}x^{23}(y^2) + \begin{pmatrix} 24 \\ 2 \end{pmatrix}x^{22}(y^2)^2 + ... =$

$x^{24} + 24x^{23}y^2 + 276x^{22}y^4 + ...$

16. $\displaystyle\sum_{i=0}^{30} \begin{pmatrix} 30 \\ i \end{pmatrix}m^{30-i}y^i$

17. Note, $P(\text{sum is } 7) = \dfrac{6}{36} = \dfrac{1}{6}$. The odds in

favor of 7 is $\dfrac{6/36}{30/36} = \dfrac{1}{5}$, i.e., 1 to 5.

18. $C(12, 3) = 220$ selections

19. $C(12, 2) \cdot C(10, 2) = 66 \cdot 45 = 2970$ outcomes

20. $1/P(8,3) = \dfrac{1}{336}$ is the probability of getting the horses and the order of finish correctly.

21. Let $T_n : \displaystyle\sum_{i=1}^{n} \left(\dfrac{1}{2}\right)^i = 1 - 2^{-n}$.

Step 1: If $n = 1$, then $T_1 : \dfrac{1}{2} = 1 - 2^{-1}$.

So T_1 is true.

Step 2: Assume $T_k : \displaystyle\sum_{i=1}^{k} \left(\dfrac{1}{2}\right)^i = 1 - 2^{-k}$ is true.

Add $\left(\dfrac{1}{2}\right)^{k+1}$ to both sides.

$$
\begin{aligned}
\sum_{i=1}^{k} \left(\dfrac{1}{2}\right)^i + \left(\dfrac{1}{2}\right)^{k+1} &= 1 - 2^{-k} + \left(\dfrac{1}{2}\right)^{k+1} \\
&= 1 - 2 \cdot 2^{-(k+1)} + 2^{-(k+1)} \\
&= 1 - 2^{-(k+1)}
\end{aligned}
$$

Then the truth of T_k implies the truth of T_{k+1}. Thus, T_n is true for every positive integer n.

Since $1 - 2^{-n} < 1$ for all positive integers n,

then by transitivity we obtain $\displaystyle\sum_{i=1}^{n} \left(\dfrac{1}{2}\right)^i < 1$

for all positive integers n.